SPILLOVER

ALSO BY DAVID QUAMMEN

NONFICTION

The Reluctant Mr. Darwin

Monster of God

The Song of the Dodo

ESSAYS

Natural Acts

The Boilerplate Rhino

Wild Thoughts from Wild Places

The Flight of the Iguana

FICTION

Blood Line

The Soul of Viktor Tronko

The Zolta Configuration

To Walk the Line

EDITED

On the Origin of Species by Charles Darwin: The Illustrated Edition

The Best American Science and Nature Writing 2000, with Burkhard Bilger

SPILLOVER

Animal Infections and the Next Human Pandemic

DAVID QUAMMEN

W. W. NORTON & COMPANY
NEW YORK · LONDON

Small portions of this book have appeared previously, in different form,
in *National Geographic* and *Outside* magazines.

Maps by Daphne Gillam, www.handcraftedmaps.com.

Printed in the United States of America
First Edition

Manufacturing by RR Donnelley, Harrisonburg
Book design by Iris Weinstein
Production manager: Julia Druskin

Library of Congress Cataloging-in-Publication Data

ISBN: 978-0-393-06680-7

W. W. Norton & Company, Inc.
500 Fifth Avenue, New York, N.Y. 10110
www.wwnorton.com

W. W. Norton & Company Ltd.
Castle House, 75/76 Wells Street, London W1T 3QT

1 2 3 4 5 6 7 8 9 0

again and ever,
to Betsy

And I looked, and behold a pale horse: and his name that sat on him was Death, and Hell followed with him. And power was given unto them over the fourth part of the earth, to kill with sword, and with hunger, and with death, and with the beasts of the earth.

—REVELATION 6:8

CONTENTS

I

PALE HORSE

1

The virus now known as Hendra wasn't the first of the scary new bugs. It wasn't the worst. Compared to some others, it seems relatively minor. Its mortal impact, in numerical terms, was small at the start and has remained small; its geographical scope was narrowly local and later episodes haven't carried it much more widely. It made its debut near Brisbane, Australia, in 1994. Initially there were two cases, only one of them fatal. No, wait, correction: There were two *human* cases, one *human* fatality. Other victims suffered and died too, more than a dozen—equine victims—and their story is part of this story. The subject of animal disease and the subject of human disease are, as we'll see, strands of one braided cord.

The original emergence of Hendra virus didn't seem very dire or newsworthy unless you happened to live in eastern Australia. It couldn't match an earthquake, a war, a schoolboy gun massacre, a tsunami. But it was peculiar. It was spooky. Slightly better known now, at least among disease scientists and Australians, and therefore slightly less spooky, Hendra virus still seems peculiar. It's a paradoxical thing: marginal, sporadic, but in some larger sense representative. For exactly that reason, it marks a good point from which to begin toward understanding the emergence of certain virulent new realities on this planet—realities that include the death of more than 30 million people since 1981. Those realities involve a phenomenon called zoonosis.

A *zoonosis* is an animal infection transmissible to humans. There are more such diseases than you might expect. AIDS is one. Influenza is a whole category of others. Pondering them as a group tends to reaffirm the old Darwinian truth (the darkest of his truths, well known and persistently forgotten) that humanity *is* a kind of animal, inextricably connected with other animals: in origin and in descent, in sickness and in health. Pondering them individually—for starters, this relatively obscure case from Australia—provides a salubrious reminder that everything, including pestilence, comes from somewhere.

2

In September 1994, a violent distress erupted among horses in a suburb at the north fringe of Brisbane. These were thoroughbred racehorses, pampered and sleek animals bred to run. The place itself was called Hendra. It was a quiet old neighborhood filled with racecourses, racing people, weatherboard houses whose backyards had been converted to stables, newsstands that sold tip sheets, corner cafes with names like The Feed Bin. The first victim was a bay mare named Drama Series, retired from racing and now heavily in foal—that is, pregnant and well along. Drama Series started showing signs of trouble in a spelling paddock, a ragged meadow several miles southeast of Hendra, where racehorses were sent to rest between outings. She had been placed there as a brood mare and would have stayed until late in her pregnancy, if she hadn't gotten sick. There was nothing drastically wrong with her—so it seemed, at this point. She just didn't look good, and her trainer thought she should come in. The trainer was a savvy little man named Vic Rail, with a forceful charm, swept-back brown hair, and a reputation for sharp practice in the local racing world. He was "tough as nails, but a lovable rogue," Vickie was, by one judgment. Some people resented him but no one denied he knew horses.

It was Rail's girlfriend, Lisa Symons, who took a horse trailer out to collect Drama Series. The mare was reluctant to move. She seemed to have sore feet. There were swellings around her lips, her eyelids, her jaw. Back at Rail's modest stable in Hendra, Drama Series sweated profusely and remained sluggish. Hoping to nourish her and save the foal, he tried to force feed her with grated carrot and molasses but she wouldn't eat. After the attempt, Vic Rail washed his hands and his arms, though in hindsight perhaps not thoroughly enough.

That was September 7, 1994, a Wednesday. Rail called his veterinarian, a tall man named Peter Reid, sober and professional, who came and looked the mare over. She was now in her own box at the stable, a cinderblock stall with a floor of sand, close amid Rail's other horses. Dr. Reid saw no discharges from her nose or eyes, and no signs of pain, but she seemed a pale image of her robust former self. "Depressed," was his word, meaning (in veterinary parlance) a physical not a psychological condition. Her temperature and her heart rate were both high. Reid noticed the facial swelling. Opening her mouth to examine her gums, he noticed remnants of the carrot shreds that she hadn't bothered or been able to swallow, and he gave her injections of antibiotic and analgesic. Then he went home. Sometime after four the next morning, he got a call. Drama Series had gotten out of her stall, collapsed in the yard, and was dying.

By the time Reid rushed back to the stables, she was dead. It had been quick and ugly. Growing agitated as her condition got worse, she had staggered out while the stall door was open, fallen down several times, gouged her leg to the bone, stood up, fallen again in the front yard, and been pinned to the ground for her own protection by a stable hand. She freed herself desperately, crashed into a pile of bricks, and then was pinned again by joint effort of the stable hand and Rail, who wiped a frothy discharge away from her nostrils—trying to help her breathe—just before she died. Reid inspected the body, noticing a trace of clear froth still at the nostrils, but did not perform a necropsy because Vic Rail couldn't afford to be so curious and, more generally, because no one foresaw a disease emergency in which every bit of such data would be crucial. Drama Series's carcass was unceremoniously carted away, by the usual contract hauler, to the dump where dead Brisbane horses routinely go.

Her cause of death remained uncertain. Had she been bitten by a snake? Had she eaten some poisonous weeds out in that scrubby, derelict meadow? Those hypotheses crumbled abruptly, thirteen days later, when her stable mates began falling ill. They went down like dominoes. This wasn't snakebite or toxic fodder. It was something contagious.

The other horses suffered fever, respiratory distress, bloodshot eyes, spasms, and clumsiness; in some, bloody froth surged from the nostrils and mouth; a few had facial swelling. Reid found one horse frantically rinsing its mouth in a water bucket. Another banged its head against the concrete wall as though maddened. Despite heroic efforts by Reid and others, twelve more animals died within the next several days, either expiring horrifically or euthanized. Reid later said that "the speed with which it went through those horses was unbelievable," but in these early moments no one had identified "it." *Something* went through those horses. At the height of the crisis, seven animals succumbed to their agonies or required euthanasia within just twelve hours. Seven dead horses in twelve hours—that's carnage, even for a casehardened veterinarian. One of them, a mare named Celestial Charm, died thrashing and gasping so desperately that Reid couldn't get close enough to give her the merciful needle. Another horse, a five-year-old gelding, had been sent from Rail's place to another spelling paddock up north, where it was sick on arrival and soon had to be put down. A vet up there necropsied the gelding and found hemorrhages throughout its organs. And in a neighbor's stable on the corner beside Rail's place in Hendra, at the same time, still another gelding went afoul with similar clinical signs and also had to be euthanized.

What was causing this mayhem? How was it spreading from one horse to another, or anyway getting into so many of them simultaneously? One possibility was a toxic contaminant in the feed supply. Or maybe poison, maliciously introduced. Alternatively, Reid began wondering whether there might be an exotic virus at work, such as the one responsible for African horse sickness (AHS), a disease carried by biting midges in sub-Saharan Africa. AHS virus affects mules, donkeys, and zebras as well as horses, but it hasn't been reported in Australia, and it isn't directly contagious from horse to horse. Furthermore, Queensland's pestiferous midges don't generally come biting in September, when the weather is cool. So AHS was not quite a fit. Then maybe another strange germ? "I'd never seen a virus do anything like that before," Reid said. A man of understatement, he recalled it as "a pretty traumatic time." He had continued to treat the suffering animals with what means and options he had, given the inconclusive diagnosis—antibiotics, fluids, antishock medicine.

Meanwhile, Vic Rail himself had taken sick. So had the stable hand. It seemed at first that they each had a touch of flu—a bad flu. Rail went into the hospital, worsened there, and, after a week of intensive care, died. His organs had failed and he couldn't breathe. Autopsy showed that his lungs were full of blood, other fluid, and (upon examination by electron microscopy) some sort of virus. The stable hand, a big-hearted man named Ray Unwin, who merely went home to endure his fever in private, survived. Peter Reid, though he had been working on the same suffering horses amid the same bloody froth, stayed healthy. He and Unwin told me their stories when I found them, years later, by asking around Hendra and making a few calls.

At The Feed Bin, for instance, someone said: Ray Unwin, yeah, most likely he'll be at Bob Bradshaw's. I followed directions to Bradshaw's stable and there on the driveway was a man who turned out to be Unwin, carrying grain in a bucket. At that point he was a middle-aged working bloke with a sandy red ponytail and a weary sadness in his eyes. He was a little shy about attention from a stranger; he'd had enough of that already from doctors, public health officials, and local reporters. Once we sat down to chat, he professed that he wasn't a "whinger" (complainer) but that his health had been "crook" (not right) since it happened.

As the horse deaths came to crescendo, the government of Queensland had intervened, in the form of veterinarians and other personnel from the Department of Primary Industries (responsible for livestock, wildlife, and agriculture throughout the state) and field officers from Queensland Health. The DPI veterinarians began doing necropsies—that is, cutting up horses, looking for clues—right in Vic Rail's little yard. Before long there were horse heads lying around, severed limbs, blood and other fluids flowing down the gutter, suspect organs and tissues going into bags. Another neighbor of Rail's, a fellow horse man named Peter Hulbert, recollected the gruesome pageant that had transpired next door, while serving me instant coffee in his kitchen. As the kettle came to a boil, Hulbert recalled the garbage containers used by DPI. "These street wheelie bins here, there was horses' legs and heads . . . —do you have sugar?"

No thanks, I said, black.

". . . horses' legs and heads and guts and everything, going into these wheelie bins. *It—was—horrendous.*" By midafternoon that day, he added, word had spread and the TV stations showed up with their news cameras. "Agh. It was bloody terrible, mate." Then the police arrived too and threw a tape cordon around Rail's place, treating it as a crime scene. Had one of his enemies done this? The racing world had its underbelly, like any business, and probably more so than most. Peter Hulbert even faced pointed questioning about whether Vic might have poisoned his own horses and then himself.

While the police wondered about sabotage or insurance scam, the health officials had other hypotheses to concern them. One was hantavirus—which is actually a group of viruses, long known to virologists following outbreaks in Russia, Scandinavia, and elsewhere but newly conspicuous since a year earlier, 1993, when a new hantavirus emerged dramatically and killed ten people around the Four Corners area of the American Southwest. Australia is justifiably wary of exotic diseases invading its borders, and hantavirus in the country would be even worse news (except for horses) than African horse sickness. So the DPI vets packed up samples of blood and tissue from the dead horses and sent them on ice to the Australian Animal Health Laboratory, a high-security institution known by its acronym, AAHL, pronounced "*aahl*," in a town called Geelong, south of Melbourne. A team of microbiologists and veterinarians there put the sample material through a series of tests, attempting to culture and identify a microbe, and to confirm that the microbe made horses sick.

They found a virus. It wasn't a hantavirus. It wasn't AHS virus. It was something new, something the AAHL microscopist hadn't seen before but which, from its size and its shape, resembled members of a particular virus group, the paramyxoviruses. This new virus differed from known paramyxoviruses in that each particle carried a double fringe of spikes. Other AAHL researchers sequenced a stretch of the viral genome and, submitting that sequence into a vast viral database, found a weak match to one subgroup of these viruses. That seemed to confirm the visual judgment of the microscopist. The matching subgroup was the morbilliviruses, which include rinderpest virus and canine distemper virus (infecting nonhuman animals) and measles (in humans). So the creature from Hendra was classified and given a name, based on those provisional identifications: equine morbillivirus (EMV). Roughly, horse measles.

About the same time, the AAHL researchers tested a sample of tissue that had been taken from Vic Rail's kidney during his autopsy. That sample also yielded a virus, identical to the virus from the horses, confirming that this equine morbillivirus didn't afflict only equines. Later, when the degree of its uniqueness became better appreciated, the label "EMV" was dropped and the virus was renamed after its place of emergence: Hendra.

Identifying the new virus was only step one in solving the immediate mystery of Hendra, let alone understanding the disease in a wider context. Step two would involve tracking that virus to its hiding place. Where did it exist when it wasn't killing horses and people? Step three would entail asking a further cluster of questions: How did the virus emerge from its secret refuge, and why here, and why now?

After our first conversation, at a café in Hendra, Peter Reid drove me several miles southeast, across the Brisbane River, to the site where Drama Series took sick. It was in an area called Cannon

Hill, formerly pastoral land surrounded by city, now a booming suburb just off the M1 motorway. Tract houses on prim lanes had been built over the original paddock. Not much of the old landscape remained. But toward the end of one street was a circle, called Calliope Circuit, in the middle of which stood a single mature tree, a Moreton Bay fig, beneath which the mare would have found shelter from eastern Australia's fierce subtropical sun.

"That's it," Reid said. "That's the bloody tree." That's where the bats gathered, he meant.

3

Infectious disease is all around us. Infectious disease is a kind of natural mortar binding one creature to another, one species to another, within the elaborate biophysical edifices we call ecosystems. It's one of the basic processes that ecologists study, including also predation, competition, decomposition, and photosynthesis. Predators are relatively big beasts that eat their prey from outside. Pathogens (disease-causing agents, such as viruses) are relatively small beasts that eat their prey from within. Although infectious disease can seem grisly and dreadful, under ordinary conditions it's every bit as natural as what lions do to wildebeests and zebras, or what owls do to mice.

But conditions aren't always ordinary.

Just as predators have their accustomed prey, their favored targets, so do pathogens. And just as a lion might occasionally depart from its normal behavior—to kill a cow instead of a wildebeest, a human instead of a zebra—so can a pathogen shift to a new target. Accidents happen. Aberrations occur. Circumstances change and, with them, exigencies and opportunities change too. When a pathogen leaps from some nonhuman animal into a person, and succeeds there in establishing itself as an infectious presence, sometimes causing illness or death, the result is a zoonosis.

It's a mildly technical term, zoonosis, unfamiliar to most people, but it helps clarify the biological complexities behind the ominous headlines about swine flu, bird flu, SARS, emerging diseases in general, and the threat of a global pandemic. It helps us comprehend why medical science and public health campaigns have been able to conquer some horrific diseases, such as smallpox and polio, but unable to conquer other horrific diseases, such as dengue and yellow fever. It says something essential about the origins of AIDS. It's a word of the future, destined for heavy use in the twenty-first century.

Ebola is a zoonosis. So is bubonic plague. So was the so-called Spanish influenza of 1918–1919, which had its ultimate source in a wild aquatic bird and, after passing through some combination of domesticated animals (a duck in southern China, a sow in Iowa?) emerged to kill as many as 50 million people before receding into obscurity. All of the human influenzas are zoonoses. So are monkeypox, bovine tuberculosis, Lyme disease, West Nile fever, Marburg virus disease, rabies, hantavirus pulmonary syndrome, anthrax, Lassa fever, Rift Valley fever, ocular larva migrans, scrub typhus, Bolivian hemorrhagic fever, Kyasanur forest disease, and a strange new affliction called Nipah encephalitis, which has killed pigs and pig farmers in Malaysia. Each of them reflects the action of a pathogen that can cross into people from other animals. AIDS is a disease of zoonotic origin caused by a virus that, having reached humans through just a few accidental events in western and central Africa, now passes human-to-human by the millions. This form of interspecies leap is common, not rare; about 60 percent of all human infectious diseases currently known either cross routinely or have recently crossed between other animals and us. Some of those—notably rabies—are familiar, widespread, and still horrendously lethal, killing humans by the thousands despite centuries of efforts at coping with their effects, concerted international attempts to eradicate or control them, and a pretty clear scientific understanding of how they work. Others are new and inexplicably sporadic, claiming a few victims (as Hendra does) or a few hundred (Ebola) in this place or that, and then disappearing for years.

Smallpox, to take one counterexample, is not a zoonosis. It's caused by variola virus, which under natural conditions infects only humans. (Laboratory conditions are another matter; the virus has sometimes been inflicted experimentally on nonhuman primates or other animals, usually for vaccine research.) That helps explain why a global campaign mounted by the World Health Organization (WHO) to eradicate smallpox was, as of 1980, successful. Smallpox could be eradicated because that virus, lacking ability to reside and reproduce anywhere but in a human body (or a carefully watched lab animal), couldn't hide. Likewise poliomyelitis, a viral disease that has afflicted humans for millennia but that (for counterintuitive reasons involving improved hygiene and delayed exposure of children to the virus) became a fearsome epidemic threat during the first half of the twentieth century, especially in Europe and North America. In the United States, the polio problem peaked in 1952 with an outbreak that killed more than three thousand victims, many of them children, and left twenty-one thousand at least partially paralyzed. Soon afterward, vaccines developed by Jonas Salk, Albert Sabin, and a virologist named Hilary Koprowski (about whose controversial career, more later) came into wide use, eventually eliminating poliomyelitis throughout most of the world. In 1988, WHO and several partner institutions launched an international effort toward eradication, which has succeeded so far in reducing polio case numbers by 99 percent. The Americas have been declared polio-free, as have Europe and Australia. Only five countries, as of latest reports in 2011, still seemed to have a minor, sputtering presence of polio: Nigeria, India, Pakistan, Afghanistan, and China. The eradication campaign for poliomyelitis, unlike other well-meant and expensive global health initiatives, may succeed. Why? Because vaccinating humans by the millions is inexpensive, easy, and permanently effective, and because apart from infecting humans, the poliovirus has nowhere to hide. It's not zoonotic.

Zoonotic pathogens can hide. That's what makes them so interesting, so complicated, and so problematic.

Monkeypox is a disease similar to smallpox, caused by a virus closely related to variola. It's a continuing threat to people in central and western Africa. Monkeypox differs from smallpox in one crucial way: the ability of its virus to infect nonhuman primates (hence the name) and some mammals of other sorts, including rats, mice, squirrels, rabbits, and American prairie dogs. Yellow fever, also infectious to both monkeys and humans, results from a virus that passes from victim to victim, and sometimes from monkey to human, in the bite of certain mosquitoes. This is a more complex situation. One result of the complexity is that yellow fever will probably continue to occur in humans—unless WHO kills every mosquito vector or every susceptible monkey in tropical Africa and South America. The Lyme disease agent, a type of bacterium, hides effectively in white-footed mice and other small mammals. These pathogens aren't *consciously* hiding, of course. They reside where they do and transmit as they do because those happenstance options have worked for them in the past, yielding opportunities for survival and reproduction. By the cold Darwinian logic of natural selection, evolution codifies happenstance into strategy.

The least conspicuous strategy of all is to lurk within what's called a reservoir host. A reservoir host (some scientists prefer "natural host") is a living organism that carries the pathogen, harbors it chronically, while suffering little or no illness. When a disease seems to disappear between outbreaks (again, as Hendra did after 1994), its causative agent has got to be *somewhere*, yes? Well, maybe it vanished entirely from planet Earth—but probably not. Maybe it died off throughout the region and will only reappear when the winds and the fates bring it back from elsewhere. Or maybe it's still lingering nearby, all around, within some reservoir host. A rodent? A bird? A butterfly? A bat? To reside undetected within a reservoir host is probably easiest wherever biological diversity is high and the ecosystem is relatively undisturbed. The converse is also true: Ecological disturbance causes diseases to emerge. Shake a tree, and things fall out.

Nearly all zoonotic diseases result from infection by one of six kinds of pathogen: viruses, bacteria, fungi, protists (a group of small, complex creatures such as amoebae, formerly but misleadingly known as protozoans), prions, and worms. Mad cow disease is caused by a prion, a weirdly folded protein molecule that triggers weird folding in other molecules, like Kurt Vonnegut's infectious form of water, ice-nine, in his great early novel *Cat's Cradle*. Sleeping sickness results from infection by a protist called *Trypanosoma brucei*, carried by tsetse flies among wild mammals, livestock, and people in sub-Saharan Africa. Anthrax is caused by a bacterium that can live dormant in soil for years and then, when scuffed out, infect humans by way of their grazing animals. Toxocariasis is a mild zoonosis caused by roundworms; you can get it from your dog. But fortunately, like your dog, you can be wormed.

Viruses are the most problematic. They evolve quickly, they are unaffected by antibiotics, they can be elusive, they can be versatile, they can inflict extremely high rates of fatality, and they are fiendishly simple, at least relative to other living or quasi-living creatures. Ebola, West Nile, Marburg, the SARS bug, monkeypox, rabies, Machupo, dengue, the yellow fever agent, Nipah, Hendra, Hantaan (the namesake of the hantaviruses, first identified in Korea), chikungunya, Junin, Borna, the influenzas, and the HIVs (HIV-1, which mainly accounts for the AIDS pandemic, and HIV-2, which is less widespread) are all viruses. The full list is much longer. There is a thing known by the vivid name "simian foamy virus" (SFV) that infects monkeys and humans in Asia, crossing between them by way of the venues (such as Buddhist and Hindu temples) where people and half-tame macaques come into close contact. Among the people visiting those temples, feeding handouts to those macaques, exposing themselves to SFV, are international tourists. Some carry away more than photos and memories. "Viruses have no locomotion," according to the eminent virologist Stephen S. Morse, "yet many of them have traveled around the world." They can't run, they can't walk, they can't swim, they can't crawl. They ride.

4

Isolating the Hendra bug had been a task for virologists, working in their high-security labs down at AAHL. "Isolating," in this sense of the word, means finding some of the virus and growing more. The isolate becomes a live, captive population of virus, potentially dangerous if any were to escape but useful for ongoing research. Virus particles are so tiny they can't be seen, except by electron microscopy, which involves killing them, so their presence during isolation must be detected indirectly. You start with a small bit of tissue, a drop of blood, or some other sample from an infected victim. Your hope is that it contains the virus. You add that inoculum, like a dash of yeast, to a culture of living cells in a nutrient medium. Then you incubate, you wait, you watch. Often, nothing happens. If you're lucky, something does. You know you've succeeded when the virus replicates abundantly and asserts itself sufficiently to cause visible damage to the cultured cells. Ideally it forms plaques, large holes in the culture, each hole representing a locus of virus-caused devastation. The process demands patience, experience, expensively exact bench tools, plus meticulous precautions against contamination (which can falsify results) or accidental release (which can infect you, endanger your co-workers, and maybe panic a town). Laboratory virologists are not generally knockabout people. You don't meet them in bars, waving their arms and bragging lustily about the perils of their métier. They tend to be focused, neat, and still, like nuclear engineers.

Discovering where a virus lives in the wild is work of a very different sort. It's an outdoor job that entails a somewhat less controllable level of risk, like trapping grizzly bears for relocation. Now, the people who look for wild viruses aren't rowdy and careless, no more so than the lab specialists; they can't afford to be. But they labor in a noisier, more cluttered, more unpredictable environment: the world. If there is reason to suspect that a certain new virus infecting humans is zoonotic (as most such viruses are), the search may lead into forests, swamps, crop fields, old buildings, sewers, caves, or the occasional horse paddock. The virus hunter is a field biologist, possibly with advanced training in human medicine, veterinary medicine, ecology, or some combination of those three—a person who finds fascination in questions that must be answered by catching and handling animals. That profile fits a lanky, soft-spoken man named Hume Field, midthirtyish at the time he became involved with Hendra.

Field grew up in the provincial towns of coastal Queensland, from Cairns to Rockhampton, a nature-loving kid who climbed trees, hiked in the bush, and spent school holidays on his uncle's dairy farm. His father was a police detective, which seems only too prefigurative of the son's later role as a viral sleuth. Young Field earned an undergraduate degree in veterinary science at the University of Queensland, in greater Brisbane, and volunteered at an animal refuge on the side, helping to rehabilitate injured wildlife. After graduation in 1976, he worked in a mixed veterinary practice in Brisbane for some years and then as a temporary fill-in (the Australians call it "doing locums") all over the state. During that time, he doctored a lot of horses. But he became increasingly aware that his deepest interest was wildlife, not livestock and pets, so in the early 1990s Field returned to the University of Queensland, this time for a doctorate in ecology.

He focused on wildlife conservation and, in due time, needed a dissertation project. Because feral cats (domestic cats gone wild on the landscape) cause considerable damage to native Australian wildlife, killing small marsupials and birds and acting as a source of disease, he undertook a study of feral cat populations and their impact. He was trapping cats, fitting them with radio collars to track how they lived, when the outbreak occurred at Vic Rail's stable. One of Field's doctoral mentors, a scientist who worked with the Department of Primary Industries, asked Field whether he would be interested in changing projects. The department needed someone to investigate the ecological side of this new disease. "So I forgot my feral cats," Field told me, when I visited him long afterward at the Animal Research Institute, a DPI facility near Brisbane, "and started off looking for wildlife reservoirs of Hendra virus."

He began his search by going back to the index case—the first equine victim, its history and locale. That was Drama Series, the pregnant mare, fallen ill in the paddock at Cannon Hill. The only clues he had were that this virus was a paramyxovirus and that another Queensland researcher had found a novel paramyxovirus in a rodent some years earlier. So Field established a trapping regime at the paddock, catching every small and medium-sized vertebrate he could—rodents, possums, bandicoots, reptiles, amphibians, birds, the odd feral cat—and drawing blood from each, with a particularly suspicious eye to the rodents. He sent the blood samples to the DPI lab to be screened for antibodies against Hendra.

Screening for antibodies is distinct from isolating virus, just as a footprint is distinct from a shoe. Antibodies are molecules manufactured by the immune system of a host in response to the presence of a biological intruder. They are custom-shaped to merge with and disable that particular virus, or bacterium, or other bug. Their specificity, and the fact that they remain in the bloodstream even after the intruder has been conquered, make them valuable as evidence of present or past infection. That's the evidence Hume Field was hoping to find. But the rodents from Cannon Hill had no antibodies to Hendra virus. Neither did anything else, leaving him to wonder why. Either he was looking in the wrong place, or in the right place in the wrong way, or at the wrong time. Bad timing might indeed be the problem, he thought. Drama Series had sickened in September, half a year had passed, and here he was searching in March, April, May. He suspected that "there could be some sort of seasonal presence of either the virus or the host" at the Cannon Hill paddock, and that maybe now it was out of season. Screening the cats, dogs, and rats around Rail's stable yielded no positives either.

Seasonal presence of the virus was one possibility. Coming and going on a shorter time scale was another. Bats, for instance, fed in large numbers at the Cannon Hill paddock by night but returned to their roosts, elsewhere, to sleep out the day. Peter Reid heard a Cannon Hill resident say that, during hours of darkness in the neighborhood, "flying foxes were as thick as the stars in the sky."

Reid had therefore suggested to AAHL that the bats should be looked at, but his suggestion evidently wasn't passed along. Hume Field and his co-workers on the reservoir hunt remained stumped until the following October, 1995, when an unfortunate event gave them a helpful new lead.

A young cane farmer named Mark Preston, who lived near the town of Mackay, about six hundred miles north of Brisbane, suffered a spate of seizures. His wife got him to a hospital. Preston's symptoms were especially alarming because they signaled a second health crisis for him in barely more than a year. Back in August 1994, he had endured a mysterious illness—headache, vomiting, stiff neck, then a provisional diagnosis of meningitis, cause unspecified—from which he had recovered. Or had seemingly recovered. Meningitis is a term applicable to any inflammation of the membranes that cover the brain and the spinal cord; it might be caused by a bacterium, a virus, even a reaction to a drug, and it might go away as inexplicably as it appeared. Preston continued to live a robust life on the farm with his wife Margaret, a veterinarian who based her practice there amid the sugar cane and the stud horses.

Did Mark Preston's seizures now indicate a recurrence of his indeterminate meningitis? Admitted to the hospital, he sunk into severe encephalitis—that is, brain inflammation, cause still unknown. Medication controlled his seizures but the doctors could watch storms of distress flickering on the electroencephalograph. "He remained deeply unconscious with persisting fever," according to a later medical report, "and died 25 days after admission."

Blood serum taken during Preston's final illness tested positive for antibodies to Hendra virus. So did his serum from a year earlier, which had been taken during the first episode, stored, and was now tested in retrospect. His immune system had been fighting the thing back then. Postmortem examination of his brain tissue, as well as other tests, confirmed the presence of Hendra. Evidently it had attacked once, subsided, lingered in latent form for a year, and then reared up and killed him. That was scary in a whole new way.

Where had he gotten it? Investigators, working backward to assemble the story, learned that in August 1994 two horses had died on the Preston farm. Mark Preston helped his wife care for them during their sudden, fatal illnesses and assisted her, at least marginally, when she performed the necropsies. Preserved tissue that Margaret Preston had drawn from both horses now also tested positive for Hendra. Despite her own exposure, though, Margaret Preston remained healthy—just as Peter Reid would remain healthy despite his exposure weeks later at Vic Rail's place. The good health of the two veterinarians raised the question of just how infectious this new virus might be. And the Preston case, at such distance from the first outbreak, caused the experts to wonder—to worry—about how far it might already have spread. Take the mileage from Hendra to Mackay as a radius of potential distribution, draw circles with that radius around the site of each outbreak, and you would circumscribe about 10 million people, nearly half the population of Australia.

How big was the problem? How widely was the virus dispersed? One group of researchers, led by an infectious diseases man named Joseph McCormack, based at the Brisbane hospital where Vic Rail had died, took a broad look. They screened serum from five thousand Queensland horses—every horse they could put a needle in, evidently—and from 298 humans, each of whom had had some level of contact with a Hendra case. None of the horses contained Hendra antibodies, nor did any of the humans. Those negatives, we can assume, brought sighs of relief from the health authorities and deepened the puzzled scowls on the faces of the scientists. "It seems," McCormack's group concluded, "that very close contact is required for transmission of infection to occur from horses to humans." But they were whistling in the dark. To say that "very close contact is required" didn't explain why Margaret Preston had outlived her husband. The reality was this: that very close contact, plus bad luck, plus maybe one or two other factors were necessary for a person to become infected, and nobody knew what the other factors were.

But the Mark Preston case gave Hume Field valuable clues—a second point on the map, a second point in time. Hendra virus in Mackay, August 1994; Hendra virus at the Cannon Hill paddock and in Rail's stable, September 1994. So Field went up to Mackay and repeated his method, trapping animals, drawing blood, sending serum to be tested for antibodies. And again he found nothing. He also drew samples from injured or otherwise debilitated wildlife of various types, creatures being nurtured in captivity until they could be released (if possible) back to the wild. The people who do such nurturing, a loose network of good-hearted amateurs, are known in Australian parlance as wildlife "carers." They tend to specialize by zoological category. There are kangaroo carers, bird carers, possum carers, and bat carers. Hume Field knew of them from his years of veterinary practice; he had virtually been one of them, during his student days at the animal refuge. Now he sampled some of the animals in their care.

But damn it: still no trace of Hendra.

In January 1996, with the search for a reservoir host at impasse, Field took part in a brainstorming session of agency officials and researchers, called by his supervisor at DPI. What were they doing wrong? How could they better target their efforts? Where would Hendra strike next? Queensland's racing industry stood in jeopardy of multi-million-dollar losses, and human lives were at risk. It was an urgent problem of governance and public relations, not just a medical riddle. One useful line of thought was explored at the meeting: biogeography. It seemed obvious that the reservoir host (or hosts), whatever type of animal it was (or they were), must exist both at Mackay and at Cannon Hill—exist there for at least part of each year, anyway, including August and September. This pointed toward animals that were either broadly distributed in Queensland or else *traveled* broadly across the state. The brainstormers (partly guided by genetic evidence suggesting there was no localization of distinct viral strains—that is, the *virus* was moving and mixing) leaned toward the second of those two possibilities: that the reservoir host was quite mobile, an animal capable of traveling hundreds of miles up and down the Queensland coast. That in turn directed suspicion at birds and . . . at bats.

Provisionally, Field and his colleagues dismissed the bird hypothesis, on two counts. First, they were unaware of any other paramyxovirus that spills over from birds into humans. Second, a mammalian reservoir simply seemed more likely, given that the virus infects humans and horses. Similarity of one kind of host animal to another is a significant indicator of the likelihood that a pathogen can make the leap. Bats are mammals, of course. And bats get around. Furthermore, bats famously harbor at least one fearful virus, rabies, although Australia at that time was considered rabies-free. (Many other bat-virus-human connections would be discovered soon afterward, including some in Australia; but at this time, 1996, the link seemed less obvious.) From the meeting, Field took away a new mandate: Look at bats.

Easily said. But catching bats on the wing, or even at their roosting sites, isn't so simple as trapping rodents or possums in a meadow. The most conspicuous and far-ranging bats native to Queensland are the so-called flying foxes, which belong to four different species within the genus *Pteropus,* each one a magnificent, fruit-eating megabat with a wingspan of three feet or more. Flying foxes customarily roost in mangroves, in paperbark swamps, or high in the limbs of rainforest trees. Special trapping tools and methods would be required. Short of gearing up immediately, Field returned first to the "carer" network. These people already had bats in captivity. At a facility in Rockhampton, up the coast toward Mackay, he found that the wounded animals under care included black flying foxes (*Pteropus alecto*). Bingo: Blood drawn from a black flying fox had antibodies to Hendra.

But one bingo moment wasn't sufficient for a scientist so fastidious as Hume Field. That datum proved that black flying foxes could be infected with Hendra, yes, but not necessarily that they were a reservoir—let alone *the* reservoir—from which horses became infected. He and his colleagues kept looking. Within a few weeks, Hendra antibodies turned up in all three other kinds, the grey-headed flying fox, the spectacled flying fox, and the little red flying fox. The DPI team also tested old samples from flying foxes, which had been archived for more than a dozen years. Again, they found telltale molecular tracks of Hendra. This showed that the bat population had been exposed to Hendra virus long before it struck Vic Rail's horses. And then, in September 1996, two years after the Rail outbreak, a pregnant grey-headed flying fox got herself snagged on a wire fence.

She miscarried twin fetuses and was euthanized. Not only did she test positive for antibodies; she also made possible the first isolation of Hendra virus from a bat. A sample of her uterine fluids yielded live virus, and that virus proved indistinguishable from Hendra as found in horses and humans. It was enough, even within scientific bounds of caution, to identify flying foxes as the "probable" reservoir hosts of Hendra.

The more that Field and his colleagues looked, the more evidence of Hendra they found. After the early bat surveys, about 15 percent of their flying foxes had tested positive for Hendra antibodies. This parameter—the percentage of sampled individuals showing some history of infection, either present or past—is called *seroprevalence*. It constitutes an estimate, based on finite sampling, of what the percentage throughout an entire population might be. As the team continued testing, the seroprevalence rose. At the end of two years, having sampled 1,043 flying foxes, Field and company reported Hendra seroprevalence at 47 percent. In plain words: Nearly half of the big bats flying around eastern Australia were present or former carriers. It almost seemed as though Hendra virus should have been raining down from the sky.

While the scientists published their findings in periodicals such as *Journal of General Virology* and *The Lancet*, some of this stuff got into the newspapers. One headline read: BAT VIRUS FEAR, RACING INDUSTRY ON ALERT. The crime-scene tape and the dismembered horses at Rail's place had been an irresistible starting point for television crews, and their interest continued. A few of those journalistic reports were accurate and sensible, but not all, and none were soothing. People became concerned. The identification of flying foxes as reservoir hosts, plus the high levels of seroprevalence within those bat populations, caused public-image trouble for a group of animals that had a legacy of such trouble already. Approval ratings for bats are never high. Now in Australia they went lower.

One eminent racehorse trainer gave me his view of the matter at a track in Hendra on a sunny Saturday during an interlude between races. *Hendra virus!* This man exploded at the mention. They shouldn't *allow* it! "They" were unspecified governmental authorities. They should get *rid* of the bats! Those bats *cause* the disease! They hang upside down and *shit* on themselves! (Can that be true? I wondered. Seemed biologically unlikely.) And they shit on *people*! It's *backwards*—let the people shit on *them*! What *good* are they? *Get rid* of them! Why doesn't that *happen*? Because the *sentimental Greenies* won't have it! he groused. We were in the Members Bar, a social sanctum for track professionals, to which I had been admitted in company with Peter Reid. The government should *protect* people! Should protect *vets*, like our friend Peter here! Harrumph, harrumph, and furthermore harrumph! et cetera. This trainer, a legendary figure in Australian racing, was a short, bantam-cocky octogenarian with gray hair combed back in dandy waves. I was a guest in his clubhouse and owed him a little respect—or anyway, a little slack. (In fairness, too, he was speaking not long after still another human victim, a Queensland veterinarian named Dr. Ben Cunneen, had died of Hendra contracted while treating sick horses. The mortal risk to horse people, and the economic risk to the entire Australian racing industry, were undeniably large.) When I showed genial interest in quoting this trainer on the record, he spoke more temperately but the gist was the same.

Among the "sentimental Greenies," he would have included bat carers. But even some of those softhearted activists, the carers, grew concerned as evidence piled up. They had two worries, uneasily counterbalanced: that the virus would make bats even more unpopular, leading to calls (like the trainer's) for bat extermination, and that they themselves might become infected in the course of their well-meaning work. The second was a new sort of anxiety. It must have caused some reexamination of commitment. They were *bat* lovers, after all, not *virus* lovers. Does a virus constitute *wildlife*? Not in most people's minds. Several such carers asked to be screened for antibodies, which opened doors for a broad survey, quickly organized and led by a young epidemiologist from the University of Queensland named Linda Selvey.

Selvey tapped into the wildlife-carer networks in southeastern Australia, eventually finding 128 bat carers willing or eager to be tested. She and her field team drew the blood and asked each participant to complete a questionnaire. The questionnaires revealed that many of these people had had prolonged and close contact with flying foxes—feeding them, handling them, not infrequently getting scratched or nipped. One carer had been bitten deeply on the hand by a Hendra-positive bat. The most unexpected finding of Selvey's survey was the percentage of those 128 carers who tested positive for antibodies: zero. Despite months and years of nurturing, despite scratches and bites and cuddling and drool and blood, not one person showed immunological evidence of having been infected with Hendra virus.

Selvey's report appeared in October 1996. She was a grad student at the time. Later she became head of the Communicable Diseases Branch of Queensland Health. Still later, as we sat over coffee in a noisy Brisbane café, I asked her: Who *are* these bat carers?

"I don't know how to describe them," Selvey answered. "People with a passion for animals, I guess." Both women and men? "Predominantly women," she said, speculating gently that women without kids might have more time and more desire for such surrogacy. Generally they do the caring in their own homes, equipped with a sizable, comfortable cage where the bats can roost when not being handled. It seemed mystifying to me that such intimate bat-human relations, combined with such a high level of bat seroprevalence, had yielded not a single case of human infection to be detected by Selvey's study. Not a single antibody-positive person out of 128 carers. What did that tell you, I asked her, about the nature of this virus?

"That it needed some sort of amplifier," she said. She was alluding to the horse.

5

Let's think about foot-and-mouth disease for a moment. Everybody has heard of it. Everybody has seen *Hud.* Most people aren't aware that, at least tenuously, it's a zoonosis. The virus that causes foot-and-mouth disease (FMD) belongs to the picornaviruses, the same group that includes poliovirus and some viruses similar to those that cause the common cold. But infection with FMD virus is a rare misfortune in humans, seldom causing worse than a rash on the hands, the feet, or the mouth lining. More frequently and consequentially, it afflicts cloven-hoofed domestic animals such as cattle, sheep, goats, and pigs. (Cloven-hoofed wildlife such as deer, elk, and antelope are also susceptible.) The main clinical signs are fever, lameness, and vesicles (little blisters) in the mouth, on the snout, on the feet. In a lactating female, the teats sometimes become blistered and then, as the blisters break, ulcerated. Bad for the mother, bad for the calf. Lethality from FMD is relatively low but the morbidity (incidence of the disease within a population) tends to be high, meaning that the disease is very contagious, making livestock ill, putting them off their feed, and causing losses of productivity that, in big-volume operations with narrow profit margins, are considered disastrous. Because of such losses, plus the swiftness of contagion, it's often treated as a terminal condition in commercial terms: Infected herds are slaughtered to prevent the virus from getting around. Nobody wants to buy stock that might be carriers, and the export trade drops to zilch. Cows, sheep, and pigs become worthless—less than worthless, an expensive liability. "Economically, it is the most important disease of animals in the world," according to one authority, who reports that "an FMD outbreak in the US could cost $27 billion in lost trade and markets." The virus spreads through direct contact, and in feces, and in milk, and is even capable of transmission by aerosol. It can travel from one farm to another on a humid breeze.

Impacts of FMD differ from one kind of animal to another.

Sheep tend to carry the infection without showing symptoms. Cattle suffer openly and pass the virus to one another by direct contact (say, muzzle to muzzle) or vertically (cow to calf) by suckling. Pigs are special: They excrete far more of the virus than other livestock, and over a longer period of time, broadcasting it prodigiously in their respiratory exhalations. They sneeze it, they chuff it, they oink it, they wheeze it and burp it and cough it into the air. One experimental study found that pig breath carried thirty times as much FMD virus as the breath of an infected cow or sheep, and that once airborne it could spread for miles. That's why pigs are considered an amplifier host of this virus.

An amplifier host is a creature in which a virus or other pathogen replicates—and from which it spews—with extraordinary abundance. Some aspect of the host's physiology, or its immune system, or its particular history of interaction with the bug, or who knows what, accounts for this especially hospitable role. The amplifier host becomes an intermediate link between a reservoir host and some other unfortunate animal, some other sort of victim—a victim requiring higher doses or closer contact before the infection can take hold. You can understand this in terms of thresholds. The amplifier host has a relatively low threshold for becoming infected, yet it produces a vast output of virus, vast enough to overcome the higher threshold in another animal.

Not every zoonotic pathogen requires an amplifier host for successful infection of humans, but some evidently do. Which ones, and how does the process work? The disease scientists are exploring those questions, among many others. Meanwhile, the concept is a hypothetical tool. Linda Selvey didn't mention the FMD paradigm when she used the word "amplifier" in our conversation about Hendra virus, but I knew what she meant.

Still . . . why horses? Why not kangaroos or wombats or koalas or potoroos? If the horse fills that amplifying role, one obvious fact deserves fresh attention: Horses aren't native to Australia. They are exotic, first brought there by European settlers barely more than two centuries ago. Hendra is probably an old virus, according to the runic evidence of its genome, as read by molecular evolutionists. Distantly diverged from its morbillivirus cousins, it may have abided unobtrusively in Australia for a very long time. Bats too are an ancient part of the native fauna; the fossil record in Queensland shows that small bats have been there for at least 55 million years, and flying foxes may have evolved in the region during the early Miocene, about 20 million years ago. Human presence is more recent, dating back only tens of millennia. More precisely, humans have inhabited Australia since the pioneering ancestors of Australian aboriginal peoples first made their way, island hopping daringly in simple wooden boats, from southeastern Asia by way of the South China Sea and the Lesser Sunda Islands to the northwestern coast of the island continent. That was at least forty thousand years ago, possibly much earlier. So three of the four principals in this complex interaction—flying foxes, Hendra virus, and people—have probably coexisted in Australia since the Pleistocene era. Horses arrived in January 1788.

It was a small change on the landscape, compared to all that would follow. Those earliest horses came aboard ships of the First Fleet, under command of Captain Arthur Phillip, who had sailed out from Britain to establish a convict colony in New South Wales. After five months of navigating the Atlantic, Phillip stopped at a Dutch settlement near the Cape of Good Hope to take on provisions and livestock before continuing eastward from Africa. He rounded Van Diemen's Land (now Tasmania) and sailed north along mainland Australia's east coast. Captain James Cook had already come and gone, "discovering" the place, but Phillip's group would be the first European settlers. At a spot near what is now Sydney, within the fine natural harbor there, his penal arks put ashore 736 convicts, 74 pigs, 29 sheep, 19 goats, 5 rabbits, and 9 horses. The horses included two stallions, four mares, and three foals. Until that day there was no record, either fossil or historic, of members of the genus *Equus* in Australia. Nor were there any oral traditions (none shared with the world so far, anyway) of Hendra virus outbreaks among aboriginal Australians.

As of January 27, 1788, then, the elements were almost certainly gathered in place—the virus, the reservoir hosts, the amplifier host, plus susceptible humans. And now another riddle presents itself. From the horses of Captain Arthur Phillip to the horses of Vic Rail is a gap of 206 years. Why did the virus wait so long to emerge? Or had it indeed emerged previously, maybe often, and never been recognized for what it is? How many past cases of Hendra, over two centuries or more, have been misdiagnosed as snakebite?

Answer from the scientists: We don't know but we're working on it.

6

Hendra virus in 1994 was just one thump in a drumbeat of bad news. The drumbeat has been sounding ever more loudly, more insistently, more rapidly over the past fifty years. When and where did it start, this modern era of emerging zoonotic diseases?

To choose one point is a little artificial, but a good candidate would be the emergence of Machupo virus among Bolivian villagers between 1959 and 1963. Machupo wasn't called Machupo at the start, of course, nor even recognized as a virus. Machupo is the name of a small river draining the northeastern Bolivian lowlands. The first recorded case of the disease came and went, almost unnoticed, as a bad but nonfatal fever afflicting a local farmer. This was during the wet season of 1959. More such illnesses, and worse, occurred in the same region over the following three years. Symptoms included fever and chills, nausea and vomiting, body aches, nosebleeds, and bleeding gums. It became known as El Tifu Negro (the Black Typhus, for the color of vomit and stool), and by late 1961 had struck 245 people, with a case fatality rate of 40 percent. It continued killing until the virus was isolated, its reservoir identified, and its dynamics of transmission understood well enough to be interrupted by preventive measures. Mouse trapping helped enormously. Most of the scientific work was done under difficult field conditions by a patched-together team of Americans and Bolivians, including an intense young scientist named Karl Johnson, pungently candid with his opinions, deeply enthralled by the dangerous beauty of viruses, who caught the disease himself and nearly died of it. This was before the Centers for Disease Control and Prevention (CDC) in Atlanta sent out well-equipped squads; Johnson and his colleagues invented their methods and tools as they went. Having struggled through his fever at a hospital in Panama, Karl Johnson would play a large and influential role in the longer saga of emerging pathogens.

If you assembled a short list of the highlights and high anxieties of that saga within recent decades, it could include not just Machupo but also Marburg (1967), Lassa (1969), Ebola (1976, with Karl Johnson again prominently involved), HIV-1 (inferred in 1981, first isolated in 1983), HIV-2 (1986), Sin Nombre (1993), Hendra (1994), avian flu (1997), Nipah (1998), West Nile (1999), SARS (2003), and the much feared but anticlimactic swine flu of 2009. That's a drama series more glutted and seething with virus than even Vic Rail's poor mare.

A person might construe this list as a sequence of dire but unrelated events—independent misfortunes that have happened to us, to humans, for one unfathomable reason and another. Seen that way, Machupo and the HIVs and SARS and the others are "acts of God" in the figurative (or literal) sense, grievous mishaps of a kind with earthquakes and volcanic eruptions and meteor impacts, which can be lamented and ameliorated but not avoided. That's a passive, almost stoical way of viewing them. It's also the wrong way.

Make no mistake, they are connected, these disease outbreaks coming one after another. And they are not simply *happening* to us; they represent the unintended results of things we are *doing*. They reflect the convergence of two forms of crisis on our planet. The first crisis is ecological, the second is medical. As the two intersect, their joint consequences appear as a pattern of weird and terrible new diseases, emerging from unexpected sources and raising deep concern, deep foreboding, among the scientists who study them. How do such diseases leap from nonhuman animals into people, and why do they seem to be leaping more frequently in recent years? To put the matter in its starkest form: Human-caused ecological pressures and disruptions are bringing animal pathogens ever more into contact with human populations, while human technology and behavior are spreading those pathogens ever more widely and quickly. There are three elements to the situation.

One: Mankind's activities are causing the disintegration (a word chosen carefully) of natural ecosystems at a cataclysmic rate. We all know the rough outlines of that problem. By way of logging, road building, slash-and-burn agriculture, hunting and eating of wild animals (when Africans do that we call it "bushmeat" and impute a negative onus, though in America it's merely "game"), clearing forest to create cattle pasture, mineral extraction, urban settlement, suburban sprawl, chemical pollution, nutrient runoff to the oceans, mining the oceans unsustainably for seafood, climate change, international marketing of the exported goods whose production requires any of the above, and other "civilizing" incursions upon natural landscape—by all such means, we are tearing ecosystems apart. This much isn't new. Humans have been practicing most of those activities, using simple tools, for a very long time. But now, with 7 billion people alive and modern technology in their hands, the cumulative impacts are becoming critical. Tropical forests aren't the only jeopardized ecosystems, but they're the richest and most intricately structured. Within such ecosystems live millions of kinds of creatures, most of them unknown to science, unclassified into a species, or else barely identified and poorly understood.

Two: Those millions of unknown creatures include viruses, bacteria, fungi, protists, and other organisms, many of which are parasitic. Students of virology now speak of the "virosphere," a vast realm of organisms that probably dwarfs every other group. Many viruses, for instance, inhabit the forests of Central Africa, each parasitic upon a kind of bacterium or animal or fungus or protist or plant, all embedded within ecological relationships that limit their abundance and their geographical range. Ebola and Marburg and Lassa and monkeypox and the precursors of the human immunodeficiency viruses represent just a minuscule sample of what's there, of the myriad other viruses as yet undiscovered, within hosts that in many cases are as yet undiscovered themselves. Viruses can only replicate inside the living cells of some other organism. Commonly they inhabit one kind of animal or plant, with whom their relations are intimate, ancient, and often (but not always) commensal. That is to say, dependent but benign. They don't live independently. They don't cause commotion. They might kill some monkeys or birds once in a while, but those carcasses are quickly absorbed by the forest. We humans seldom have occasion to notice.

Three: But now the disruption of natural ecosystems seems more and more to be unloosing such microbes into a wider world. When the trees fall and the native animals are slaughtered, the native germs fly like dust from a demolished warehouse. A parasitic microbe, thus jostled, evicted, deprived of its habitual host, has two options—to find a new host, a new *kind* of host . . . or to go extinct. It's not that they target us especially. It's that we are so obtrusively, abundantly available. "If you look at the world from the point of view of a hungry virus," the historian William H. McNeill has noted, "or even a bacterium—we offer a magnificent feeding ground with all our billions of human bodies, where, in the very recent past, there were only half as many people. In some 25 or 27 years, we have doubled in number. A marvelous target for any organism that can adapt itself to invading us." Viruses, especially those of a certain sort—those whose genomes consist of RNA rather than DNA, leaving them more prone to mutation—are highly and rapidly adaptive.

All these factors have yielded not just novel infections and dramatic little outbreaks but also new epidemics and pandemics, of which the most gruesome, catastrophic, and infamous is the one caused by a lineage of virus known to scientists as HIV-1 group M. That's the lineage of HIV (among twelve different sorts) that accounts for most of the worldwide AIDS pandemic. It has already killed 30 million humans since the disease was noticed three decades ago; roughly 34 million other humans are presently infected. Despite the breadth of its impact, most people are unaware of the fateful combination of circumstances that brought

HIV-1 group M out of one remote region of African forest, where its precursor lurked as a seemingly harmless infection of chimpanzees, into human history. Most people don't know that the real, full story of AIDS doesn't begin among American homosexuals in 1981, or in a few big African cities during the early 1960s, but at the headwaters of a jungle river called the Sangha, in southeastern Cameroon, half a century earlier. Even fewer people have caught wind of the startling discoveries that, just within the past several years, have added detail and transformative insight to that story. Those discoveries will get their place later ("The Chimp and the River") in this account. For now I'll just note that, even if the subject of zoonotic spillover addressed nothing but the happenstance of AIDS, it would obviously command serious attention. But as mentioned already, the subject addresses much more—other pandemics and catastrophic diseases of the past (plague, influenza), of the present (malaria, influenza), and of the future.

Diseases of the future, needless to say, are a matter of high concern to public health officials and scientists. There's no reason to assume that AIDS will stand unique, in our time, as the only such global disaster caused by a strange microbe emerging from some other animal. Some knowledgeable and gloomy prognosticators even speak of the Next Big One as an inevitability. (If you're a seismologist in California, the Next Big One is an earthquake that drops San Francisco into the sea, but in this realm of discourse it's a vastly lethal pandemic.) Will the Next Big One be caused by a virus? Will the Next Big One come out of a rainforest or a market in southern China? Will the Next Big One kill 30 or 40 million people? The concept by now is so codified, in fact, that we could think of it as the NBO. The chief difference between HIV-1 and the NBO may turn out to be that HIV-1 does its killing so slowly. Most other new viruses work fast.

I've been using the words "emergence" and "emerging" as though they are everyday language, and maybe they are. Among the experts, they're certainly common parlance. There's even a journal dedicated to the subject, *Emerging Infectious Diseases*, published monthly by the CDC. But a precise definition of "emergence" might be useful here. Several have been offered in the scientific literature. The one I prefer simply says that an emerging disease is "an infectious disease whose incidence is increasing following its first introduction into a new host population." The key words, of course, are "infectious," "increasing," and "new host." A re-emerging disease is one "whose incidence is increasing in an existing host population as a result of long-term changes in its underlying epidemiology." Tuberculosis is re-emerging as a severe problem, especially in Africa, as the TB bacterium exploits a new opportunity: infecting AIDS patients whose immune systems are disabled. Yellow fever re-emerges among humans wherever *Aedes aegypti* mosquitoes are allowed to resume carrying the virus between infected monkeys and uninfected people. Dengue, also dependent on mosquito bites for transmission and native monkeys as reservoirs, re-emerged in Southeast Asia after World War II due at least partly to increased urbanization, wider travel, lax wastewater management, inefficient mosquito control, and other factors.

Emergence and spillover are distinct concepts but interconnected. "Spillover" is the term used by disease ecologists (it has a different use for economists) to denote the moment when a pathogen passes from members of one species, as host, into members of another. It's a focused event. Hendra virus spilled over into Drama Series (from bats) and then into Vic Rail (from horses) in September 1994. Emergence is a process, a trend. AIDS emerged during the late twentieth century. (Or was it the *early* twentieth century? I'll return to that question.) Spillover leads to emergence when an alien bug, having infected some members of a new host species, thrives in that species and spreads among it. In this sense, the strict sense, Hendra hasn't emerged into the human population, not yet, not quite. It is merely a candidate.

Not all emerging diseases are zoonotic, but most are. From where else might a pathogen emerge, if not from another organism? Well, granted, some novel pathogens do seem to emerge from the environment itself, without need for shelter in a reservoir host. Case in point: The bacterium now called *Legionella pneumophila* emerged from the cooling tower of an air-conditioning system at a hotel in Philadelphia, in 1976, to create the first-known outbreak of Legionnaires' disease and kill thirty-four people. But that scenario is far less typical than the zoonotic one. Microbes that infect living creatures of one kind are the most likely candidates to infect living creatures of another kind. This has been borne out statistically by several review studies in recent years. One of them, published by two scientists at the University of Edinburgh in 2005, looked at 1,407 recognized species of human pathogen and found that zoonotic bugs account for 58 percent. Of the full total, 1,407, just 177 can be considered emerging or re-emerging. Three-fourths of those emergent pathogens are zoonotic. In plain words: Show me a strange new disease and, most likely, I can show you a zoonosis.

A parallel survey, from a team led by Kate E. Jones of the Zoological Society of London, appeared in the journal *Nature* in 2008. This group considered more than three hundred "events" of emerging infectious disease (EIDs, in their shorthand) that occurred between 1940 and 2004. They wondered about changing trends as well as discernible patterns. Although their list of events was independent of the Edinburgh researchers' list of pathogens, Jones and her colleagues found almost the same portion (60.3 percent) to be zoonotic. "Furthermore, 71.8% of these zoonotic EID events were caused by pathogens with a wildlife origin," as distinct from domestic animals. They cited Nipah in Malaysia and SARS in southern China. Further still, the increment of disease events associated with wildlife, as opposed to livestock, seems to be increasing over time. "Zoonoses from wildlife represent the most significant, growing threat to global health of all EIDs," these authors concluded. "Our findings highlight the critical need for health monitoring and identification of new, potentially zoonotic pathogens in wildlife populations, as a forecast measure for EIDs." That sounds reasonable: *Let's keep an eye on wild creatures. As we besiege them, as we corner them, as we exterminate them and eat them, we're getting their diseases.* It even sounds reassuringly doable. But to highlight the need for monitoring and forecasting is also to highlight the urgency of the problem and the discomfiting reality of how much remains unknown.

For instance: Why did Drama Series, the original mare, fall sick in that paddock when she did? Was it because she shaded herself beneath a fig tree and munched some grass besmeared with bat urine containing the virus? How did Drama Series pass her infection to the other horses at Vic Rail's stable? Why did Rail and Ray Unwin get infected but not the devoted veterinarian, Peter Reid? Why did Mark Preston get sick but not Margaret Preston? Why did the outbreaks at Hendra and Mackay occur in August and September 1994, close in time though distant geographically? Why did all those bat carers remain uninfected, despite their months and years of fondling flying foxes?

These local riddles about Hendra are just small forms of big questions that scientists such as Kate Jones and her team, and the Edinburgh researchers, and Hume Field, and many others around the world are asking. Why do strange new diseases emerge when they do, where they do, as they do, and not elsewhere, other ways, at other times? Is it happening more now than in the past? If so, how are we bringing these afflictions upon ourselves? Can we reverse or mitigate the trends before we're hit with another devastating pandemic? Can we do that without inflicting fearful punishment on all those other kinds of infected animals with which we share the planet? The dynamics are complicated, the possibilities are many, and while science does its work slowly, we all want a fast response to the biggest question: What sort of nasty bug, with what unforeseen origins and what inexorable impacts, will emerge next?

7

During one trip to Australia I stopped in Cairns, a balmy resort city about a thousand miles north of Brisbane, for a conversation with a young veterinarian there. I can't recall how I located her, because she was wary of publicity and didn't want her name used in print. But she agreed to talk to me about her experience with Hendra. Although her experience had been brief, it included two roles: as doctor, as patient. At that time she was the only known Hendra survivor in Australia, besides the stable hand Ray Unwin, who had also suffered infection with the virus and lived. We spoke in the office of a small veterinary clinic where she worked.

She was an ebullient woman, twenty-six years old, with pale blue eyes and hennaed brunette hair pulled back in a tight bun. She wore silver earrings, shorts, and a red short-sleeve shirt with a clinic logo. While an earnest border collie kept us company, nudging my hands for affection as I tried to write notes, the vet described a night in October 2004 when she had gone out to attend to a suffering horse. The owners were concerned because this particular animal, a ten-year-old gelding, seemed "off color."

The horse was named Brownie, she remembered that. He lived on a family farm down at Little Mulgrave, about twenty miles south of Cairns. She remembered it all, in fact, a night full of vivid impressions. Brownie was a quarterhorse-thoroughbred cross. Not a racer, no, a pet. The family included a teenage daughter; Brownie was her special favorite. At eight o'clock that evening the horse seemed normal, but then something went suddenly wrong. The family suspected colic, bad stomach—maybe he had eaten some toxic greens. Around eleven o'clock they phoned for help and got the young vet, who was on call that night. She jumped in her car, and when she arrived Brownie was in desperate condition, panting heavily, feverish, down on the ground. "I found the horse had a heart rate through the roof, temperature through the roof," she told me, "and there was bloody red froth coming out the nose." Giving him a quick look, taking his vitals, she came close to the horse and, when he snorted, "I got quite a degree of bloody sort of red mucousy froth on my arms." The teenage girl and her mother were already smeared with blood from having tried to comfort Brownie. Now he could barely lift his head. The vet, a fiercely caring professional, told them the horse was dying. Knowing her duty, she said: "I want to euthanize it." She ran back to her car, got the euthanasia solution and tools, but by the time she returned Brownie was dead.

In his last agonal gasps, he had brought up more bubbly red froth through his nostrils and mouth.

Were you wearing gloves? I asked.

No. The protocol was to use gloves for a postmortem, but not for live animals. Then the one situation led so swiftly to the other. "I was wearing exactly what I'm wearing now. A pair of shoes, short socks, blue shorts, and short sleeves."

A surgical mask?

No, no mask. "You know, in the laboratory all those precautions are easy to take. When it's twelve at night and it's pouring down rain and you're out in the middle of the dark and you're operating via the car headlights with a hysterical family in the background, it's not always easy to take the proper precautions. And the other thing was, that I just didn't know." Didn't know what she was confronting in Brownie's case, she meant. "I wasn't really thinking infectious disease." She was defensive on these points because there had been second-guessing of her procedures, an investigation, questions about negligence. She had been exonerated—in fact, she made her own complaint about having not been properly warned—but it couldn't have been helpful to her career, and that's presumably why she wanted anonymity. She had a story to tell, yet she also wished to put it behind her.

In the minutes after Brownie's death, she had changed into boots, long pants, and shoulder-length gloves and begun the postmortem exam. The owners were keen to know whether Brownie had eaten some sort of poisonous grass that might threaten their other horses too. The vet sliced opened Brownie's abdomen and found his guts looking normal. No sign of twisted bowel or other blockage that might cause colic. In the process, "I got a couple of splashes of abdominal fluid on my leg." You can't do a postmortem on a horse without getting smeared, she explained. Next she looked into the chest, by way of a modest incision between the fourth and fifth ribs. If it wasn't colic it was probably cardiac trouble, she suspected, and saw that hunch immediately confirmed. "The heart was massively enlarged. The lungs were wet and full of bloody fluid and there was just fluid right through the chest cavity. So he died of congestive heart failure. That was all I could conclude. I couldn't conclude whether it was infectious or not." She offered to take samples for lab testing, but the owners declined. Enough information, enough expense, too bad about Brownie, and they would simply bury the carcass with a bulldozer.

Were there bats around this property? I asked.

"There's bats everywhere." Everywhere throughout northern Queensland, she meant, not just at Little Mulgrave. "If you walk out the back here, you'll see a couple hundred bats." The entire area of Cairns and its environs: warm climate, plenty of fruit trees, plenty of fruit-eating bats. But the subsequent inquiry turned up nothing about Brownie's situation that seemed to have closely exposed him to bats. "They couldn't say, other than random chance, why this particular horse got infected." Buried beneath ten feet of dirt, having left behind no samples of blood or tissue, he couldn't even be labeled "infected" except by later inference.

Immediately after the postmortem, the vet washed her hands and arms thoroughly, wiped down her legs, and then went home to take a Betadine shower. She keeps a large supply of Betadine, the professional antiseptic of choice, for such occasions. She gave herself a good surgical scrub and got into bed, after a hard but not too unusual night. It wasn't until nine or ten days later that she started feeling headachy and sick. Her doctor suspected the flu, or a cold, or maybe tonsillitis. "I get tonsillitis a lot," she said. He gave her some antibiotics and sent her home.

She missed a week's work, languishing with symptoms that felt like influenza or bronchitis: mild pneumonia, sore throat, a bad cough, muscle weakness, fatigue. At one point a senior colleague asked whether she had considered the possibility that the dead horse had infected her with Hendra virus. The young vet, trained in Melbourne (way down in temperate Australia) before she moved up to tropical Cairns, had scarcely heard Hendra virus mentioned in veterinary school. It was too obscure, too new, and not an issue in the Melbourne area. Only two of the four kinds of reservoir bats range that far south, and evidently they had yet to cause concern. Now she went to the hospital for a blood test, then another, and yes indeed: She had antibodies to Hendra virus. By that time she was back on her feet, working again. She had been infected and shaken it off.

When I met her, more than a year later, she was feeling fine, apart from a little weariness and more than a little anxiety. She knew well that the case of Mark Preston—his infection during a horse postmortem, his recovery, his interlude of good health, then his relapse—cautioned against complacency that the virus had left her forever. State health officials were tracking her case; if the headaches returned, if she felt dizzy or suffered a seizure, if her nerves tingled, if she started coughing or sneezing, they wanted to know it. "I still go and see the infectious disease control specialists," she said. "I get weighed by the Department of Primary Industries on a regular basis." From blood tests they charted her antibody levels, which continued to fluctuate peculiarly down and up. Lately the numbers were back up. Did that portend a relapse, or did it just reflect her robust acquired immunity?

The scariest part, she told me, was the uncertainty. "It's the fact that this disease has been around for so little that they can't tell me whether there's going to be any future health risk." How would she be in seven years, ten years? How high was the chance of recrudescence? Mark Preston died suddenly after a year. Ray Unwin said his health was still "crook." The young vet in Cairns only wanted to know, in her own case, the same thing we all want to know: What next?

II

THIRTEEN GORILLAS

8

Not many months after the events at Vic Rail's stables, another spillover occurred, this one in Central Africa. Along the upper Ivindo River in northeastern Gabon, near the border with the Republic of the Congo, lies a small village called Mayibout 2, a sort of satellite settlement just a mile upriver from the village of Mayibout. In early February 1996, eighteen people in Mayibout 2 became suddenly sick after they participated in the butchering and eating of a chimpanzee.

Their symptoms included fever, headache, vomiting, bloodshot eyes, bleeding from the gums, hiccupping, muscle pain, sore throat, and bloody diarrhea. All eighteen were evacuated downriver to a hospital in the district capital, a town called Makokou, by decision of the village chief. It's less than fifty miles as the crow flies from Mayibout 2 to Makokou, but by pirogue on the sinuous Ivindo, a journey of seven hours. The boat wound back and forth between walls of forest along the banks. Four of the evacuees were moribund when they arrived and dead within two days. The four bodies, returned to Mayibout 2, were buried according to traditional ceremonial practice, with no special precautions against the transmission of whatever had killed them. A fifth victim escaped from the hospital, straggled back to the village, and died there. Secondary cases soon broke out among people infected while caring for the first victims—their loved ones or friends—or in handling the dead bodies. Eventually thirty-one people got sick, of whom twenty-one died: a case fatality rate of almost 68 percent.

Those facts and numbers were collected by a team of medical researchers, some Gabonese, some French, who reached Mayibout 2 during the outbreak. Among them was an energetic Frenchman named Eric M. Leroy, a Paris-trained veterinarian and virologist then based at the Centre International de Recherches Médicales de Franceville (CIRMF), in Franceville, a modest city in southeastern Gabon. Leroy and his colleagues found evidence of Ebola virus in samples from some patients, and they deduced that the butchered chimpanzee had been infected with Ebola. "The chimpanzee seems to have been the index case for infecting 18 primary human cases," they wrote. Their investigation also turned up the fact that the chimp hadn't been killed by village hunters; it had been found dead in the forest and scavenged.

Four years later, I sat at a campfire near the upper Ivindo River with a dozen local men who were working as forest crew for a long overland trek. These men, most of them from villages in northeastern Gabon, had been walking for weeks before I joined them on the march. Their job involved carrying heavy bags through the jungle and building a simple camp each night for the biologist, one Mike Fay, whose obsessive sense of mission drove the whole enterprise forward. Fay is an unusual man, even by the standards of tropical field biologists: physically tough, obdurate, free-spirited, smart, and fiercely committed to conservation. His enterprise, which he labeled the Megatransect, was a two-thousand-mile biological survey, on foot, through the wildest remaining forest areas of Central Africa. He took data every step of the way, recording elephant dung piles and leopard tracks and chimpanzee sightings and botanical identifications, tiny notations by the thousands, all going into his waterproof yellow notebooks in scratchy left-handed print, while the crewmen strung out behind him toted his computers, his satellite phone, his special instruments and extra batteries, as well as tents and food and medical supplies enough for both him and themselves.

Fay had already been walking for 290 days by the time he reached this part of northeastern Gabon. He had crossed the Republic of the Congo with a field crew of forest-tough Congo men, mostly Bambendjellés (one ethnic group of the short-statured peoples sometimes termed Pygmies), but those fellows had been disallowed entry at the Gabonese border. So Fay had been forced to raise a new team in Gabon. He recruited them largely from a cluster of gold-mining camps along the upper Ivindo River. The hard, stumbling work he demanded, cutting trail, schlepping bags, was evidently preferable to digging for gold in equatorial mud. One man served as cook as well as porter, stirring up massive amounts of rice or *fufu* (a starchy staple made from manioc flour, like an edible wallpaper paste) at each evening's campfire, and adorning it with some sort of indeterminate brown sauce. The ingredients for that variously included tomato sauce, dried fish, canned sardines, peanut butter, freeze-dried beef, and *pili-pili* (hot pepper), all deemed mutually compatible and combined at the whim of the chef. No one complained. Everyone was always hungry. The only thing worse than a big portion of such stuff, at the end of an exhausting day of stumbling through the jungle, was a small portion. My role amid this gang, on assignment for *National Geographic*, was to walk in Fay's footsteps and produce a series of stories describing the work and the journey. I would accompany him for ten days here, two weeks there, and then escape back to the United States, let my feet heal (we wore river sandals), and write an installment.

Each time I rejoined Fay and his team, there was a different logistical arrangement for our rendezvous, depending on the remoteness of his location and the urgency of his need to be resupplied. He never diverted from the zigzag line of his march. It was up to me to get to him. Sometimes I went in by bush plane and motorized dugout, along with Fay's trusted logistics man and quartermaster, a Japanese ecologist named Tomo Nishihara. Tomo and I would pile ourselves into the canoe amid whatever stuff he was bringing for the next leg of Fay's trek: fresh bags of fufu and rice and dried fish, crates of sardines, oil and peanut butter and pili-pili and double-A batteries. But even a dugout canoe couldn't always reach the spot where Fay and his crew, famished and bedraggled, would be waiting. On this occasion, with the trekkers crossing a big forest block called Minkébé, Tomo and I roared out of the sky in a Bell 412 helicopter, a massive 13-seater, chartered expensively from the Gabonese army. The forest canopy, elsewhere thick and unbroken, was punctuated here by several large granite gumdrops that rose above everything, hundreds of feet high, like El Capitan standing out of a green ground fog. Atop one of those inselbergs was the landing zone to which Fay had directed us. It was forty miles due west of Mayibout 2.

That day had been a relatively easy one for the crew—no swamps crossed, no thickets of skin-slicing vegetation, no charging elephants provoked by Fay's desire to take video at close range. They were bivouacked, awaiting the helicopter. Now the supplies had arrived—including even some beer! This allowed for a relaxed, genial atmosphere around the campfire. Quickly I learned that two of the crewmen, Thony M'Both and Sophiano Etouck, had roots in Mayibout 2. They were present when Ebola virus struck the village.

Thony, an extrovert, slim in build and far more voluble than the other fellow, was willing to talk about it. He spoke in French while Sophiano, a shy man with a body-builder's physique, an earnest scowl, a goatee, and a nervous stutter, sat silent. Sophiano, by Thony's account, had watched his brother and most of his brother's family die.

Having just met these two men, I couldn't decently press for more information that evening. Two days later we set off on the next leg of Fay's hike, across the Minkébé forest, heading southward away from the inselbergs. We got busy and distracted with the physical challenges of foot travel through trackless jungle terrain, and were exhausted (especially they, working harder than I) by nightfall. Halfway along, though, after a week of difficult walking, common miseries, and shared meals, Thony loosened enough to tell me more. His memories agreed generally with the report of the CIRMF team from Franceville, apart from small differences on some numbers and details. But his perspective was more personal.

Thony called it *l'épidémie*, the epidemic. This happened in 1996, yes, he said, around the same time some French soldiers came up to Mayibout 2 in a Zodiac raft and camped near the village. It was unclear whether the soldiers had a serious purpose—rebuilding an old airstrip?—or were just there to amuse themselves. They shot off their rifles. Maybe, Thony guessed, they also possessed some sort of chemical weaponry. He mentioned these details because he thought they might have relevance to the epidemic. One day some boys from the village went out hunting with their dogs. The intended prey was porcupines. Instead of porcupines they got a chimp—not killed by the dogs, no. A chimp found dead. They brought it back. The chimp was rotten, Thony said, its stomach putrid and swollen. Never mind, people were glad and eager for meat. They butchered the chimp and ate it. Then quickly, within two days, everyone who had touched the meat started getting sick.

They vomited; they suffered diarrhea. Some went downriver by motorboat to the hospital at Makokou. But there wasn't enough fuel to transport every sick person. Too many victims, not enough boat. Eleven people died at Makokou. Another eighteen died in the village. The special doctors quickly came up from Franceville, yes, Thony said, wearing their white suits and helmets, but they didn't save anyone. Sophiano lost six family members. One of those, one of his nieces—he was holding her as she died. Yet Sophiano himself never got sick. No, nor did I, said Thony. The cause of the illnesses was a matter of uncertainty and dark rumor. Thony suspected that the French soldiers, with their chemical weapons, had killed the chimpanzee and carelessly left its meat to poison the villagers. Anyway, his fellow survivors had learned their lesson. To this day, he said, no one in Mayibout 2 eats chimpanzee.

I asked about the boys who went hunting. Them, all the boys, they died, Thony said. The dogs did not die. Had he ever before seen such a disease, such an epidemic? "*No,*" Thony answered. "*C'etait le premier fois.*" Never.

How did they cook the chimp? I pried. In a normal African sauce, Thony said, as though that were a silly question. I imagined chimpanzee hocks in a peanutty gravy, with pili-pili, ladled over fufu.

Apart from the chimpanzee stew, one other stark detail lingered

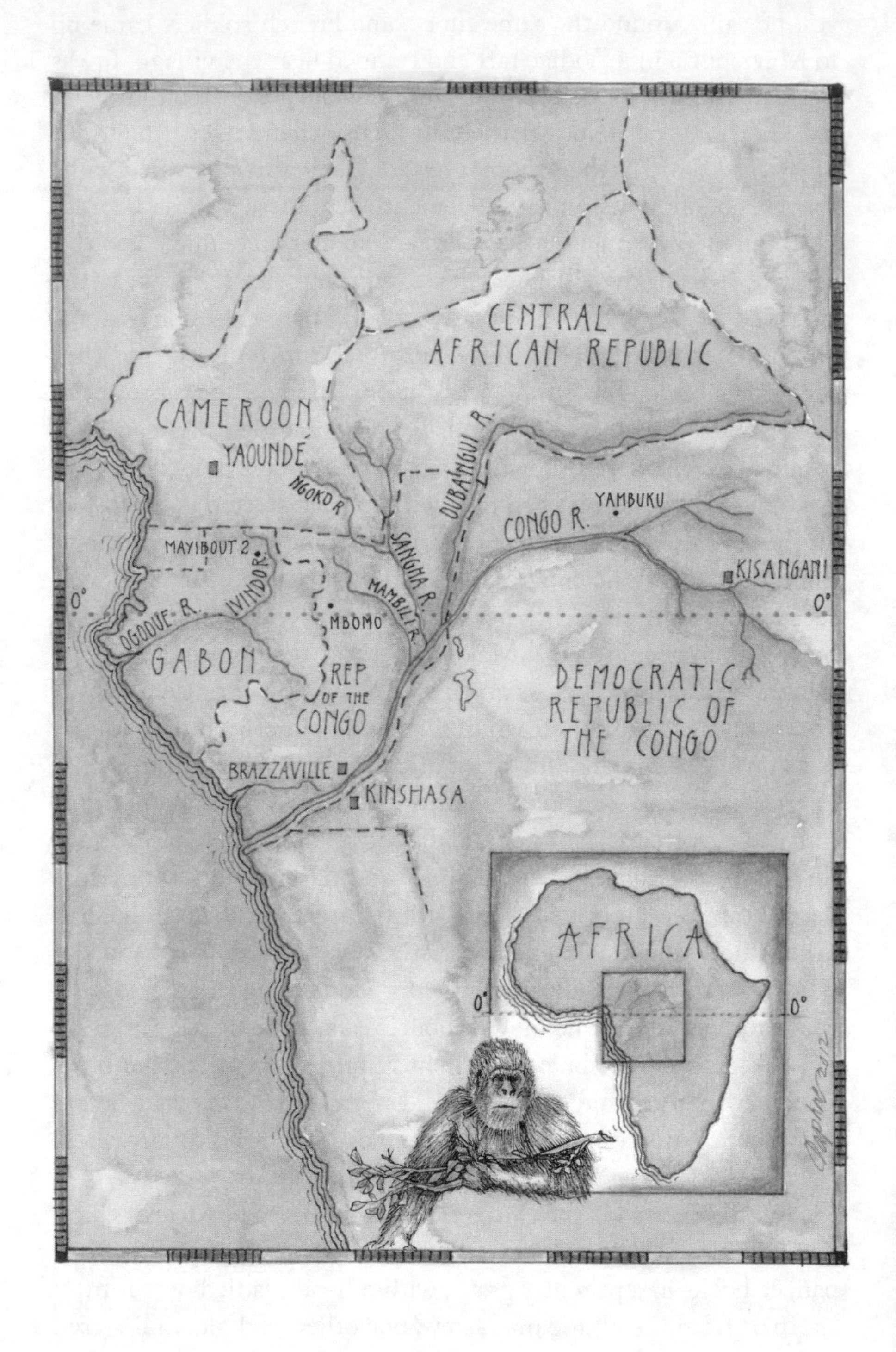
CENTRAL AFRICAN REPUBLIC
CAMEROON
YAOUNDÉ
NGOKO R.
DUBANGUI R.
SANGHA R.
YAMBUKU
CONGO R.
MAYIBOUT 2
IVINDO R.
KISANGANI
0°
0°
MAMBILI R.
MBOMO
OGOOUÉ R.
GABON
REP OF THE CONGO
DEMOCRATIC REPUBLIC OF THE CONGO
BRAZZAVILLE
KINSHASA
AFRICA
0°
0°

in my mind. It was something Thony had mentioned during our earlier conversation. Amid the chaos and horror in the village, Thony told me, he and Sophiano had seen something bizarre: a pile of thirteen gorillas, all dead, lying nearby in the forest.

Thirteen gorillas? I hadn't asked about dead wildlife. This was volunteered information. Of course, anecdotal testimony tends to be shimmery, inexact, sometimes utterly false, even when it comes from eyewitnesses. To say *thirteen dead gorillas* might actually mean a dozen, or fifteen, or simply lots—too many for an anguished brain to count. People were dying. Memories blur. To say *I saw them* might mean exactly that or possibly less. *My friend saw them, he's a close friend, I trust him like I trust my eyes.* Or maybe: *I heard about it on pretty good authority.* Thony's testimony, it seemed to me, belonged in the first epistemological category: reliable if not necessarily precise. I believed he saw these dead gorillas, roughly thirteen, in a group if not a pile; he may even have counted them. The image of thirteen gorilla carcasses strewn on the leaf litter was lurid but plausible. Subsequent evidence indicates that gorillas are highly susceptible to Ebola.

Scientific data are another matter, very different from anecdotal testimony. Scientific data don't shimmer with poetic hyperbole and ambivalence. They are particulate, quantifiable, firm. Fastidiously gathered, rigorously sorted, they can reveal emergent meanings. That's why Mike Fay was walking across Central Africa with his yellow notebooks: to search for big patterns that might emerge from masses of small data.

The next day we continued on through the forest. We were still more than a week from the nearest road. It was excellent gorilla habitat, well structured, rich with their favorite plant foods, and nearly untouched by humans: no trails, no camps, no evidence of hunters. It should have been full of gorillas. And once, in the recent past, it had been: A census of Gabon's ape populations done two decades earlier, by a pair of scientists from CIRMF, had yielded an estimate of 4,171 gorillas within the Minkébé forest bloc. Nevertheless, during our weeks of bushwhacking, we saw none. There was an odd absence of gorillas and gorilla sign—so odd that, for

Fay, it seemed dramatic. This was exactly the sort of pattern, positive or negative, that his methodology was meant to illuminate. During the course of his entire Megatransect he recorded in his notebook every gorilla nest he saw, every mound of gorilla dung, every stem fed upon by gorilla teeth—as well as elephant dung, leopard tracks, and similar traces of other animals. At the end of our Minkébé leg, he subtotaled his data. This took him hours, holed away in his tent, collating the latest harvest of observations on his laptop. Then he emerged.

Over the past fourteen days, Fay informed me, we had stepped across 997 piles of elephant dung and not one dollop from a gorilla. We had passed amid millions of stems of big herbaceous plants, including some kinds (belonging to the family *Marantaceae*) with nutritious pith that gorillas devour like celery; but not one of those stems, so far as he'd noticed, had shown gorilla tooth marks. We had heard zero gorilla chest-beat displays, seen zero gorilla nests. It was like the curious incident of the dog in the nighttime—a silent pooch, speaking eloquently to Sherlock Holmes with negative evidence that something wasn't right. Minkébé's gorillas, once abundant, had disappeared. The inescapable inference was that something had killed them off.

9

The spillover at Mayibout 2 was no isolated event. It was part of a pattern of disease outbreaks across Central Africa—a pattern of which the meaning is still a matter of puzzlement and debate. The disease in question, once known as Ebola hemorrhagic fever, is now simply called Ebola virus disease. The pattern stretches from 1976 (the first recorded emergence of Ebola virus) to the present, and from one side of the continent (Côte d'Ivoire) to the other (Sudan and Uganda). The four major lineages of virus that showed themselves during those emergence events are collectively known as ebolaviruses. On a smaller scale, within Gabon alone, there has been a tight clustering of Ebola incidents: three in less than two years, and all three rather closely localized in space. Mayibout 2 was the middle episode of that cluster.

An earlier outbreak began during December 1994 in the gold-mining camps on the upper Ivindo, the same area from which Mike Fay later recruited his Gabonese crew. These camps lie about twenty-five miles upstream from Mayibout 2. At least thirty-two people got sick, showing the usual range of symptoms (fever, headache, vomiting, diarrhea, and some bleeding) that suggest Ebola virus disease. The source was hard to pinpoint, though one patient told of having killed a chimpanzee that had wandered into his camp and acted strangely. Maybe that animal was infected, inadvertently bringing the contagion to hungry humans. According to another account, the first case was a man who had come across a dead gorilla, took parts of it back to his camp, and shared. He died and so did others who touched the meat. Around the same time came some reports of chimps, as well as gorillas, seen dead in the forest. More generally, the miners (and their families—these camps were essentially villages) by their very presence, their needs for food, shelter, and fuel, had caused disturbance to the forest canopy and the creatures that lived in it.

From the mining camps, those victims in 1994 were transferred downriver (as they would be again from Mayibout 2) to Makokou General Hospital. Then arose a wave of secondary cases, focused around the hospital or in villages nearby. In one of those villages was a *nganga*, a traditional healer, whose house may have been a point of transmission between a certain mining-camp victim of the outbreak, seeking folk medicine, and an unlucky local person visiting the healer about something less dire than Ebola. Possibly the virus was passed by the healer's own hands. Anyway, by the time this sequence ended, forty-nine cases had been diagnosed, with twenty-nine deaths, for a case fatality rate of almost 60 percent.

A year later came the outbreak at Mayibout 2, second in the series. Eight months after that, the CIRMF scientists and others responded to a third outbreak, this one near the town of Booué in central Gabon.

The Booué situation had probably begun three months earlier, in July 1996, with the death of a hunter at a timber camp known as SHM, about forty miles north of Booué. In retrospect, this hunter's fatal symptoms were recognized as matching Ebola virus disease, though his case hadn't triggered alarm at the time. Another hunter died mysteriously in the same logging camp six weeks later. Then a third. What sort of meat were they supplying to the camp? Probably a wide range of wild species, including monkeys, duikers, bush pigs, porcupines, possibly even (despite legal restrictions) apes. And again there were reports of chimpanzees seen dead in the forest—fallen dead, that is, not shot dead. The three early human cases seem to have been independent of one another, as though each hunter contracted the virus from the wild. Then the third hunter broadened the problem, making himself a transmitter as well as a victim.

He was hospitalized briefly at Booué but left that facility, eluded medical authorities, went to a nearby village, and sought help there from another nganga. Despite the healer's ministrations the hunter died—and then so did the nganga and the nganga's nephew. A cascade had begun. During October and into succeeding months there was a wider incidence of cases in and around Booué, suggesting more person-to-person transmission. Several patients were transferred to hospitals in Libreville, Gabon's capital, and died there. A Gabonese doctor, having performed a procedure on one of those patients, fell sick himself and, showing little confidence in his own country's health care, flew to Johannesburg for treatment. That doctor seems to have survived, but a South African nurse who looked after him sickened and died. Ebola virus had thereby emerged from Central Africa into the continent at large. The eventual tally from this third outbreak, encompassing Booué, Libreville, and Johannesburg, was sixty cases, of which forty-five were fatal. Rate of lethality? For that one, you can do the math in your head.

Amid this welter of cases and details, a few common factors stand out: forest disruption at the site of the outbreak, dead apes as well as dead humans, secondary cases linked to hospital exposure or traditional healers, and a high case fatality rate, ranging from 60 to 75 percent. Sixty percent is extremely high for any infectious disease (except rabies); it's probably higher, for instance, than mortalities from bubonic plague in medieval France at the worst moments of the Black Death.

In the years since 1996, other outbreaks of Ebola virus disease have struck both people and gorillas within the region surrounding Mayibout 2. One area hit hard lies along the Mambili River, just over the Gabon border in northwestern Congo, another zone of dense forest encompassing several villages, a national park, and a recently created reserve known as the Lossi Gorilla Sanctuary. Mike Fay and I had walked through that area also, in March 2000, just four months before my rendezvous with him at the Minkébé inselbergs. In stark contrast to the emptiness of Minkébé, gorillas had been abundant within the Mambili drainage when we saw it. But two years later, in 2002, a team of researchers at Lossi began finding gorilla carcasses, some of which tested positive for antibodies to Ebola virus. (A positive test for antibodies is less compelling evidence than a find of live virus, but still suggestive.) Within a few months, 90 percent of the individual gorillas they had been studying (130 of 143 animals) had vanished. How many had simply run away? How many were dead? Extrapolating rather loosely from confirmed deaths and disappearances to overall toll throughout their study area, the researchers published a paper in *Science* under the forceful (but overconfident) headline: EBOLA OUTBREAK KILLED 5000 GORILLAS.

10

In 2006 I returned to the Mambili River, this time with a team led by William B. (Billy) Karesh, then director of the Field Veterinary Program for the Wildlife Conservation Society (WCS) of New York and now filling a similar role at the EcoHealth Alliance. Billy Karesh is a veterinarian and an authority on zoonoses. He's a peripatetic field man, raised in Charleston, South Carolina, nourished on Marlin Perkins, whose usual working uniform is a blue scrub shirt, a gimme cap, and a beard. An empiricist by disposition, he speaks quietly, barely moving his mouth, and avoids categorical pronouncements as though they might hurt his teeth. Often he wears a sly smile, suggesting amusement at the wonders of the world and the varied spectacle of human folly. But there was nothing amusing about his mission to the Mambili. He had come to shoot gorillas—not with bullets but with tranquilizer darts. He meant to draw blood samples and test them for antibodies to Ebola virus.

Our destination was a site known as the Moba Bai complex, a group of natural clearings near the east bank of the upper Mambili, not far from the Lossi sanctuary. A *bai* in Francophone Africa is a marshy meadow, often featuring a salt lick, and surrounded by forest like a secret garden. In addition to Moba Bai, the namesake of this complex, there were three or four others nearby. Gorillas (and other wildlife) frequent such bais, which are waterlogged and sunny, because of the sodium-rich sedges and asters that grow beneath the open sky. We arrived at Moba, coming upstream on the Mambili, in an overloaded dugout pushed by a 40-horse outboard.

The boat carried eleven of us and a formidable pile of gear. We had a gas-powered refrigerator, two liquid-nitrogen freezer tanks (for preserving samples), carefully packaged syringes and needles and vials and instruments, medical gloves, hazmat suits, tents and tarps, rice, fufu, canned tuna, canned peas, several boxes of bad red wine, numerous bottles of water, a couple of folding tables, and seven stackable white plastic chairs. With these tools and luxurious provisions we established a field camp across the river from

Moba. Our team included an expert tracker named Prosper Balo, plus other wildlife veterinarians, other forest guides, and a cook. Prosper had worked at Lossi before and during the outbreak. With his guidance, we would prowl the complex of bais, all full of succulent vegetation and previously famed for the dozens of gorillas that came there daily to eat and relax.

Billy Karesh had visited the same area twice previously, before Ebola struck, seeking baseline data on gorilla health. During a 1999 trip, he had seen sixty-two gorillas here in one day. In 2000 he returned to try darting a few. "Every day," he told me, "every bai had at least a family group." Not wanting to be too disruptive, he had tranquilized only four animals, weighed them and examined them for obvious diseases (such as yaws, a bacterial skin infection), and taken blood samples. All four apes had tested negative for Ebola antibodies. This time things were different. He wanted blood serum from survivors of the 2002 die-off. So we began, with high expectations. Days passed. As far as we could see, there *were* no survivors.

Precious few, anyway—not enough to make gorilla-darting (which is always a parlous enterprise, with some risk for both the darter and the dartees) productive of data. Our stakeout at Moba lasted more than a week. Early each morning we crossed the river, walked quietly to one bai or another, concealed ourselves in thick vegetation along the edge, and waited patiently for gorillas to appear. None did. Often we hunkered in the rain. When it was sunny, I read a thick book or dozed on the ground. Karesh stood ready with his air rifle, the darts loaded full of tilletamine and zolazepam, drugs of choice for tranquilizing a gorilla. Or else we hiked through the forest, following closely behind Prosper Balo as he searched for gorilla sign and found none.

On the morning of day 2, along a swampy trail to the bais, we saw leopard tracks, elephant tracks, buffalo tracks, and chimpanzee sign, but no evidence of gorillas. On day 3, with still no gorillas, Karesh said: "I think they're dead. Ebola went through here." He figured that only a lucky few, uninfected by the disease or else resistant enough to survive it, remained. Then again, he said, "those are the ones we're interested in," because they, if any, might carry antibodies. On day 4, separating from the rest of us, Karesh and Balo managed to locate a single, distraught male gorilla from the sound of his chest beats and screaming barks, and to crawl within ten yards of him in the thick underbrush. Suddenly the animal stood, only his head visible, in front of them. "I could have killed him," Karesh said later. "Pitted him." Drilled him between the eyes, that is, but not immobilized him with a safe shot to the flank. So Karesh held his fire. The gorilla let out another bark and ran off.

My notes from day 6 include the entry: "Nada nada nary gorilla nada." On our final chance, day 7, Balo and Karesh tracked another couple of animals for hours through the boggy forest without getting so much as a good glimpse. Gorillas had become desperately scarce, round about Moba Bai, and the stragglers were fearfully shy. Meanwhile the rain continued, the tents grew muddy, and the river rose.

When we weren't in the forest, I spent time in camp talking with Karesh and the three Africa-based WCS veterinarians on his team. One was Alain Ondzie, a lanky and bashful Congolese, trained in Cuba, fluent in Spanish as well as French and several Central African languages, with a likable tendency to dip his head and giggle joyously whenever he was teased or amused. Ondzie's main job was to respond to reports of dead chimps or gorillas anywhere in the country, getting to the site as quickly as possible and taking tissue samples to be tested for Ebola virus. He described to me the tools and procedures for such a task, with the carcass invariably putrefied by the time he reached it and the presumption (until otherwise proven) that it might be seething with Ebola. His working costume was a hazmat suit with a vented hood, rubber boots, a splash apron, and three pairs of gloves, duct taped at the wrists. Making the first incision for sampling was dicey because the carcass might have become bloated with gas; it could explode. In any case the dead ape was usually covered with scavenging insects—ants, tiny flies, even bees. Ondzie told of one occasion when three bees from a carcass ran up his arms, under his hood flap, down across his bare body, and commenced to sting him as he worked on the samples. Can Ebola virus travel on the stinger of a bee? No one knows.

Does this work frighten you? I asked Ondzie. Not anymore, he said. Why do you do it? I asked. Why do you love it? (as he clearly did). "*Ca, c'est une bonne question,*" he said, with the characteristic bob and giggle. Then he added, more soberly: Because it allows me to apply what I've learned, and to keep learning, and it might save some lives.

Another member of the team was Patricia (Trish) Reed, who had come out to Africa as a biologist fifteen years earlier, studied Lassa fever and then AIDS, hired on with CIRMF in Franceville, gotten some field experience in Ethiopia, and then collected a DVM from the veterinary school at Tufts University in Boston. She was back at CIRMF, doing research on a monkey virus, when the WCS field vet working out here was killed in a plane crash coming into a backcountry Gabonese airstrip. Karesh hired Reed as the dead woman's replacement.

The scope of her work, Reed told me, encompassed a range of infectious diseases that threaten gorilla health, of which Ebola is only the most exotic. The others were largely human diseases of more conventional flavor, to which gorillas are susceptible because of their close genetic similarity to us: TB, poliomyelitis, measles, pneumonia, chickenpox, et cetera. Gorillas can be exposed to such infections wherever unhealthy people are walking, coughing, sneezing, and crapping in the forest. Any such spillover in the reverse direction—from humans to a nonhuman species—is known as an *anthroponosis.* The famous mountain gorillas, for instance, have been threatened by anthroponotic infections such as measles, carried by ecotourists who come to dote upon them. (Mountain gorillas constitute a severely endangered subspecies of the eastern gorilla, confined to the steep hillsides of the Virunga Volcanoes in Rwanda and neighboring lands. The western gorilla of Central African forests, a purely lowland species, is more numerous but far from secure.) Combined with destruction of their habitat by logging operations, and the hunting of them for bushmeat to be consumed locally or sold into markets, infectious diseases could push western gorillas from their current levels of relative abundance (maybe a hundred thousand in total) to a situation in which small, isolated populations survive tenuously, like the mountain gorillas, or go locally extinct.

But the forests of Central Africa are still relatively vast, compared to the small Virunga hillsides that harbor mountain gorillas; and the western gorilla doesn't face many ecotourists in its uncomfortable, nearly impenetrable home terrain. So measles and TB aren't the worst of its problems. "I would say that, without a doubt, Ebola is the biggest threat" to the western species, Reed said.

What makes Ebola virus among gorillas so difficult, she explained, is not just its ferocity but also the lack of data. "We don't know if it was here before. We don't know if they survive it. But we need to know how it passes through groups. We need to know *where* it *is*." And the question of *where* has two dimensions. How broadly is Ebola virus distributed across Central Africa? Within what reservoir species does it lurk?

On the eighth day, we packed up, reloaded the boats, and departed downstream on the Mambili, taking away no blood samples to add to the body of data. Our mission had been thwarted by the very factor that made it relevant: a notable absence of gorillas. Here was the curious incident of the dog in the nighttime again. Billy Karesh had seen one gorilla at close range but been unable to dart it, and had tracked two others with the help of Prosper Balo's keen eye for spoor. The rest, the many dozens that formerly frequented these bais, had either dispersed to parts unknown or they were . . . dead? Anyway, once gorillas had been abundant hereabouts, and now they were gone.

The virus seemed to be gone too. But we knew it was only hiding.

11

Hiding where? For almost four decades, the identity of Ebola's reservoir host has been one of the darkest little mysteries in the world of infectious disease. That mystery, along with efforts to solve it, dates back to the first recognized emergence of Ebola virus disease, in 1976.

Two outbreaks occurred in Africa that year, independently but almost simultaneously: one in the north of Zaire (now the Democratic Republic of the Congo), one in southwestern Sudan (in an area that today lies within the Republic of South Sudan), the two separated by three hundred miles. Although the Sudan situation began slightly earlier, the Zaire event is the more famous, partly because a small waterway there, the Ebola River, eventually gave its name to the virus.

The focal point of the Zaire outbreak was a small Catholic mission hospital in a village called Yambuku, within the district known as Bumba Zone. In mid-September, a Zairian doctor there reported two dozen cases of a dramatic new illness—not the usual malarial fevers but something more grisly, more red, characterized by bloody vomiting, nosebleeds, and bloody diarrhea. Fourteen of the patients had died, as of the doctor's cabled alert to authorities in Kinshasa, Zaire's capital, and others were in danger. By the start of October, Yambuku Mission Hospital had closed, for the grim reason that most of its staff members were dead. An international response team of scientists and physicians converged on the area several weeks later, under the direction of the Zairian Minister of Health, to do a crash study of the unknown disease and give advice toward controlling it. This group, consisting of members from France, Belgium, Canada, Zaire, South Africa, and the United States, including nine from the CDC in Atlanta, became known as the International Commission. Their leader was Karl Johnson, the same American physician and virologist who had worked on Machupo virus in Bolivia back in 1963, barely surviving his own infection with that disease. Thirteen years later, still intense, still dedicated, and not noticeably mellowed by near-death experience or professional ascent, he was head of the Special Pathogens Branch at the CDC.

Johnson had helped solve the Machupo crisis by his attention to the ecological dimension—that is, where did the virus live when it wasn't killing Bolivian villagers? The reservoir question had been tractable, in that case, and the answer had quickly been found: A native mouse was carrying Machupo into human households and granaries. Trapping out the mouse effectively ended the outbreak. Now, amid the desperate and befuddling days of October and November 1976, in northern Zaire, confronting a different invisible and unidentified killer, as the death toll rose into the hundreds, Johnson and his fellow researchers found time to wonder about Ebola virus as he had wondered about Machupo virus: Where did this thing come from?

By then they knew that the pathogen *was* a virus. That knowledge derived from isolations performed quickly on clinical samples shipped to overseas laboratories, including the CDC. (Johnson, before flying to Zaire, had led the CDC isolation effort himself.) They knew that this virus was similar to Marburg virus, another lethal agent, identified nine years before; the electron micrographs showed that it was equally filamentous and twisty, like an anguished tapeworm. But the lab tests also revealed Ebola virus as distinct enough from Marburg virus to constitute something new. Eventually these two wormy viruses, Ebola and Marburg, would be classified within a new family, *Filoviridae*: the filoviruses.

Johnson's group knew also that the new agent, Ebola virus, must reside in some living animal—something other than humans—where it could exist less disruptively and maintain a continuous presence. But the question of its reservoir was less urgent than other concerns, such as how to break the chain of person-to-person transmission, how to keep patients alive, how to end the outbreak. "Only limited ecological investigations were made," the team reported later, and the results of those investigations were all negative. No sign of Ebola virus appeared anywhere except in humans. But the negative data are interesting in retrospect, at least as a record of where these early researchers looked. They pureed 818 bedbugs collected from Ebola-affected villages, finding no evidence of the virus in any. They considered mosquitoes. Nothing. They drew blood from ten pigs and one cow—all of which proved Ebola-free. They caught 123 rodents, including 69 mice, 30 rats, and 8 squirrels, not one of which was a viral carrier. They read the entrails of six monkeys, two duikers, and seven bats. These animals also were clean.

The International Commission members were chastened by what they had seen. "No more dramatic or potentially explosive epidemic of a new acute viral disease has occurred in the world in the past 30 years," their report warned. The case fatality rate of 88 percent, they noted, was higher than any on record, apart from the rate for rabies (almost 100 percent among patients not treated before they show symptoms). The Commission made six urgent recommendations to Zairian officialdom, among which were health measures at the local level and nationwide surveillance. But the identification of Ebola's reservoir wasn't mentioned. That was a scientific matter, slightly more abstract than the action items offered to President Mobutu's government. It would have to wait.

The wait has continued.

Three years after Yambuku, Karl Johnson and several other members of the Commission were still wondering about the reservoir question. They decided to try again. Lacking funds to mount an expedition devoted solely to finding Ebola's hideout, they hitched their effort to an ongoing research program on monkeypox in Zaire, coordinated by the World Health Organization. Monkeypox is a severe affliction, though not so dramatic as Ebola virus disease, and also caused by a virus that lurks in a reservoir host or hosts, at that time still unidentified. So it seemed natural and economical to do a combined search, using two sets of analytical tools to screen a single harvest of specimens. Again the field team collected animals from villages and surrounding forest in Bumba Zone, as well as in other areas of northern Zaire and southeastern Cameroon. This time their trapping and hunting efforts, plus the bounties they paid for creatures delivered alive by villagers, yielded more than fifteen hundred animals representing 117 species. There were monkeys, rats, mice, bats, mongooses, squirrels, pangolins, shrews, porcupines, duikers, birds, tortoises, and snakes. Blood was taken from each, and then snips of liver, kidney, and spleen. All these samples, deep-frozen in individual vials, were shipped back to the CDC for analysis. Could any live virus be grown from the sampled tissues? Could any Ebola antibodies be detected in blood serum? The bottom line, reported with candor by Johnson and coauthors in the pages of *The Journal of Infectious Diseases*, was negatory: "No evidence of Ebola virus infection was found."

One factor making the hunt for Ebola's reservoir especially difficult, especially hard to focus, is the transitory nature of the disease within human populations. It disappears entirely for years at a time. This is a mercy for public health but a constraint for science. Viral ecologists can look for Ebola anywhere, in any creature of any species, in any African forest, but those are big haystacks and the viral needle is small. The most promising search targets, in space and in time, are wherever and whenever people are dying of Ebola virus disease. And for a long interlude, no one *was* dying of that illness—no one whose death came to the attention of medical authorities, anyway.

After the Yambuku outbreak of 1976, and then two small episodes in Zaire and Sudan between 1977 and 1979, ebolaviruses barely showed themselves anywhere in Africa for fifteen years. There may have been some scattered cases during the early 1980s, retrospectively suspected, but there was no confirmed outbreak that evoked emergency response; and in each of those minor instances the chain of infection seemed to have burned itself out. Burning out is a concept with special relevance to such highly lethal and moderately contagious pathogens. It means that a few people died, a few more got infected, a fraction of those also died but others recovered, and the pathogen didn't continue to propagate. The incident expired on its own before shock troops from WHO, the CDC, and other centers of expertise had to be mustered. Then, after an interval, it returned—with the outbreaks at Mayibout 2 and elsewhere in Gabon, and even more alarmingly at a place called Kikwit.

Kikwit, in Zaire, lay about three hundred miles east of Kinshasa. It differed from Yambuku, and Mayibout 2, and the timber camp outside Booué in one crucial way: It was a city of two hundred thousand people. It contained several hospitals. It was connected to the wider world in a way that those other outbreak sites weren't. But like them it was surrounded by forest.

The first identified case in the Kikwit outbreak was a forty-two-year-old man who worked in or near that forest and probably, to some small extent, disturbed it. He farmed several patches of cleared land, planting corn and cassava, and made charcoal from timber, all at a spot five miles southeast of the city. How did he get his wood supply, how did he clear daylight for his gardens? Presumably by cutting trees. This man fell sick on January 6, 1995, and died of a hemorrhagic fever a week later.

By that time he had directly infected at least three members of his family, all fatally, and launched the infection into his wider circle of social contacts, ten more of whom died within coming weeks. Some of those contacts evidently carried the virus into the city's maternity hospital, where it infected a laboratory technician, and from there into Kikwit General Hospital. The technician, while being treated at Kikwit General, infected several doctors and nurses who did surgery on him (suspecting a gut perforation related to typhoid, they cut open his abdomen), as well as two Italian nuns who helped with his care. The technician died, the nuns died, and local officials hypothesized that this was epidemic dysentery, a misdiagnosis that allowed the virus to spread further among patients and staff at other hospitals in the Kikwit region.

Not everyone accepted the dysentery hypothesis. One doctor at the Ministry of Health thought it looked instead like a viral hemorrhagic fever, which suggested Ebola. That good guess was confirmed quickly from blood specimens received by the CDC, in Atlanta, on May 9. Yes indeed: It was Ebola virus. By the end of the outbreak, in August, 245 people had died, including 60 hospital staff members. Performing abdominal surgery on Ebola patients, when you thought they were suffering from something else (such as gastrointestinal bleeding from ulcers), was risky work.

Meanwhile, another international team came out to search for the reservoir, converging on Kikwit in early June. This group consisted of people from the CDC, from a Zairian university, from the US Army Medical Research Institute of Infectious Diseases (USAMRIID, formerly a bioweapons lab but now committed to disease research and biodefense) in Maryland, and one fellow from the Danish Pest Infestation Laboratory, who presumably knew a lot about rodents. They began work at the site to which the spillover seemed traceable—that is, at the charcoal pit and crop fields of the unlucky forty-two-year-old man, the first victim, southeast of the city. From that site and others, over the following three months, they trapped and netted thousands of animals. Mostly those were small mammals and birds, plus a few reptiles and amphibians. All the traps were set within forest or savanna areas outside the city limits. Within Kikwit itself, the team netted bats at a Sacred Heart mission. They killed each captured animal, drew blood, and dissected out the spleen (in some cases other organs too, such as a liver or a kidney), which went into frozen storage. They also took blood from some dogs, cows, and pet monkeys. The total yield included 3,066 blood samples and 2,730 spleens, all shipped back to the CDC for analysis. The blood samples, after having been irradiated to kill any virus, were tested for Ebola virus antibodies, using the best available molecular method of the time. The spleens were transferred to a biosafety level 4 (BSL-4) laboratory, a new sort of facility since Karl Johnson's early work (and of which he was one of the pioneering designers), with multiple seals, negative air pressure, elaborate filters, and lab personnel working in spacesuits—a containment zone in which Ebola virus could be handled without risk (theoretically) of accidental release. No one knew whether any of these Zairian spleens contained the virus but each had to be treated as though it did. From the spleen material, minced finely and added to cell cultures, the lab people tried to grow the virus.

None grew. The cell cultures remained blithely unspotted by viral blooms. And the antibody tests yielded no positive hits either. Once again, Ebola virus had spilled over, caused havoc, and then disappeared without showing itself anywhere but in the sick and dying human victims. It was Zorro, it was the Swamp Fox, it was Jack the Ripper—dangerous, invisible, gone.

This three-month, big-team effort at Kikwit shouldn't be considered a total failure; even negative results from a well-designed study tend to reduce the universe of possibilities. But it was another hard try ending in frustration. Maybe the Kikwit team had gotten there too late, five months after the charcoal maker fell ill. Maybe the shift from wet season to dry season had caused the reservoir, whatever it is, to migrate or hide or decrease in abundance. Maybe the virus itself had declined to a minimal population, a tenuous remnant, undetectable even within its reservoir during the off season. The Kikwit team couldn't say. The most notable aspect of their eventual report, apart from its long list of animals that *didn't* contain Ebola virus, was its clear statement of three key assumptions that had guided their search.

First, they suspected (based on earlier studies) that the reservoir is a mammal. Second, they noted that Ebola virus disease outbreaks in Africa had always been linked to forests. (Even the urban epidemic at Kikwit had begun with that charcoal-maker out amid the woods.) It seemed safe to assume, therefore, that the reservoir is a forest creature. Third, they noted also that Ebola outbreaks had been sporadic in time—with years sometimes passing between one episode and the next. Those gaps implied that infection of humans from the reservoir is a rare occurrence. Rarity of spillover in turn suggested two possibilities: that either the reservoir itself is a rare animal or that it's an animal only rarely in contact with people.

Beyond that, the Kikwit team couldn't say. They published their paper in 1999 (among a whole series of reports on Ebola, in a special supplement of the *Journal of Infectious Diseases*), authoritatively documenting a negative conclusion. After twenty-three years, the reservoir still hadn't been found.

12

"We need to know where it is," Trish Reed had said. She was alluding to the two unanswered questions about Ebola virus and its location in space. The first question is ecological: In what living creature does it hide? That's the matter of reservoir. The second question is geographical: What's its distribution across the African landscape? The second may be impossible to answer until the reservoir is identified and *its* distribution traced. In the meantime, the only data reflecting Ebola virus's whereabouts are the plotted points of human outbreaks on a map.

Let's glance across that map. In 1976 Ebola virus made its debut, as I've mentioned, with the dramatic events in Yambuku and the slightly smaller crisis in southwestern Sudan, which was nonetheless large enough to account for 151 deaths. The Sudanese outbreak centered at a town near the Zairian border, five hundred miles northeast of Yambuku. It began among employees of a cotton factory, in the rafters of which roosted bats and on the floor of which skittered rats. The lethality was lower than in Zaire, "only" 53 percent, and laboratory analysis revealed that the Sudanese virus was genetically distinct enough from the virus in Zaire to be classified in a separate species. That species later became known, in careful taxonomic parlance, as *Sudan ebolavirus*. The official common name is simply Sudan virus, which lacks the frisson of the word "Ebola" but nonetheless denotes a dangerous, blazing killer. The version Karl Johnson found at Yambuku, originally and still called Ebola virus, belongs to the species *Zaire ebolavirus*. This may seem confusing, but the accurate, up-to-date labels are important for keeping things straight. Eventually there would be five recognized species.

In 1977 a young girl died of hemorrhagic fever at a mission hospital in a village called Tandala, in northwestern Zaire. A blood sample taken after her death and sent unrefrigerated to the CDC yielded Ebola virus, not in cell cultures but only after inoculating live guinea pigs and then finding the virus replicating in their organs. (These were early days still in the modern field campaign against emerging viruses, and methodology was being extemporized to compensate for difficulties, such as keeping live virus frozen under rough field conditions in the tropics.) Karl Johnson again was part of the laboratory team; this seemed a logical extension of his work on the first outbreak, just a year earlier and two hundred miles east. But the nine-year-old girl, dead in Tandala, was an isolated case. Her family and friends remained uninfected. There was not even a hypothesis as to how she got sick. The later published report, with Johnson again as coauthor, only noted suggestively, in describing the girl's native area: "Contact with nature is intimate, with villages located in clearings of the dense rain forest or along the rivers of the savannah." Had she touched a dead chimpanzee, breathed rodent urine in a dusty shed, or pressed her lips to the wrong forest flower?

Two years later Sudan virus also resurfaced, infecting a worker at the same cotton factory where it had originally emerged. The worker was hospitalized, upon which he infected another patient, and by the time the virus finished ricocheting through that hospital, twenty-two people were dead. The case fatality rate was again high (65 percent), though lower than for Ebola virus. Sudan virus seemed to be not quite so lethal.

Then another decade passed before filoviruses made their next appearance, in another shape, in an unexpected place: Reston, Virginia.

You know about this if you've read *The Hot Zone*, Richard Preston's account of a 1989 outbreak of an Ebola-like virus among captive Asian monkeys at a lab-animal quarantine facility in suburban Reston, just across the Potomac from Washington, DC. Filovirus experts express mixed opinions about Preston's book, but there's no question that it did more than any journal article or newspaper story to make ebolaviruses infamous and terrifying to the general public. It also led to "a shower of funding," one expert told me, for virologists "who before didn't see a dime for their work on these exotic agents!" If this virus could massacre primates in their cages within a nondescript building in a Virginia office park, couldn't it go anywhere and kill anyone?

The facility in question was known as the Reston Primate Quarantine Unit and owned by a company called Hazelton Research Products, which was a division of Corning. The unfortunate monkeys were long-tailed macaques (*Macaca fascicularis*), an animal much used in medical research. They had arrived in an air shipment from the Philippines. Evidently they brought their filovirus with them, a lethal stowaway, like smallpox virus making its way through the crew of a sailing ship. Two macaques were dead on arrival, which wasn't unusual after such a stressful journey; but over the following weeks, within the building, many more died, which *was* unusual. Eventually the situation triggered alarm and the infective agent was recognized as an ebolavirus—*some* sort of ebolavirus, as yet unspecified. A team from USAMRIID came in, like a SWAT team in hazmat suits, to kill all the remaining macaques. Then they sealed the Reston Primate Quarantine Unit and sterilized it with formaldehyde gas. You can read Preston for the chilling details. There was great anxiety among the experts because this ebolavirus seemed to be traveling from monkey to monkey in airborne droplets; a leak from the building might therefore send it wafting out into Washington-area traffic. Was it lethal to humans as well as to macaques? Several staff members of the Quarantine Unit eventually tested positive for antibodies but—sigh of relief—those people showed no symptoms. Laboratory work revealed that the virus was similar to Ebola virus yet, like Sudan virus, different enough to be classified in a new species. It came to be known as Reston virus.

Notwithstanding that name, Reston virus seems to be native to the Philippines, not to suburban Virginia. Subsequent investigation of monkey-export houses near Manila, on the island of Luzon, found a sizable die-off of animals, most of which were infected with Reston virus, plus twelve people with antibodies to the virus. But none of the dozen Filipinos got sick. So the good news about Reston virus, derived both from the 1989 US scare and from retrospective research on Luzon, is that it doesn't seem to cause illness in humans, only in monkeys. The bad news is that no one understands why.

Apart from Reston virus, ebolaviruses in the wild remain an African phenomenon. But the next emergence, in November 1992, added yet another point to the African map. Chimpanzees began dying at a forest refuge in Côte d'Ivoire, West Africa. The refuge, Taï National Park, lying near Côte d'Ivoire's border with Liberia, encompassed one of the last remaining areas of primary rainforest in that part of Africa. It harbored a rich diversity of animals, including several thousand chimpanzees.

One community of those chimps had been followed and studied for thirteen years by a Swiss biologist named Christophe Boesch. During the 1992 episode, Boesch and his colleagues noticed a sudden drop in the population—some chimps died, others disappeared—but the scientists didn't detect a cause. Then, in late 1994, eight more carcasses turned up over a short period of time, and again other animals went missing. Two of the chimp bodies, only moderately decayed, were cut open and examined by researchers at Taï. One of those proved to be teeming with an Ebola-like agent, though that wasn't apparent at the time. During the necropsy, a thirty-four-year-old female Swiss graduate student, wearing gloves but no gown, no mask, became infected. Infected how? There wasn't any obvious moment of fateful exposure, no slip of the scalpel, no needlestick mishap. Probably she got chimp blood onto a broken patch of skin—a small scratch?—or caught a gentle splash of droplets in the face. Eight days later, the woman started shivering.

She took a dose of malaria medicine. That didn't help. She was moved to a clinic in Abidjan, Côte d'Ivoire's capital, and there treated again for malaria. Her fever continued. On day 5 came vomiting and diarrhea, plus a rash that spread over her whole body. On day 7 she was carried aboard an ambulance jet and flown to Switzerland. Now she *was* wearing a mask, and so were the doctor and the nurse in attendance. But no one knew what ailed her. Dengue fever, hantavirus infection, and typhoid were being considered, and malaria still hadn't been ruled out. (Ebola wasn't at the top of the list because it had never been seen in Côte d'Ivoire.) In Switzerland, hospitalized within a double-door isolation room with negative air pressure, she was tested for a whole menu of nasty things, including

Lassa fever, Crimean-Congo hemorrhagic fever, chikungunya, yellow fever, Marburg virus disease, and now, yes, Ebola virus disease. The last of those possibilities was investigated using three kinds of assays, each one specific: for Ebola virus, for Sudan virus, for Reston virus. No positive results. The antibodies in those assays didn't recognize the virus, whatever it was, in her blood.

The laboratory sleuths persisted, designing a fourth assay that was more generalized—comprehensive for the whole group of ebolaviruses. Applied to her serum, that one glowed, a positive, announcing the presence of antibodies to an ebolavirus of some sort. So the Swiss woman was the world's first identified victim of what became known as Taï Forest virus. The chimpanzee she had necropsied, its tissues tested later, was the second victim, recognized posthumously.

Unlike the chimp, she survived. After another week, she left the hospital. She had lost thirteen pounds and her hair later fell out, but otherwise she was okay. Besides being the initial case of Taï Forest virus infection, the Swiss woman holds one other distinction: She is the first person known to have carried an ebolavirus infection off the African continent. There is no reason to assume that she will be the last.

13

Ebolavirus spillovers continued throughout the 1990s and into the twenty-first century, sporadic and scattered enough to make field research difficult, frequent enough to keep some scientists focused and some public health officials worried. In 1995, soon after the Côte d'Ivoire episode, it was Ebola virus in Kikwit, about which you've read. Six months after that outbreak, as you'll also recall, the new one began at Mayibout 2. What I haven't yet mentioned about Mayibout 2 is that, though the village lies in Gabon, the virus was Ebola as known originally from Zaire, which seems to be the most broadly distributed of the group. At the timber camp near Booué, Gabon, it was Ebola virus.

Also that year, 1996, Reston virus reentered the United States by way of another shipment of Philippine macaques. Sent from the same export house near Manila that had shipped the original sick monkeys to Reston, Virginia, these went to a commercial quarantine facility in Alice, Texas, near Corpus Christi. One animal died and, after it tested positive for Reston virus, forty-nine others housed in the same room were euthanized as a precaution. (Most of those, tested posthumously, were negative.) Ten employees who had helped unload and handle the monkeys were also screened for infection, and they also tested negative, but none of them were euthanized.

Uganda became the next known locus of the virus in Africa, with an outbreak of Sudan virus that began near the northern town of Gulu in August 2000. Northern Uganda shares a border with what in those days was southern Sudan, and it wasn't surprising that Sudan virus might cross or straddle that border. Cross it how, straddle it how? By way of the individual movements or the collective distribution of the reservoir species, identity unknown. This is a pointed example of why solving the reservoir mystery is important: If you know which animal harbors a certain virus and where that animal lives—and conversely, where it *doesn't* live—you know where the virus may next spill over, and where it probably won't. That provides some basis for focusing your vigilance. If the reservoir is a rodent that lives in the forests of southwestern Sudan but not in the deserts of Niger, the goat herders of Niger can relax. They have other things to worry about.

In Uganda, unfortunately, the 2000 spillover led to an epidemic of Sudan virus infections that spread from village to village, from hospital to hospital, from the north of the country to the southwest, killing 224 people.

The case fatality rate was again "only" 53 percent, exactly what it had been in the first Sudanese outbreak, back in 1976. This precise coincidence seems to reflect a significant difference in virulence between Sudan and Ebola viruses. Their difference, in turn, might reflect different evolutionary adjustments to humans as a secondary host (though random happenstance is also a possible explanation). Many factors contribute to the case fatality rate during an outbreak, including diet, economic conditions, public health in general, and the medical care available in the location where an outbreak occurs. It's hard to isolate the inherent ferocity of a virus from those contextual factors. What can be said, though, is that Ebola virus *appears* to be the meanest of the four ebolaviruses you've heard about, as gauged by its effect on human populations. Taï Forest virus can't reliably be placed on that spectrum at all, not yet—for lack of evidence. Having infected just one known human (or possibly two, counting an unconfirmed later case) and killed none, Taï Forest virus may be less prone to spillover. It may or may not be less lethal; one case, like one roll of the dice, proves nothing about what's likely to emerge as numbers grow larger. Then again, Taï Forest virus might also be spilling more frequently but inconsequentially—infecting people yet not causing notable illness. No one has screened the populace of Côte d'Ivoire to exclude that possibility.

The role of evolution in making Taï Forest virus (or any virus) less virulent in humans is a complicated matter, not easily deduced from simple comparison of case fatality rates. Sheer lethality may be irrelevant to the virus's reproductive success and long-term survival, the measures by which evolution keeps score. Remember, the human body isn't the primary habitat of ebolaviruses. The reservoir host is.

Like other zoonotic viruses, ebolaviruses have probably adapted to living tranquilly within their reservoir (or reservoirs), replicating steadily but not abundantly and causing little or no trouble. Spilling over into humans, they encounter a new environment, a new set of circumstances, often causing fatal devastation. And one human can infect another, through direct contact with bodily fluids or other sources of virus. But the chain of ebolavirus infection, at least so far, has never continued through many successive cases, great distances, or long stretches of time. Some scientists use the term "dead-end host," as distinct from "reservoir host," to describe humanity's role in the lives and adventures of ebolaviruses. What the term implies is this: Outbreaks have been contained and terminated; in each situation the virus has come to a dead end, leaving no offspring. Not the virus in toto throughout its range, of course, but that *lineage* of virus, the one that has spilled over, betting everything on this gambit—it's gone, kaput. It's an evolutionary loser. It hasn't caught hold to become an endemic disease within human populations. It hasn't caused a huge epidemic. Ebolaviruses, judged by experience so far, fit that pattern. Careful medical procedures (such as barrier nursing by way of isolation wards, latex gloves, gowns, masks, and disposable needles and syringes) usually stop them. Sometimes simpler methods can bring a local spillover to a dead end too. This has probably happened more times than we'll ever know. Advisory: If your husband catches an ebolavirus, give him food and water and love and maybe prayers but keep your distance, wait patiently, hope for the best—and, if he dies, don't clean out his bowels by hand. Better to step back, blow a kiss, and burn the hut.

This business about dead-end hosts is the conventional wisdom. It applies to the ordinary course of events. But there's another perspective to consider. Zoonoses by definition involve events beyond the ordinary, and the scope of their consequences can be extraordinary too. Every spillover is like a sweepstakes ticket, bought by the pathogen, for the prize of a new and more grandiose existence. It's a long-shot chance to transcend the dead end. To go where it hasn't gone and be what it hasn't been. Sometimes the bettor wins big. Think of HIV.

14

In late 2007 a fifth ebolavirus emerged, this time in western Uganda.

On November 5, 2007, the Ugandan Ministry of Health received a report of twenty mysterious deaths in Bundibugyo, a remote district along the mountainous border with the Democratic Republic of the Congo (the new name, as of 1997, for what had been Zaire). An acute infection of some unknown sort had killed those twenty people, abruptly, and put others at risk. Was it a rickettsial bacterium, such as the one that causes typhus? An ebolavirus was another possibility, but considered less likely at first, because few of the patients hemorrhaged. Blood samples were gathered quickly, flown to the CDC in Atlanta, and tested there, using both a generalized assay that might detect any form of ebolavirus and specific assays for each of the known four. Although the specific tests were all negative, the general test rang up some positives. So on November 28, the CDC informed Ugandan officials: It's an ebolavirus, all right, but not one we've ever seen.

Further laboratory work established that this new virus was at least 32 percent different genetically from any of the other four. It became Bundibugyo virus. Soon a CDC field team arrived in Uganda to help respond to the outbreak. As usual in such situations, their efforts along with those of the national health authorities involved three tasks: caring for patients, trying to prevent further spread, and investigating the nature of the disease. The eventual tally was 116 people infected, of whom 39 died.

Also as usual, the scientific team later published a journal article, in this case announcing the discovery of a new ebolavirus. First author on the paper was Jonathan S. Towner, a molecular virologist at the CDC with field experience in the search for reservoirs. Besides guiding the lab work, he went to Uganda and did a stint with the response team. The Towner paper contained a very interesting statement, as an aside, concerning the five ebolaviruses: "Viruses of each species have genomes that are at least 30–40% divergent from one another, a level of diversity that presumably reflects differences in the ecologic niche they occupy and in their evolutionary history." Towner and company suggested that some of the crucial differences between one ebolavirus and another—including the differences in lethality—might be related to where and how they live, where and how they *have* lived, within their reservoir hosts.

The events in Bundibugyo left many Ugandans uneasy. And they were entitled to their uneasiness: Uganda now held a sorry distinction as the only country on Earth that had suffered outbreaks of two different ebolaviruses (Sudan virus at Gulu in 2000, Bundibugyo virus in 2007), as well as outbreaks of both Ebola virus disease and Marburg virus disease, caused by another filovirus, within a single year. (The creepy circumstances of the Marburg spillover, at a gold mine called Kitaka in June 2007, are part of a story I'll come to in its turn.) Given such national ill fortune, it's not surprising that there were rumors, stories, and anxieties circulating among Ugandans, in late 2007, that made tracing genuine ebolavirus leads all the more difficult.

A pregnant woman, showing signs of hemorrhagic fever, delivered her baby and then died. The baby, left in the care of a grandmother, soon died too. That was sad but not peculiar; orphaned infants often die in the hard conditions of a village. More notable was that the grandmother also died. An ape (chimp or gorilla?) reportedly bit a domestic goat, infecting it; the goat was slaughtered in due course, skinned by a thirteen-year-old boy, and then the boy's family began falling ill. No, a dead monkey was eaten. No, bats were eaten. Mostly these tales couldn't be substantiated, but their currency and their general themes reflected a widespread, intuitive comprehension of zoonoses: Relations between humans and other animals, wild or domestic, must somehow lie at the root of the disease troubles. In early December, and then again in January 2008, came reports of suspicious animal deaths (monkeys and pigs) in outlying regions of the country. One of those reports also involved dogs that died after being bitten by the sickened monkeys. Was it an epidemic of rabies? Was it Ebola? The Ministry of Health sent people to collect specimens and investigate.

"Then there was a new epidemic—of fear," said Dr. Sam Okware,

Commissioner of Health Services, when I visited him in Kampala a month later. Among Dr. Okware's other duties, he served as chairman of the national Ebola virus task force. "That was the most difficult to contain," he said. "There was a new epidemic—of panic."

These are remote places, he explained. Villages, settlements, small towns surrounded by forest. The people feed themselves mostly on wildlife. During the Bundibugyo outbreak, residents of that area were shunned. Their economy froze. Outsiders wouldn't accept their money, scared that it carried infection. Population drained from the major town. The bank closed. When patients recovered (if they were lucky enough to recover) and went home from the hospital, "again they were shunned. Their houses were burned." Dr. Okware was a thin, middle-aged man with a trim mustache and long, gesticulant hands that moved through the air as he spoke of Uganda's traumatic year. The Bundibugyo outbreak, he said, was "insidious" more than dramatic, smoldering ambiguously while health officials struggled to comprehend it. There were still five questions pending, he said, and he began to list them: (1) Why were only half of the members of each household affected? (2) Why were so few hospital workers affected, compared to other Ebola outbreaks? (3) Why did the disease strike so spottily within the Bundibugyo district, hitting some villages but not others? (4) Was the infection transmitted by sexual contact? After those four he paused, momentarily unable to recall his fifth pending question.

"The reservoir?" I suggested. Yes, that's it, he said: *What's the reservoir?*

Bundibugyo virus in Uganda, 2007, completes the outline sketch of ebolavirus classification and distribution as presently known. Four different ebolaviruses are scattered variously across Central Africa and have emerged from their reservoir hosts to cause human disease (as well as gorilla and chimpanzee deaths) in six different countries: South Sudan, Gabon, Uganda, Côte d'Ivoire, the Republic of the Congo, and the Democratic Republic of the Congo. A fifth ebolavirus seems to be endemic to the Philippines, and to have traveled from there several times to the United States in infected macaques. But how did it get to the Philippines, if the ancestral origin of ebolaviruses is equatorial Africa? Could it have arrived there in one soaring leap, leaving no traces in between? From southwestern Sudan to Manila is almost seven thousand miles as the bat flies. But no bat can fly that far without roosting. Are ebolaviruses more broadly distributed than we suspect? Should scientists start looking for them in India, Thailand, and Vietnam? Or did Reston virus get to the Philippines the same way Taï Forest virus got to Switzerland and Johannesburg—by airplane?

If you contemplate all this from the perspective of biogeography (the study of which creatures live where on planet Earth) and phylogeny (the study of evolving lineages), one thing becomes evident: The current scientific understanding of ebolaviruses constitutes pinpricks of light against a dark background.

15

People in the villages where Ebola struck—the survivors, the bereaved, the scared but lucky ones not directly affected—had their own ways of understanding this phenomenon, and one way was in terms of malevolent spirits. In a single word, which loosely encompasses the variety of beliefs and practices seen among different ethnic and language groups and is often used to explain rapid death of adults: sorcery.

The village of Mékouka, on the upper Ivindo River in northeastern Gabon, offers an instance. Mékouka was one of the gold camps in which the outbreak of 1994 got its start. Three years later, a medical anthropologist named Barry Hewlett, an American, visited there to learn from the villagers themselves how they had thought about and responded to the outbreak. Many local people told him, using a term from their Bakola language, that this Ebola thing was *ezanga,* meaning some sort of vampirism or evil spirit. Asked to elaborate, one villager explained that ezanga are "bad human-like spirits that cause illness in people" as retribution for accumulating material goods and not sharing. (This wouldn't seem to apply to that man on the upper Ivindo, in 1994, who reportedly shared his tainted gorilla meat before he died.) Ezanga could even be summoned and targeted at a victim, like casting a hex. Neighbors or acquaintances, envious of the wealth or power someone has amassed, could send ezanga to gnaw at the person's internal organs, making him sick unto death. That's why gold miners and timber-company employees suffered such high risk of Ebola, Hewlett was told. They were envied and they didn't share.

Barry Hewlett had investigated the Mékouka outbreak in retrospect, months after the events occurred. Still fascinated by the subject, and concerned that an important dimension was being omitted by the more clinical methods of research and response, he got himself to the scene in Gulu, Uganda, in late 2000, while that outbreak was still going on. He found that the predominant ethnic group there, the Acholi, were also inclined to attribute Ebola virus disease to supernatural forces. They believed in a form of malign spirit, called *gemo*, that sometimes swept in like the wind to cause waves of sickness and death. Ebola wasn't their first gemo. The Acholi previously suffered epidemics of measles and smallpox, Hewlett learned, and those were likewise explained. Several elders told Hewlett that disrespect for the spirits of nature could bring on a gemo.

Once a true gemo was recognized, as distinct from a lesser spate of illness in the community, Acholi cultural knowledge dictated a program of special behaviors, some of which were quite appropriate for controlling infectious disease, whether you believed it was caused by spirits or by a virus. These behaviors included quarantining each patient in a house apart from other houses; relying on a survivor of the epidemic (if there were any) to provide care to each patient; limiting movement of people between the affected village and others; abstaining from sexual relations; not eating rotten or smoked meat; and suspending the ordinary burial practices, which would involve an open casket and a final "love touch" of the deceased by each mourner, filing up for that purpose. Dancing was also prohibited. Such traditional Acholi strictures (along with intervention by the Uganda Ministry of Health and support from the CDC, Médecins Sans Frontières, and WHO) may have helped suppress the Gulu outbreak.

"We have a lot to learn from these people," Barry Hewlett told me, one day in Gabon, "as to how they've responded to these epidemics over time." Modern society has lost that sort of ancient, painfully acquired accumulation of cultural knowledge, he said. Instead we depend on the disease scientists. Molecular biology and epidemiology are useful, but other traditions of knowledge are useful too. "Let's listen to what people are saying here. Let's find out what's going on. They've been living with epidemics for a long time."

Hewlett is a gentle-spirited man with a professorship at Washington State University and two decades of field experience in Central Africa. By the time I met him, at an international ebolavirus conference in Libreville, we had each visited one other village famed for suffering the disease—a place called Mbomo, in the Republic of the Congo, along the western edge of Odzala National Park. Mbomo lies not far from the Mambili River and the Moba Bai complex, where I had watched Billy Karesh trying to dart gorillas. The outbreak around Mbomo began in December 2002, probably among hunters who handled infected gorillas or duikers, and spread throughout an area that encompassed at least two other villages. A large difference between Hewlett's experience in Mbomo and mine was that he arrived during the outbreak. The grease was still flaming in the pan when he made his inquiries.

One early patient, Hewlett learned, was pulled out of the village clinic because his family disbelieved the Ebola diagnosis and preferred relying on a traditional healer. After that patient died at home, unattended by medical personnel and uncured by the healer, things got testy. The healer pronounced that this man had been poisoned by sorcery and that the perpetrator was his older brother, a successful man working in a nearby village. The older brother was a teacher who had "risen" to become a school inspector and didn't share the good fortune with his family. So again, as with ezanga among the Bakola people in northeastern Gabon, there were jealous animosities underlying the accusations of sorcery. Then another brother died, and a nephew, at which point family members burned the older brother's Mbomo house and sent a posse to kill him. They were stopped by the police. The older brother, though now taken for an evil magus, escaped vengeance. Then community relations deteriorated generally as more victims died from the invisible terror, with no cure available, no satisfactory explanation, to a point where anyone who looked out of the ordinary or above the crowd became suspect.

Another element of the dangerous brew in and around Mbomo was a mystic secret society, *La Rose Croix*, more familiar (if barely) to you and me as Rosicrucianism. It's an international organization that has existed for centuries, mostly devoted to esoteric study, but in this part of the Congo it had a bad reputation, akin to sorcery. Four teachers within one nearby village were members, or were thought to be members—and these teachers had been telling children about Ebola virus before the outbreak occurred. That led some traditional healers to suspect that the teachers had advance knowledge—supernatural knowledge—of the outbreak. Something had to be done, yes? On the day before Barry Hewlett and his wife arrived in Mbomo, the four teachers were murdered with machetes while they worked in their crop fields.

Soon afterward, the disease outbreak expanded to include so many community members that sorcery no longer seemed a plausible explanation to local people. The alternative was *opepe*, an epidemic, Mbomo's equivalent (in Kota, one of the local languages) to what Barry Hewlett had heard about, from the Acholi, as gemo. "This illness is killing everyone," one local man told the Hewletts, and therefore it couldn't be sorcery, which targets individual victims or their families. By early June 2003, there had been 143 cases in Mbomo and the surrounding area, with 128 deaths. That's a case fatality rate of 90 percent, at the top of the range even for Ebola virus.

With their deep interest in local explanations and their patient listening methods, the Hewletts heard things that wouldn't fit within the multiple-choice categories of an epidemiological questionnaire. Another of their informants, an Mbomo woman, declared: "Sorcery does not kill without reason, does not kill everybody, and does not kill gorillas or other animals." Oh, yes, again gorillas. That was another aspect of the Mbomo brew—everyone knew there were dead apes in the forest all roundabout. They had died at the Lossi sanctuary. They had died, so far as Billy Karesh could tell, at Moba Bai. Carcasses had been seen in the environs of Mbomo itself. And, as the woman said, sorcery didn't apply to gorillas.

16

When a silverback gorilla dies of Ebola, he does it beyond the eyes of science and medicine. No one is there in the forest to observe the course of his agony, with the possible exception of other gorillas. No one takes his temperature or peers down his throat. When a female gorilla succumbs to Ebola, no one measures the rate of her breathing or checks for a telltale rash. Thousands of gorillas may have been killed by the virus but no human has ever attended one of those deaths—not even Billy Karesh, not even Alain Ondzie. A small number of carcasses have been found, some of which have tested positive for Ebola antibodies. A larger number of carcasses have been seen and reported by casual witnesses, in Ebola territory at Ebola times, but because the forest is a hungry place, most of those carcasses could never be inspected and sampled by scientific researchers. The rest of what we know about Ebola's effect on gorillas is inferential: Many of them—major portions of some regional populations, such as the ones at Lossi, Odzala, and Minkébé—have disappeared. But nobody knows just how Ebola virus affects the gorilla body.

With humans it's different. The numbers I've mentioned above offer one gauge of that difference: 245 fatal cases during the outbreak at Kikwit, another 224 at Gulu, 128 in and around Mbomo, et cetera. The total of human fatalities from Ebola virus disease, since its discovery in 1976, is about fifteen hundred—not many compared to such widespread and relentless global afflictions as malaria and tuberculosis, or to the great waves of death brought by the various influenzas, but enough to generate a significant body of data. Furthermore, doctors and nurses saw many of those fifteen hundred people die. So the medical profession knows a good bit about the range of symptoms and the pathological effects produced on a human body during death by ebolavirus infection. It's not quite like you might think.

If you devoured *The Hot Zone* when it was published, as I did, or if you have been secondarily exposed to its far-reaching influence on public impressions about ebolaviruses, you may carry some wildly gruesome notions. Richard Preston is a vivid writer, a skillful writer, an industrious researcher, and it was his purpose to make a truly horrible disease seem almost preternaturally horrific. You may recall his depiction of a Sudanese hospital in which the virus "jumped from bed to bed, killing patients left and right," creating dementia and chaos, and not only killing patients but causing them to bleed profusely as they died, liquefying their organs, until "people were dissolving in their beds." You may have shuddered at Preston's statement that Ebola virus in particular "transforms virtually every part of the body into a digested slime of virus particles." You may have paused before turning the page when he told you that, after death, an Ebola-infected cadaver "suddenly deteriorates," its internal organs deliquescing in "a sort of shock-related meltdown." You may not have noticed that meltdown was a metaphor, meaning dysfunction, not actual melting. Or maybe it wasn't. At a later point, bringing another filovirus into the story, Preston mentioned a French expatriate, living in Africa, who "essentially melts down with Marburg virus while traveling on an airplane." You may remember one phrase in particular, as Preston described victims in a darkened Sudanese hut: comatose, motionless, and "bleeding out." That seemed to be so different from just "bleeding." It suggested a human body draining away in a gush. There was also the statement that Ebola causes a victim's eyeballs to fill up with blood, bringing blindness and more. "Droplets of blood stand out on the eyelids: You may weep blood. The blood runs from your eyes down your cheeks and refuses to coagulate." The mask of red death—where medical reporting meets Edgar Allan Poe.

It's my duty to advise that you need not take these descriptions quite literally—at least, not as the typical course of a fatal case of Ebola virus disease. Expert testimony, some published and some spoken, tempers Preston on several of these more lurid points, without minimizing the terribleness of Ebola in terms of real suffering and death. Pierre Rollin, for instance, deputy chief of the Special Pathogens Branch of the CDC, is one of the world's most experienced ebolavirus hands. He worked at the Pasteur Institute in Paris before moving to Atlanta, and has been a member of response teams to many Ebola and Marburg outbreaks over the past fifteen years, including those at Kikwit and Gulu. When I asked him, during an interview in his office, about the public perception that this disease is extraordinarily bloody, Rollin interrupted me genially to say: "—which is bullshit." When I mentioned the descriptions in Preston's book, Rollin mockingly said, "They melt, splash on the wall," and gave a frustrated shrug. Mr. Preston could write what he pleased, Rollin added, so long as the product was labeled fiction. "But if you say it's a true story, you have to speak to the true story, and he didn't. Because it was much more exciting to have blood everywhere and scaring everywhere." A few patients do bleed to death, Rollin said, but "they don't explode, and they don't melt." In fact, he said, the often-used term "Ebola hemorrhagic fever" is itself a misnomer for Ebola virus disease, because more than half the patients don't bleed at all. They die of other causes, such as respiratory distress and shutdown (but not dissolution) of internal organs.

Karl Johnson, one of the pioneers of Ebola outbreak response, whose credentials I've already sketched, offered a similar but even more pointed reaction, expressed with his usual candor. We were talking—in my own office, this time—during one of his periodic trips to Montana for fly-fishing. We had become friends and he had coached me a bit, informally, on how to think about zoonotic viruses. Finally I got him to sit for an interview, and *The Hot Zone* inescapably came up. Waxing serious, Karl said: "Bloody tears is bullshit. Nobody has ever had bloody tears." Furthermore, Karl noted, "People who die are not formless bags of slime." Johnson also concurred with Pierre Rollin that the bloodiness angle has been oversold. If you want a really bloody disease, he said, look at Crimean-Congo hemorrhagic fever. Ebola is bad and lethal, sure, but not bad and lethal precisely that way.

In the real world, as described in the scientific literature, the list of major symptoms of Ebola virus disease goes like this: abdominal pain, fever, headache, sore throat, nausea and vomiting, loss of appetite, arthralgia (joint pain), myalgia (muscle pain), asthenia (weakness), tachypnea (rapid breathing), conjunctival injection, and diarrhea. Conjunctival injection means pink eye, not bloody tears. All these symptoms tend to show up in many or most fatal cases. Additional symptoms including chest pain, hematemesis (vomiting of blood), bleeding from the gums, bloody stools, bleeding from needle-puncture sites, anuria (inability to pee), rash, hiccups, and ringing in the ears have appeared in a smaller fraction of cases. During the Kikwit outbreak, 59 percent of all patients didn't bleed noticeably at all, and bleeding in general was no indicator of who would or wouldn't survive. Rapid breathing, urine retention, and hiccups, on the other hand, were ominous signals that death would probably come soon. Among those patients who did bleed, blood loss never seemed massive, except among pregnant women who spontaneously aborted their fetuses. Most of the nonsurvivors died stuporous and in shock. Which is to say: Ebola virus generally killed with a whimper, not with a bang or a splash.

Despite all these data, gathered amid woeful and dangerous conditions while the primary mission was not science but saving lives, even the experts aren't sure exactly *how* the virus typically causes death. "We don't know the mechanism," Pierre Rollin told me. He could point to liver failure, to kidney failure, to breathing difficulties, to diarrhea, and in the end it often seemed that multiple causes were converging in an unstoppable cascade. Karl Johnson voiced similar uncertainty, but mentioned that the virus "really goes after the immune system," shutting down production of interferon, a class of proteins essential to immune response, so that "nothing stops the continued replication of the virus."

This idea of immune suppression by ebolaviruses has also appeared lately in the literature, along with speculation that it might allow catastrophic overgrowth of a patient's natural populations of bacteria, normally resident in the gut and elsewhere, as well as unhindered replication of the virus itself. Runaway bacterial growth might in turn put blood into the urine and feces, and even lead to "intestinal destruction," according to one source. Maybe that's what Preston had in mind when he wrote about liquefied organs and people dissolving in their beds. If so, he was blurring the distinction between what Ebola virus does and what garden-variety bacteria can do in the absence of a healthy immune system keeping them cropped. But, hey, don't we all like a dramatic story better than a complicated one?

Still another aspect of the pathology of Ebola virus disease is a phenomenon called disseminated intravascular coagulation, familiar to the medical community as DIC. It's also known as consumptive coagulopathy (if that helps you), because it involves consumption of too much of the blood's coagulating capacity in a misdirected way. Billy Karesh had told me about DIC as we boated down the Mambili River after our gorilla stakeout. Disseminated intravascular coagulation, he explained, is a form of pathological blood sludge, in which the normal clotting factors (coagulation proteins and platelets) are pulled out to form tiny clots along the insides of blood vessels throughout the victim's body, leaving little or no coagulation capacity to prevent leakage elsewhere. As a result, blood may seep from capillaries into a person's skin, forming bruiselike purple marks (hematomas); it may dribble from a needle puncture that seems never to heal, or it may leak into the gastrointestinal tract or the urine. Still worse, the mass aggregation of small clots in the vessels may block blood flow to the kidneys or the liver, causing organ failure as often seen with Ebola.

At least that was the understanding of DIC's role in Ebola virus disease at the time Karesh alerted me to it. More recently, Karl

Johnson and others have begun questioning whether the immune-shutdown effect that the virus somehow achieves, and the consequent blossoms of bacteria, might better explain some of the damage formerly blamed on DIC. "When it was first discovered, DIC, da da da, was the key to everything in hemorrhagic fever," Johnson told me, again cheerily dismissive of conventional wisdom. Now, he said, he was reading a hell of a lot less about DIC in the literature.

Ebola virus is still an inscrutable bug in more ways than one, and Ebola virus disease is still a mystifying affliction as well as a ghastly, incurable one—with or without DIC, with or without melting organs and bloody tears. "I mean, it's awful," Johnson stressed. "It really, really is." He had seen it almost before anyone else, under especially mystifying conditions—in Zaire, 1976, before the virus even had a name. But the thing hasn't changed, he said. "And frankly, everybody in the world is much too afraid of it, including the medical fraternity worldwide, to really want to try and study it." To study its effect on a living, struggling human body, he meant. To do that, you would need the right combination of hospital facilities, BSL-4 facilities, dedicated and expert professionals, and circumstances. You couldn't do it during the next outbreak at a mission clinic in an African village. You would need to bring Ebola virus into captivity—into a research situation, under highly controlled scrutiny—and not just in the form of frozen samples. You would need to study a raging infection inside somebody's body.

That isn't easy to arrange. He added: "We haven't had an Ebola patient yet in the US." But for everything that happens, there is a first time.

17

England had its first case of Ebola virus disease in 1976. Russia had its first case (that we know of) in 1996. Unlike the Swiss woman who did the chimp necropsy in Côte d'Ivoire, these two unfortunate people didn't pick up their infections during African fieldwork and come home prostrate in an ambulance jet. Their exposure derived from laboratory accidents. Each of them suffered a small, fateful, self-inflicted injury while doing research.

The English accident occurred at Britain's Microbiological Research Establishment, a discreetly expert institution within a high-security government compound known as Porton Down, not far from Stonehenge in the rolling green countryside southwest of London. Think of Los Alamos, but tucked into the boonies of pastoral England instead of the mountains of New Mexico, and with bacteria and viruses in place of uranium and plutonium as the strategic materials of interest. In its early years, beginning in 1916, Porton Down was an experiment station for the development of chemical weapons such as mustard gas; during World War II, its scientists worked also on biological weapons derived from anthrax and botulin bacteria. But eventually, at Porton Down as at USAMRIID, with changing political circumstances and government scruples, the emphasis shifted to defense—that is, research on countermeasures against biological and chemical weapons. That work involved high-containment facilities and techniques for studying dangerous new viruses, and therefore qualified Porton Down to offer assistance in 1976, when WHO assembled a field team to investigate a mysterious disease outbreak in southwestern Sudan. Deep-frozen blood samples from desperately ill Sudanese patients arrived for analysis—at about the same time, during that fretful autumn, as blood samples from Yambuku went to the CDC. The field people were asking the laboratory people to help answer a question: What *is* this thing? It hadn't yet been given a name.

One of the lab people at Porton Down was Geoffrey S. Platt. On November 5, 1976, in the course of an experiment, Platt filled a syringe with homogenized liver from a guinea pig that had been infected with the Sudanese virus. Presumably he intended to inject that fluid into another test animal. Something went amiss, and instead he jabbed himself in the thumb.

Platt didn't know exactly what pathogen he had just exposed himself to, but he knew it wasn't good. The fatality rate from this unidentified virus, as he must have been aware, was upwards of 50 percent. Immediately he peeled off his medical glove, plunged his thumb into a hypochlorite solution (bleach, which kills virus) and tried to squeeze out a drop or two of blood. None came. He couldn't even see a puncture. That was a good sign if it meant there *was* no puncture, a bad sign if it meant a little hole sealed tight. The tininess of Platt's wound, in light of subsequent events, testifies that even a minuscule dose of an ebolavirus is enough to cause infection, at least if that dose gets directly into a person's bloodstream. Not every pathogen is so potent. Some require a more sizable foothold. Ebolaviruses have force but not reach. You can't catch one by breathing shared air, but if a smidgen of the virus gets through a break in your skin (and there are always tiny breaks), God help you. In the terms used by the scientists: It's not very contagious but it's highly infectious. Six days after the needle prick, Geoffrey Platt got sick.

At the start he merely felt nauseous and exhausted, with abdominal pain. Given the circumstances, though, his malaise was taken very seriously. He was admitted to a special unit for infectious diseases at a hospital near London and, within that unit, put into a plastic-walled isolator tent under negative air pressure. The historical records don't mention it but you can be sure his nurses and doctors wore masks. He was given injections of interferon, to help stimulate his immune system, and blood serum (flown up from Africa) that had been drawn from a recovered Ebola patient to supply some borrowed antibodies. On the fourth day, Platt's temperature spiked and he vomited. This suggested the virus was thriving. For the next three days, his crisis period, he suffered more vomiting, plus diarrhea, and a spreading rash; his urine output was low; and a fungal growth in his throat hinted at immune failure. All these were gloomy signs. Meanwhile he was given more serum. Maybe it helped.

By the eighth day, Platt's vomiting and diarrhea had ended. Two days later, the rash began to fade and the fungus was under control. He had been lucky, perhaps genetically, as well as privileged to receive optimal medical care. The virus disappeared from his blood, from his urine, and from his feces (though it lingered awhile in his semen; apparently he promised doctors that he wouldn't make that a risk issue for anyone else). He was taken out of the isolator. Eventually he went home. He had lost weight, and during the long, slow convalescence much of his hair fell out. But like the Swiss woman, he survived.

The Russian researcher, in 1996, wasn't so lucky. Her name, as given in one Russian news account (but unspoken in the western medical literature), was Nadezhda Alekseevna Makovetskaya. Employed at a virological institute under the Ministry of Defense, she had been working on an experimental therapy against Ebola virus disease, derived from the blood serum of horses. Horses aren't susceptible to Ebola—not like they are to Hendra—which is why they are used to make antibodies. Testing the efficacy of this treatment required exposing additional horses. "It is difficult to describe working with a horse infected with Ebola," according to the dry, cautious statement from Russia's chief biowarfare man at the time, a lieutenant general named Valentin Yevstigneyev, in the Ministry of Defense. No doubt he was right about that. A horse can be nervous and jumpy, even if it's not suffering convulsions. Who would want to get close with a needle? "Under normal conditions this animal is difficult to manage and we had to work in special protective gear," said General Yevstigneyev. What he meant by "we" might be broadly interpreted. He was a high officer and military bureaucrat, not likely pulling the latex mitts onto his own hands. "One false step, one torn glove and the consequences would be grave." Makovetskaya had evidently taken that false step. Or maybe it wasn't her mistake so much as the twitch of a sensitive gelding. "She tore her protective gloves but concealed it from the leadership," by General Yevstigneyev's unsympathetic account, "since it happened just before the New Year holidays." Was he implying that she hadn't wanted to miss seasonal festivities while sitting in quarantine? He didn't mention a needlestick, or a scratch, or an open cut beneath the torn glove, though some such misfortune must have been involved. "As a result, by the time she turned to a doctor for help it was too late." The details of Makovetskaya's symptoms and death remain secret.

Another Russian woman stuck herself with Ebola in May 2004, and about this case slightly more is known. Antonina Presnyakova was a forty-six-year-old technician working at a high-security viral research center called Vektor (which sounds like something from Ian Fleming) in southwestern Siberia. Presnyakova's syringe carried blood from a guinea pig infected with Ebola virus. The needle went through two layers of gloves into her left palm. She immediately entered an isolation clinic, developed symptoms within a few days, and died at the end of two weeks.

These three cases reflect the inherent perils of doing laboratory research on such a lethal, infectious virus. They also suggest the context of concerns that surrounded America's closest approach to a home-grown case of Ebola. This one occurred also in 2004, just months before the death of Antonina Presnyakova.

18

Kelly L. Warfield grew up in a suburb of Frederick, Maryland, not many miles from Fort Detrick, the US Army base devoted to medical research and biodefense within which sits USAMRIID. She was a local girl, bright and curious, whose mother owned a convenience store just outside the Fort Detrick gate. Helping her mom since she was a middle-schooler, Kelly first saw and spoke with scientists from the disease-research institute when they stopped into the store to buy Diet Coke, quarts of milk,

Nicorette gum, Tylenol . . . whatever it is that top-level, Army-affiliated virologists buy. Unlike your average young convenience-store clerk, Kelly herself had a strong early aptitude for science. During high-school summers she worked in a government institute of standards and measures. After her freshman year of college and each summer until graduation, she served as a laboratory assistant at the National Cancer Institute, which had a branch on the grounds of Fort Detrick. She finished a bachelor's degree in molecular biology and considered her options for grad school. Around the same time she read *The Hot Zone*, which had recently been published.

"I'm a *Hot Zone* kid," Warfield told me much later. She couldn't vouch for the book's scientific accuracy, she added, but its effect on her then was galvanic. She was inspired by one of the main characters, Nancy Jaax, an Army major and veterinary pathologist at USAMRIID, who had been part of the response team at the infected monkey house in Reston. Warfield herself hoped to return to Fort Detrick after graduate school and join USAMRIID as a scientist—if possible, to work on Ebola virus.

She looked for a doctoral program that would teach her virology and found a good one at Baylor College of Medicine, in Houston. An entire department at Baylor was devoted to viral research, with two dozen virologists, some of whom were quite eminent, though none dealt with such high-hazard pathogens as Ebola. Warfield found a place in the lab of a mentor there and began studying a group of gastrointestinal viruses, the rotaviruses, which cause diarrhea in humans. Her dissertation project looked at immune response against rotavirus infection in mice. That was intricate and significant work (rotaviruses kill a half million children around the world every year), though not especially dramatic. She got experience in using lab animals (especially mice) as models for human immune response to viral infections, and she learned a bit about making vaccines. In particular, she gained expertise in a line of vaccine development using viruslike particles (VLPs), rather than the more conventional approach, which uses live virus attenuated by laboratory-induced evolution. VLPs are essentially the outer shells of viruses, capable of inducing antibody production (immune readiness) but empty of functional innards, and therefore incapable of replicating or causing disease. VLPs seem to hold high promise for vaccines against viruses, such as Ebola, that might be too dangerous for live-virus vaccination.

It took some time for Kelly to achieve her dream, but not much, and she wasted none. With the doctorate finished, twenty-six-year-old Dr. Warfield began work at USAMRIID in June 2002, just days after her graduation in Houston. The Army's institute had hired her, in part, for her VLP skills. Immediately she enrolled in the Special Immunizations Program, a punishing series of shots and more shots required before a new person can be cleared to enter the BSL-3 labs. (BSL-3 comprises the laboratory suites in which researchers generally work on dangerous but curable diseases, many caused by bacteria, such as anthrax and plague. BSL-4 is reserved for work on pathogens such as Ebola, Marburg, Nipah, Machupo, and Hendra, for which there are neither vaccines nor treatments.) They vaccinated her against a whole list of unsavory things that she might or might not ever face in the lab—against Rift Valley fever, against Venezuelan equine encephalitis, against smallpox, and against anthrax—all within a year.

Some of these vaccines can make a person feel pretty sick. Anthrax, for Warfield, was a particular disfavorite. "Ooof, terrible!" she recalled, during our long conversation at her current home, in a new suburb outside of Frederick. "That's a terrible vaccine." After all these challenges to her immune system, and possibly as a result, she suffered an attack of rheumatoid arthritis, which runs in her family. Rheumatoid arthritis is an immune dysfunction, and the medicine used to control it can potentially suppress normal immune responses. "So I wasn't allowed to get any vaccines anymore." Nonetheless, she was cleared to enter the BSL-3 suites, and then soon the BSL-4s. She began working with live Ebola virus.

Much of her effort went into the VLP research, though she also helped on other projects within her boss's lab. One involved testing a form of laboratory-created antibodies that might serve as a treatment against Ebola virus disease. These antibodies, developed by a private company in collaboration with USAMRIID, were designed to thwart the virus by tangling with a cellular protein involved in viral replication, not with the virus itself. It was a clever idea. Warfield again used mice as her test animals; she now had years of experience at handling and injecting them. For the experiment she infected fifty or sixty mice with Ebola virus and then, during the following days, gave them the experimental antibody treatment. Would they live, would they die? The mice were kept in clear plastic cages, like tall-sided pans, ten mice to a pan. Methodical procedures and constant attention are crucial to BSL-4 work, as Warfield well knew. Her methodical procedures for this experiment included filling a syringe full of antibody solution, enough for ten doses, and then injecting the ten mice from each pan with the same syringe, the same needle. It wasn't as though cross-infection was a concern, since they had already been dosed with the same batch of Ebola. Dosing multiple mice with a single syringe saved time, and time in a BSL-4 lab adds up toward stress and increased risk, because the physical circumstances are so difficult.

Picture those circumstances for Kelly Warfield. Customarily she worked in the BSL-4 suite known as AA-5, off a cinderblock corridor in the most secure wing of USAMRIID, behind three pressure-sealed doors and a Plexiglas window. She wore a blue vinyl protective suit (she and her colleagues simply called them "blue suits," not spacesuits or hazmats) with a fully enclosed hood, a clear face shield, and a ventilation hookup. Attached to her hookup was a yellow hose, coiling down from the ceiling to bring filtered air. She wore rubber boots and two pairs of gloves—latex gloves beneath heavier canners gloves, sealed to her suit at the wrists with electrical tape. Even with canners gloves over latex, her hands were the most vulnerable part of her body; they couldn't be protected with vinyl because they had to be delicately dexterous. Her workbench was a stainless steel cart, like a hospital cart, easy to clean, easy to move. If you didn't love the work, you wouldn't put yourself in this place.

She was alone in AA-5, under exactly those circumstances, at five thirty on the evening of February 11, 2004. She had come late to the day's tasks for the Ebola experiment because earlier hours had been filled with other demands. One pan of mice sat on her cart, along with a plastic beaker, a clipboard, and not much else in the way of materials and tools. It was the last pan of mice for the day. She filled a syringe and carefully injected nine mice, one after another—gripping each animal by the skin behind its neck, turning it belly up, inserting the needle into its abdomen deftly, quickly, adding no more discomfort than necessary to the life of each doomed and Ebola-ridden mouse. After each injection, she placed that mouse in the beaker, to keep the finished group apart from the others. One mouse to go. Maybe she was a little tired. Accidents happen. It was this very last mouse that caused the trouble. Just after being injected, it suddenly kicked away the needle, deflecting the point into the base of Kelly Warfield's left thumb.

The wound, if there was a wound, seemed to be only a very light graze. "At first, I didn't think that the needle went through the gloves," she told me. "It didn't hurt. Nothing hurt." Remaining calm by an act of discipline, she set the mouse back in his pan, put the syringe away, and then squeezed her hand. She could see blood emerging under the layers of glove. "So I knew I had stuck myself."

We were seated at her dinette table, on a mild September afternoon, as she talked me through the events of that February day. The house, which she shared with her Army-physician husband and her young son, was light and cheery with a lived-in feel; there were pieces of kid art on the refrigerator, a few toys lying around, a large green backyard, two half-poodle dogs, and a sign on the kitchen wall commanding: DO NOT ENTER WITHOUT WEARING VENTILATED SUIT. Today she was dressed in a red jacket and pearl earrings, not in blue vinyl.

She recalled her mind racing forward, from an immediate "Oh my God, I've done it" reaction to a sober consideration of just what she *had* done. She had not injected herself with live Ebola virus—or at least, not much. The syringe didn't carry Ebola virus; it carried antibodies, which would be harmless to anyone. But the needle had gone into ten Ebola-infected mice before going into her. If its point had picked up any particles of Ebola and brought them along, then she might have received a tiny dose. And she knew that a tiny dose could be enough. Quickly she unhooked her yellow hose and exited the BSL-4 suite, by way of the first of the pressurized doors, into an airlock space equipped with a chemical shower. There she showered out, dosing her blue-suit exterior with a virus-killing solution.

Then she pushed through the second door, to a locker-room area known as the Gray Side. She shed the boots, peeled off the blue suit and the gloves as fast as she could, leaving her clad only in medical scrubs. She used a wall phone to call two close friends, one of whom was Diane Negley, the BSL-4 suite supervisor. It was now suppertime or later, and Negley didn't answer at home, so Warfield left a chilling, desperate message on Negley's machine, the gist of which was: I've had an accident, stuck myself, please come back to work. The other friend, a co-worker named Lisa Hensley, who hadn't yet left the building, answered her call and said: "Start scrubbing. I'm on my way down." Warfield began scrubbing her hands with Betadine, rinsing with water and saline solution, scrubbing again. In her fervor she splashed water all over the floor. Hensley arrived quickly, joined her in the Gray Side, and started making calls to alert other people, including those in the Medical Division who handled accidents, while Warfield continued the Betadine scrub. After five or ten minutes, feeling she had done what she could on the wound site, Warfield stripped out of her medical scrubs, took a soap-and-water shower, and dressed. Hensley did likewise. But when they tried to exit the Gray Side, that pressure-sealed door wouldn't open. Its electronic lock didn't respond to their badges. Warfield, full of adrenaline, scared, with no luxury of being patient, busted open the door on manual override and alarms started ringing in other parts of the building.

Word had spread fast through the institute and, by now, a small crowd had gathered in the corridor. Warfield passed amid their stares and their questions, headed for the Medical Division. There she was ushered into a small room, questioned about her accident by the doctor on duty, a civilian woman, and given a "physical exam," through the whole course of which the doctor never touched her. "It was like she was afraid that I already had Ebola," Warfield recalled. The incubation period for Ebola virus is measured in days, not hours or minutes. It takes at least two days and usually more than a week for the virus to establish itself, replicate abundantly, and make a person symptomatic or infectious. But the civilian doctor didn't seem to know that, or to care. "She acted like I was a leper already." That doctor went off to confer with others, after which the head of the Medical Division took Warfield into his office, sat her down, and gently told her the recommended next step. They wanted to put her in the Slammer.

The Slammer at USAMRIID is a medical containment suite, designed for care of a person infected with any dangerous pathogen and—equally—for protecting against the spread of that infection to others. It consists of two hospital-style rooms set behind more pressure-sealed doors and another chemical shower. Earlier on the day of our conversation, having gotten me clearance for a tour of USAMRIID, Warfield had shown me through the Slammer, explaining its features with a trace of mordant pride. On the outside, a wide main door is labeled: CONTAINMENT ROOM. AUTHORIZED PERSONNEL ONLY. That's door number 537 within USAMRIID's labyrinthine corridors. It's the door through which a new patient enters the suite and, if things go well, through which the same patient eventually walks out. If things don't go well, the patient exits under other circumstances, not walking and not via door 537. All other human traffic—the flow of medical caregivers and faithful, intrepid friends—must pass through a smaller door into a change room, where piles of scrub suits sit folded and ready on shelves, and then through a pressurized steel door into an airlock shower. On the other side of the shower stall is another steel door. The two pressurized steel doors are never both open at once. So long as the patient shows no signs of infection, approved visitors are admitted to the Slammer wearing scrubs, gowns, masks, and gloves. If the patient proves to be infected, the suite becomes an active BSL-4 zone, in which doctors and nursing staff (no visitors now) must wear full blue suits. In that situation, the medical people shower thoroughly on the way out, leaving their scrub clothing behind in a bag to be autoclaved.

Warfield led me. We could pass through the shower stall in street clothes because the containment suite was unoccupied. When she slammed the first steel door behind us, triggering pressurization, I heard a *voosh* and felt the change in my ears. She said: "There's why it's called the Slammer."

She had entered the suite around noon on February 12, 2004, the day following her accident, after having drawn up a will and an advance directive (stipulating end-of-life medical decisions) with help from an Army lawyer. Her husband was in Texas for advanced military training and she had apprised him of the situation by phone. In fact, she had stayed on the phone with him much of the previous night, helped through the hours of terror and dread by his long-distance support. At some point she told him: "If I get sick, please *please* give me a lot of morphine. I've seen this disease"—she had watched it kill monkeys in the lab, though never a human—"and I know it *hurts*." On the first weekend, he managed to fly up from Texas and they spent Valentine's Day in the suite holding hands through his latex gloves. There was no kissing through his mask.

The incubation period for Ebola virus disease, as I've mentioned, is reckoned to be at least two days; it can be longer than three weeks. Individual case histories differ, of course, but at that time twenty-one days seemed to be the outer limit. Expert opinion held that, if an exposed person hasn't shown the disease within that length of time, she wouldn't. Kelly Warfield was therefore sentenced to twenty-one days in the Slammer. "It was like prison," she told me. Then she amended her statement: "It's like prison *and* you're gonna die."

Another difference from prison is that there were more blood tests. Each morning her friend Diane Negley, who happened to be a certified phlebotomist and who knew enough about Ebola to be cognizant of the risk to herself, tapped a vein and took away some of Warfield's blood. In exchange, she brought a donut and a latte. Negley's morning visit was the highlight of Warfield's day. During the first week or so, Negley took fifty milliliters of blood daily, a sizable volume (more than three tablespoons) that allowed for multiple tests plus a bit extra to put in frozen storage. One test, using the PCR (polymerase chain reaction) technique that's familiar to all molecular biologists, looked for sections of Ebola RNA (the virus's genetic molecule, equivalent to human DNA) in her blood. That test, which can ring a loud alarm but is sometimes unreliable, delivering a false positive, was routinely performed twice on each sample. Another test screened for interferon, the presence of which might signal a viral infection of any sort. Still another test targeted changes in blood coagulation, for an early alert in case of disseminated intravascular coagulation, the catastrophic clotting phenomenon that makes blood ooze out where it shouldn't. Warfield encouraged the medical people to take all the blood they desired. She recalled telling them: "If I die, I want you to learn everything you can about me"—everything they could about Ebola virus disease, she meant. "Store every sample. Analyze everything you can. Please *please* take something away from this if I die. I want you to learn." She told her family the same: If the worst happens, let them autopsy me. Let them salvage all possible information.

If she did die, Warfield knew, her body wouldn't come out of the Slammer through door 537. After autopsy, it would come through the autoclave chute, a sterilizing cooker, which would leave nothing her loved ones would want to see in an open coffin.

All her test results during the first week were normal and reassuring—with a single exception. The second PCR test from one day's sample came back positive. It said she had Ebola virus in her blood.

It was wrong. The provisional result gave Warfield a fright but that mistake was soon corrected by further testing. Woops, no, sorry. Never mind.

Another kerfuffle arose when USAMRIID's leadership realized that Warfield suffered rheumatoid arthritis, the medications for which might have suppressed her immune system. "That became this huge controversy," she told me. Certain honchos of the institute's top leadership acted surprised and angry, although the condition was clearly on file in her medical records. "They had all these teleconferences with all these experts. Everybody wanted to know why someone that was immunocompromised was working in the

BSL-4 suites." There was in fact no evidence that her immune system wasn't working fine. The commander of USAMRIID never made a personal visit to see her in the Slammer, not even through the glass, but he sent her an email announcing that he was suspending her access to BSL-4 labs and impounding her badge. It was a "slap in the face," added onto her other miseries and worries, Warfield said.

After more than two weeks of vampiric blood draws and reassuring tests, Warfield began feeling guardedly confident she wouldn't die of Ebola. She was weak and weary, her veins were weary too, so she asked that the blood sampling be reduced to a daily minimum. She got another unsettling jolt one evening as she undressed, discovering red spots on her arm and wondering whether they might herald the start of Ebola's characteristic rash. She had seen similar spots on lab-infected monkeys. That night she lay awake, obsessing about the spots, but they turned out to be nothing. She had Ambien to help her sleep. She had a stationary bike in case she wanted exercise. She had TV and Internet and a phone. As the weeks passed, the terrifying element of her situation faded slowly beneath the good news and the tedium.

She stayed sane with help from her mother and a few close friends (who could visit her often), her husband (who couldn't), her father (who remained off the visitor list so he could look after her son, in case everyone else got infected and quarantined and then died), and a certain amount of nervous laughter. Her son, whose name is Christian, was just three at the time and barred by age regulations from entering USAMRIID. Warfield judged he was too young, in any case, to be burdened with knowing exactly what was going on; she and her husband explained to Christian simply that mom would be absent for three weeks doing "special work." She was given a video linkup, a sort of Slammer Cam, through which she could see and talk with her loved ones on the outside. Hi, it's me, Kelly, live from Ebolaville, how was your day? Diane Negley, besides supplying the morning donut and coffee, heroically smuggled in one beer every Friday night. Food was a problem at first, there being no cafeteria at USAMRIID, until the Army realized it had funds that could be spent on supplying a patient in the Slammer with carryout. After that, Warfield had her choice each evening among Frederick's best: Chinese, Mexican, pizza. And she could share with her visiting friends, such as Negley, who would sit in the blind spot beneath the security camera, flip up her face shield, and eat. These high-carb consolations led Warfield and her pals to invent a game: "*Ebola Makes You* . . ." and then fill in the blank. Ebola makes you fat. Ebola makes you silly. Ebola makes you diabetic from too much chocolate ice cream. Ebola makes you appreciate little joys and smiles in the moment.

On the morning of March 3, 2004, door 537 opened and Kelly Warfield walked out of the Slammer. Her mother and (by special exemption) Christian were in the waiting room down the corridor. She took her son home. That afternoon she returned to USAMRIID, where her friends and colleagues threw her a coming-out party with food, testimonials, and balloons. Several months later, after a period of suspended access, a battery of tests on her immune system, a somewhat humiliating regimen of retraining and supervision, and a bit of persistent struggle, she regained her clearance for the BSL-4 suites. She could return to tickling the tail of the dragon that might have killed her.

Did you ever consider *not* going back to Ebola? I asked.

"No," she said.

Why do you love this work so much?

"I don't know," she said, and began to ruminate. "I mean, *why* Ebola? It only kills maybe a couple hundred people a year." That is, it hasn't been a disease of massive global significance and, notwithstanding the lurid scenarios that some people evoke, it's unlikely ever to become one. But she could cite its attractions in scientific terms. She took deep interest, for instance, in the fact that such a simple organism can be so potently lethal. It contains only a tiny genome, enough to construct just ten proteins, which account for the entire structure, function, and self-replicating capacity of the thing. (A herpesvirus, by contrast, carries about ten times more genetic complexity.) Despite the minuscule genome, Ebola virus is ferocious. It can kill a person in seven days. "How can something that is so small and so simple just be so darn dangerous?" Warfield posed the question and I waited. "That's just really fascinating to me."

Her son Christian, grown to a handsome first-grader, at this point arrived home from school. Kelly Warfield had given me most of her day and now there was time for just one more question. Although she is a molecular biologist, not an ecologist, I mentioned those two unsolved mysteries of Ebola's life in the wild: the reservoir host and the spillover mechanism.

Yes, very intriguing also, she agreed. "It pops up and kills a bunch of people, and before you can get there and figure anything out, it's gone."

It disappears back into the Congo forest, I said.

"It disappears," she agreed. "Yeah. Where did it come from and where did it go?" But that was out of her area.

19

Think of a BSL-4 laboratory—not necessarily AA-5 at USAMRIID but any among a handful around the world in which this virus is studied. Think of the proximity, the orderliness, and the certitude. Ebola virus is in these mice, replicating, flooding their bloodstreams. Ebola virus is in that tube, frozen solid. Ebola virus is in the Petri dish, forming plaques among human cells. Ebola virus is in the syringe; beware its needle. Now think of a forest in northeastern Gabon, just west of the upper Ivindo River. Ebola virus is everywhere and nowhere. Ebola virus is present but unaccounted for. Ebola virus is near, probably, but no one can tell you which insect or mammal or bird or plant is its secret repository. Ebola virus is not in *your* habitat. You are in *its*.

That's how Mike Fay and I felt as we hiked through the Minkébé forest in July 2000. Six days after my helicopter fly-in we left the inselbergs area, trudging southwest on Fay's compass line through a jungle of great trees, thorny vines interwoven into torturous thickets, small streams and ponds, low ridges between the stream drainages, mud-bordered swamps dense with thorny vegetation, fallen fruits as big as bocce balls, driver ants crossing our path, groups of monkeys overhead, forest elephants in abundance, leopards, almost no signs of human visitation, and roughly a trillion cheeping frogs. The reservoir host of Ebola virus was there too, presumably, but we couldn't have recognized it for that if we'd looked it in the face. We could only take sensible precautions.

On the eleventh day of walking, one of Fay's forest crewmen spotted a crested mona monkey on the forest floor, a youngster, alive but near death, with blood dripping from its nostrils. Possibly it had missed its grip in a high tree and suffered a fatal fall. Or . . . maybe it was infected with something, such as Ebola, and came down to die. Under standing instructions from Fay, the crewman didn't touch it. Fay's crew of hardworking Bantus and Pygmies always hungered after wild meat for the evening pot, but he forbade hunting on conservation grounds—and during this stretch through Minkébé he had commanded his cook even more sternly: Do *not* feed us anything found dead on the ground. That night we ate another brownish stew, concocted from the usual freeze-dried meats and canned sauces, served over instant mashed potatoes. The dying monkey, I fervently hoped, had been left behind.

One night later, at the campfire after dinner, Fay helped me tease some direct testimony from Sophiano Etouck, the shier of the two survivors from Mayibout 2. I had heard the whole story—including the part about Sophiano's personal losses—from the voluble Thony M'Both, but Sophiano himself, burly, diffident, had never spoken up. Now finally he did. The sentences were diced cruelly by his stutter, which sometimes brought him to what seemed an impassable halt; but Sophiano pushed on, and between blockages his words came quickly.

He had been traveling to one of the gold camps. Farther upriver. And stopped in Mayibout 2 to stay with family. That night one of his nieces said she was feeling bad. Malaria, everyone thought. A routine thing. The next morning, it got worse. Then other people too. They vomited, they had diarrhea. Started dying. I lost six, Sophiano said. Thony had gotten the number right but was a little confused about the identities. An uncle, a brother, a widowed sister-in-law. Her three daughters. The men in white suits, they came to take charge. One of them, a Zairian, had seen the disease before. At Kikwit. Twenty doctors had died there at Kikwit, the Zairian told us. They told us, this thing is very infectious. If a fly lands on you after having touched one of the corpses, they said, you will die. But I held one of my nieces in my arms. She had a tube in her wrist, an IV drip. It got clogged, backed up. Her hand swelled. And then with a pop her blood sprayed all over my chest, Sophiano said. But I didn't get sick. You've got to take the remedy, the doctors told me. You've got to stay here twenty-one days under quarantine. I thought, the hell with that. I didn't take the remedy. After my family people had been buried, I left Mayibout 2. I went to Libreville and stayed with another sister, hiding, Sophiano confessed. Because I was afraid the doctors would hassle me, he said.

This was our last evening in the forest before a resupply rendezvous four or five miles onward, at a point where Fay's preplotted line of march crossed a road. That road led eastward to Makokou. Some of Fay's crew would leave him there. They were exhausted, spent, fed up. Others would stay with him because, though also exhausted, they needed the work badly, or because it was better than gold mining, or because those reasons supplemented another: the sheer fascination of being involved with an enterprise so sublimely crazed and challenging. Another half year of hard walking across forests and swamps lay between them and Fay's end point, the Atlantic Ocean.

Sophiano would stay. He had been through worse.

20

The identity of Ebola's reservoir host (or hosts) remains unknown, as of this writing, although suspects have been implicated. Several different groups of researchers have explored the question. The most authoritative, most advantageously placed, and most persistent of them is the team led by Eric M. Leroy, of CIRMF, in Franceville, Gabon. As mentioned earlier, Leroy was one of the visiting doctors dressed in mystifying white suits who took part in the response effort at Mayibout 2. Although he and his colleagues may not have saved many (or *any*, as remembered by Thony M'Both) of the Mayibout patients from death, that outbreak was transformative for Leroy himself. He trained as an immunologist as well as a veterinarian and a virologist, and until 1996 studied the effects of another kind of virus (SIV, of which much more below) on the immune systems of mandrills. Mandrills are large, baboon-like monkeys with red noses, puffy blue facial ridges, and contorted expressions, all of which give them the look of angry, dark clowns. Leroy was also curious about the immune physiology of bats. Then came Mayibout 2 and Ebola.

"It is a little bit like a fate," Leroy told me when I visited him in Franceville.

Back at CIRMF after Mayibout 2, he explored Ebola further in his lab. He and a colleague, like him an immunologist, investigated some molecular signals in blood specimens taken during the outbreak. They found evidence suggesting that the medical outcome for an individual patient—to survive and recover, or to die—might be related not to the size of the infectious dose of Ebola virus but to whether the patient's blood cells produced antibodies promptly in response to infection. If they didn't, why not? Was it because the virus itself somehow quickly decommissioned their immune systems, interrupting the normal sequence of molecular interactions involved in antibody production? Does the virus kill people (as is now widely supposed) by creating immune dysfunction before overwhelming them with viral replication, which then inflicts further devastating effects? Leroy and his immunologist colleague, with a group of additional coauthors, published this study in 1999, after which he became interested in other dimensions of Ebola: its ecology and its evolutionary history.

The ecology of Ebola virus encompasses the reservoir question: Where does it hide between outbreaks? Another ecological matter is spillover: By what route, and under what circumstances, does the virus pass from its reservoir into other animals, such as apes and humans? To ask those questions is one thing; to get data that might help answer them is more tricky. How does a scientist study the ecology of such an elusive pathogen? Leroy and his team went into the forest, near locations where Ebola-infected gorilla or chimp carcasses had recently been found, and began trapping animals wholesale. They were groping for a hypothesis. Ebola might abide in one of these creatures—but which one?

In the course of several expeditions between 2001 and 2003, into Ebola-stricken areas of Gabon and the Republic of the Congo, Leroy's group caught, killed, dissected, and took samples of blood and internal organs from more than a thousand animals. Their harvest included 222 birds of various species, 129 small terrestrial mammals (shrews and rodents), and 679 bats. Back at the lab in Franceville, they tested the samples for traces of Ebola using two different methods. One method was designed to detect Ebola-specific antibodies, which would be present in animals that had responded to infection. The other method used PCR (as it had been used on Kelly Warfield) to screen for fragments of Ebola's genetic material. Having looked so concertedly at the bat fauna, which accounted for two-thirds of his total collections, Leroy found something: evidence of Ebola virus infection in bats of three species.

These were all fruit bats, relatively big and ponderous, like the flying foxes harboring Hendra virus in Australia. One of them, the hammer-headed bat (*Hypsignathus monstrosus*), is the largest bat in Africa, as big as a crow. People hunt it for food. But in this case the evidence linking bats and virus, though significant, wasn't definitive. Sixteen bats (including four hammer-headed) had antibodies. Thirteen bats (again including some hammer-headed) had bits of the genome of Ebola virus, detectable by PCR. That amounted to twenty-nine individuals, representing a small fraction of the entire sample. And the results among even those twenty-nine seemed ambiguous, in that no individual bat tested positive by both methods. The sixteen bats with antibodies contained no Ebola RNA, and vice versa. Furthermore, Leroy and his team did not find live Ebola virus in a single bat—nor in any of the other animals they opened.

Ambiguous or not, these results seemed dramatic when they appeared in a paper by Leroy and his colleagues in late 2005. It was a brief communication, barely more than a page, but published by *Nature*, one of the world's most august scientific journals. The headline ran: FRUIT BATS AS RESERVOIRS OF EBOLA VIRUS. The text itself, more carefully tentative, said that bats of three species "may be acting as a reservoir" of the virus. Some experts reacted as though the question were now virtually settled, others reserved judgment. "The only thing missing to be sure that bats are the reservoir," Leroy told me, during our conversation ten months later, "is virus isolation. Live virus from bats." That was 2006. It still hasn't happened, so far as the world knows, though not for lack of effort on his part. "We continue to catch bats—to try to isolate the virus from their organs," he said.

But the reservoir question, Leroy emphasized, was only one aspect of Ebola that engaged him. Using the methods of molecular genetics, he was also studying its phylogeny—the ancestry and evolutionary history of the whole filovirus lineage, including Marburg virus and the various ebolaviruses. He wanted to learn too about the natural cycle of the virus, how it replicates within its reservoir (or reservoirs) and maintains itself in those populations. Finally, knowing something about the natural cycle would help in discovering how the virus is transmitted to humans: the spillover moment. Does that transmission somehow occur directly (for instance, by people eating bats), or through an intermediate host? "We don't know if there's direct transmission from bats to humans," he said. "We only know there is direct transmission from dead great apes to humans." Understanding the dynamics of transmission—including seasonal factors, the geographical pattern of outbreaks, and the circumstances that bring reservoir animals or their droppings into contact with apes or humans—might give public health authorities a chance to predict and even prevent some outbreaks. But there exists a grim circularity: Gathering more data requires more outbreaks.

Ebola is difficult to study, Leroy explained, because of the character of the virus. It strikes rarely, it progresses quickly through the course of infection, it kills or it doesn't kill within just a few days, it affects only dozens or hundreds of people in each outbreak, and those people generally live in remote areas, far from research hospitals and medical institutes—far even from his institute, CIRMF. (It takes about two days to travel, by road and river, from Franceville to Mayibout 2.) Then the outbreak exhausts itself locally, coming to a dead end, or is successfully stanched by intervention. The virus disappears like a band of jungle guerrillas. "There is nothing to do," Leroy said, expressing the momentary perplexity of an otherwise patient man. He meant, nothing to do except keep trying, keep working, keep sampling from the forest, keep responding to outbreaks as they occur. No one can predict when and where Ebola virus will next spill. "The virus seems to decide for itself."

21

The geographical pattern of Ebola outbreaks among humans is, as I've mentioned, controversial. Everyone knows what that pattern looks like but experts dispute what it means. The dispute involves Ebola virus in particular, the one among those five ebolaviruses that has emerged most frequently, in multiple locations across Africa, and therefore cries most loudly for explanation. From its first known appearance to the present, from Yambuku (1976) to Tandala (1977) to the upper Ivindo River gold camps (1994) to Kikwit (1995) to Mayibout 2 (1996) to Booué (later 1996) to the northern border region between Gabon and the Republic of the Congo (2001–2002) to the Mbomo area (2002–2003) to its recurrence at Mbomo (2005) and then to its two more recent appearances near the Kasai River in what's now the Democratic Republic of the Congo (2007–2009), Ebola virus has seemingly hopscotched its way around Central Africa. What's going on? Is that pattern random or does it have causes? If it has causes, what are they?

Two schools of thought have arisen. I think of them as the wave school and the particle school—my little parody of the classic wave-or-particle conundrum about the nature of light. Back in the seventeenth century, as your keen memory for high-school physics will tell you, Christiaan Huygens proposed that light consists of waves, whereas Isaac Newton argued that light is particulate. They each had some experimental grounds for believing as they did. It took quantum mechanics, more than two centuries later, to explain that wave-versus-particle is not a resolvable dichotomy but an ineffable duality, or at least an artifact of the limitations of different modes of observing.

The particle view of Ebola sees it as a relatively old and ubiquitous virus in Central African forests, and each human outbreak as an independent event, primarily explicable by an immediate cause. For instance: Somebody scavenges an infected chimpanzee carcass; the carcass is infected because the chimp itself scavenged a piece of fruit previously gnawed by a reservoir host. The subsequent outbreak among humans results from a local, accidental event, each outbreak therefore representing a particle, discrete from others. Eric Leroy is the leading proponent of this view. "I think the virus is present all the time, within reservoir species," he told me. "And sometimes there is transmission from reservoir species to other species."

The wave view suggests that Ebola has *not* been present throughout Central Africa for a long time—that, on the contrary, it's a rather new virus, descended from some viral ancestor, perhaps in the Yambuku area, and come lately to other sites where it has emerged. The local outbreaks are not independent events, but connected as part of a wave phenomenon. The virus has been expanding its range within recent decades, infecting new populations of reservoir in new places. Each outbreak, by this view, represents a local event primarily explicable by a larger cause—the arrival of the wave. The main proponent of the wave idea is Peter D. Walsh, an American ecologist who has worked often in Central Africa and specializes in mathematical theory about ecological facts.

"I think it's spreading from host to host in a reservoir host," Walsh said, when I asked him to explain where the virus was traveling and how. This was another conversation in Libreville, a teeming Gabonese city with pockets of quietude, through which all Ebola researchers eventually pass. "Probably a reservoir host that's got large population sizes and doesn't move very much. At least, it doesn't transmit the virus very far." Walsh didn't claim to know the identity of that reservoir, but it had to be some animal that's abundant and relatively sedentary. A rodent? A small bird? A nonmigrating bat?

The evidence on each side of this dichotomy is varied and intriguing, though inconclusive. One form of that evidence is the genetic differences among variants of Ebola virus as they have been found, or left traces of themselves, in human victims, gorillas, and other animals sampled at different times and places. Ebola virus in general seems to mutate at a rate comparable to other RNA viruses (which means relatively quickly), and the amount of variation detectable between one strain of Ebola virus and another can be a very important clue about their origins in space and time. Peter Walsh, working with two coauthors on a paper published in 2005, combined such genetic data with geographical analysis to suggest that all known variants of Ebola virus descended from an ancestor closely resembling the Yambuku virus of 1976.

Walsh's collaborators were Leslie Real, a highly respected disease ecologist and theoretician at Emory University, and a bright younger colleague named Roman Biek. Together they presented maps, graphs, and family trees illustrating strong correlations among three kinds of distance: distance in miles from Yambuku, distance in time from that 1976 event, and distance in genetic differences from the Yambuku-like common ancestor. "Taken together, our results clearly point to the conclusion that [Ebola virus] has gradually spread across central Africa from an origin near Yambuku in the mid-1970s," they wrote. Their headline, stating the thesis plainly, was WAVE-LIKE SPREAD OF EBOLA ZAIRE. It may or may not be a new pathogen—at least, new in these places. (Other evidence, published more recently, suggests that filoviruses may be millions of years old.) But maybe something happened, and happened rather recently, to reshape the virus and unleash it upon humans and apes. "Under this scenario, the distinct phylogenetic tree structure, the strong correlation between outbreak date and distance from Yambuku, and the correlation between genetic and geographic distances can be interpreted as the outcome of a consistently moving wave of [Ebola virus] infection." One consequence of the moving wave, they argued, is massive mortality among the apes. Some regional populations have been virtually exterminated—such as the gorillas of the Minkébé forest, of the Lossi sanctuary, of the area around Moba Bai—because Ebola hit them like a tsunami.

So much for the wave hypothesis. The particle hypothesis embraces much of the same data, construed differently, to arrive at a vision of independent spillovers, not a traveling wave. Eric Leroy's group also collected more data, including samples of muscle and bone from gorillas, chimps, and duikers found dead near human outbreak sites. In some of the carcasses (especially the gorillas), they detected evidence of Ebola virus infection, with small but significant genetic differences in the virus among individual animals. Likewise they looked at a number of human samples, from the outbreaks in Gabon and the Congo during 2001–2003, and identified eight different viral variants. (These were lesser degrees of difference than the gaps among the five ebolaviruses.) Such distinct viruses, they proposed, should be understood in the context that their genetic character is relatively stable. The differences among variants suggest long isolation in separate locales, not a rolling wave of newly arrived, rather uniform virus. "Thus, Ebola outbreaks probably do not occur as a single outbreak spreading throughout the Congo basin as others have proposed," Leroy's team wrote, alluding pointedly to Walsh's hypothesis, "but are due to multiple episodic infection of great apes from the reservoir."

This apparent contradiction between Leroy's particle hypothesis and Walsh's wave hypothesis reflects an argument at cross-purposes, I think. The confusion may have arisen from back-channel communications and a certain sense of competition as much as from ambiguity in their published papers. What Walsh suggested—to recapitulate in simplest form—is a wave of Ebola virus sweeping across Central Africa by newly infecting some reservoir host or hosts. From its recent establishment in the host, according to Walsh, the virus spilled over, here and there, into ape and human populations. The result of that process is manifest as a sequence of human outbreaks coinciding with clusters of dead chimps and gorillas—*almost* as though the virus were sweeping through ape populations across Central Africa. Walsh insisted during our Libreville chat, though, that he had never proposed a continental wave of dying gorillas, one group infecting another. His wave of Ebola, he explained, has been traveling mainly through the reservoir populations, not through the apes. Ape deaths have been numerous and widespread, yes, and to some degree amplified by ape-to-ape contagion, but the larger pattern reflects progressive viral establishment in some other group of animals, still unidentified, with which apes frequently come into contact. Leroy, on the other hand, has presented his particle hypothesis of "multiple independent introductions" as a diametric alternative not to Walsh's idea as here stated but to the notion of a continuous wave among the apes.

In other words, one has cried: *Apples!* The other has replied: *Not oranges, no!* Either might be right, or not, but in any case their arguments don't quite meet nose to nose.

So . . . is light a wave or a particle? The coy, modern, quantum-mechanical answer is *yes.* And is Peter Walsh correct about Ebola virus or is Eric Leroy? The best answer again may be *yes.* Walsh and Leroy eventually coauthored a paper, along with Roman Biek and Les Real as deft reconcilers, offering a logical amalgam of their respective views on the family tree of Ebola virus variants (all descended from Yambuku) and of the hammer-headed bat and those two other kinds of bats as (relatively new) reservoir hosts. But even that paper left certain questions unanswered, including this one: If the bats have just recently become infected with Ebola virus, why don't they suffer symptoms?

The four coauthors did agree on a couple other basic points. First, fruit bats might be reservoirs of Ebola virus but not necessarily the *only* reservoirs. Maybe another animal is involved—a more ancient reservoir, long since adapted to the virus. (If so, where is *that* creature hiding?) Second, they agreed that too many people have died of Ebola virus disease, but not nearly so many people as gorillas.

22

After our fruitless stakeout near Moba Bai, in northwestern Congo, Billy Karesh and I and the expert gorilla guide Prosper Balo, along with other members of the team, traveled three hours back down the Mambili River by pirogue. We carried no samples of frozen gorilla blood, but I was nevertheless glad to have had the chance to come looking. From the lower Mambili we turned upstream on one of its branches, motored to a landing, and then drove a dirt road to the town of Mbomo, central to the area where Ebola virus had killed 128 people during the 2002–2003 outbreak.

Mbomo is where Barry Hewlett, arriving just after the four teachers were hacked to death, had encountered murderous suspicions between one resident and another that the Ebola deaths resulted from sorcery. We stopped at a little hospital, a U-shaped arrangement of low concrete structures surrounding a dirt courtyard, like a barebones motel. Each of the rooms, tiny and cell-like, gave directly onto the courtyard through a louvered door. As we stood in the heat, Alain Ondzie told me that Mbomo's presiding physician, Dr. Catherine Atsangandako, had famously locked an Ebola patient into one of those cells just a year earlier, supplying him with food and water through the slats. The man was a hunter, presumably infected by handling one form or another of wild meat. He had died behind his louvered door, a lonely end, but the doctor's draconian quarantine was generally credited with having prevented a wider outbreak.

Dr. Catherine herself was out of town today. The only evidence of her firm hand was a sign, painted in stark red letters:

ATTENTION EBOLA
NE TOUCHONS JAMAIS
NE MANIPULONS JAMAIS
LES ANIMAUX TROUVES
MORTS EN FORET

Don't touch dead animals in the forest.

Mbomo had another small distinction: It was Prosper Balo's hometown. We visited his house, walking to it along a narrow byway and then a grassy path, and found its dirt courtyard neatly swept, with wooden chairs set out for us under a palm. We met his wife, Estelle, and some of his many children. His mother offered us palm whiskey. The children jostled for their father's attention; other relatives gathered to meet the strange visitors; we took group photos. Amid this cheery socializing, in response to a few gentle queries, we learned some details about how Ebola had affected Estelle and her family during that grim period in 2003, when Prosper had been away.

We learned that her sister, two brothers, and a child had all died in the outbreak, and that Estelle herself was shunned by townspeople because of her association with those fatalities. No one would sell food to her. No one would touch her money. Whether it was infection they feared, or dark magic, is uncertain. She had to hide in the forest. She would have died herself, Prosper said, if he hadn't taught her the precautions he'd learned from Dr. Leroy and the other scientists, around that time, while helping them in their search for infected animals: Sterilize everything with bleach, wash your hands, and don't touch corpses. But now the bad days were past and, with Prosper's arm around her, Estelle was a smiling, healthy young woman.

Prosper remembered the outbreak in his own way, mourning Estelle's losses and some of a different sort. He showed us a treasured book, like a family bible—except it was a botanical field guide—on the endpapers of which he had written a list of names: Apollo, Cassandra, Afrodita, Ulises, Orfeo, and almost twenty others. They were gorillas, an entire group that he had known well, that he had tracked daily and observed lovingly at Lossi. Cassandra was his favorite, Prosper said. Apollo was the silverback. "*Sont tous disparus en deux-mille trois,*" he said. All of them, gone in the 2003 outbreak. In fact, though, they hadn't entirely *disparus*: He and other trackers had followed the group's final trail and found six gorilla carcasses along the way. He didn't say which six. Cassandra, dead with others in a fly-blown pile? It was very hard, he said. He had lost his gorilla family, and also members of his human family.

For a long time Prosper stood holding the book, opened for us to see those names. He comprehended emotionally what the scientists who study zoonoses know from their careful observations, their models, their data. People and gorillas, horses and duikers and pigs, monkeys and chimps and bats and viruses: We're all in this together.

III

EVERYTHING COMES FROM SOMEWHERE

23

Ronald Ross came west from India, in 1874, at age seventeen, to study medicine at St. Bartholomew's Hospital in London. He came to the study of malaria somewhat later.

Ross was a true son of the empire. His father, General Campbell Ross, a Scottish officer with roots in the Highlands, had served in the British Indian Army through the Sepoy Rebellion and fought in fierce battles against the hill tribes. Ronald had been "home" to England before, having endured a boarding school near Southampton. He fancied the idea of becoming a poet, or a painter, or maybe a mathematician; but he was the eldest of ten children, with all attendant pressures, and his father had decided he should enter the Indian Medical Service (IMS). After a lackluster five years at St. Bartholomew's, Ross flunked the IMS qualifying exam, an inauspicious start for an eventual Nobel laureate in medicine. The two facts from his youth that do seem to have augured well and truly are that he won a schoolboy prize for mathematics and, during medical training, he diagnosed a woman as suffering from malaria. It was an unusual diagnosis, malaria being virtually unknown in England, even amid the Essex marshes where this woman lived. History doesn't record whether Ross's diagnosis was right because he scared her with talk of the deadly disease and she disappeared, presumably back into lowland Essex. Anyway, Ross tried the IMS exam again after a year, squeaked through, and was posted to duty in Madras. That's where he started noticing mosquitoes. They annoyed him because they were so abundant in his bungalow.

Ross didn't bloom early as a medical detective. He dabbled and dawdled for years, distracted with the enthusiasms of the polymath. He wrote poetry, plays, music, bad novels, and what he hoped were groundbreaking mathematical equations. His medical duties at the Madras hospital, which involved treating malarial soldiers with quinine, among other tasks, demanded only about two hours daily, which left him plenty of time for extracurricular noodling. But eventually the extracurriculars included wondering about malaria. What caused it—miasmal vapors, as the traditional view held, or some sort of infectious bug? If a bug, how was that bug transmitted? How could the disease be controlled?

After seven years of unexceptional service he returned to England on furlough, did a course in public health, learned to use a microscope, found a wife, and took her back to India. This time his post was a small hospital in Bangalore. He started looking through his microscope at blood smears from feverish soldiers. He lived an intellectually isolated life, far from scientific societies and fellow researchers, but in 1892 he learned belatedly that a French doctor and microscopist named Alphonse Laveran, working in Algeria and then Rome, had discovered tiny parasitic creatures (now known as protists) in the blood of malaria patients. Those parasites, Laveran argued, caused the disease. During another visit to London, with help from an eminent mentor there, Ross himself saw the "Laveran bodies" in malarial blood and was converted to Laveran's idea, so far as it went.

Laveran had detected the important truth that malaria is caused by microbes, not by bad air. But that still left unexplained the wider matters of how these microbes reproduced in a human body, and how they passed from one host to another. Were they carried and ingested in water, like the germ causing cholera? Or might they be transmitted in the bite of an insect?

Ronald Ross's eventual discovery of the mosquito-mediated life cycle of malarial parasites, for which he won his Nobel Prize in 1902, is famous in the annals of disease research and I won't retell it here. It's a complicated story, both because the life cycle of the parasites is so amazingly complex and because Ross, himself a complicated man, had so many influences, competitors, enemies, wrong ideas as well as right ones, and distracting disgruntlements. Two salient points are enough to suggest the connections of that story to our subject, zoonoses. First, Ross delineated the life history of malarial parasites not as he found them infecting humans but as he found them infecting birds. Bird malaria is distinct from human malaria but it served as his great analogy. Second, he came to see the disease as a subject for applied mathematics.

24

Numbers can be an important aspect of understanding infectious disease. Take measles. At first glance, it might seem nonmathematical. It's caused by a paramyxovirus and shows itself as a respiratory infection, usually accompanied by a rash. It comes and it goes. But epidemiologists have recognized that, with measles virus, as with other pathogens, there's a critical minimum size of the host population, below which it can't persist indefinitely as an endemic, circulating infection. This is known as the critical community size (CCS), an important parameter in disease dynamics. The critical community size for measles seems to be somewhere around five hundred thousand people. That number reflects characteristics specific to the disease, such as the transmission efficiency of the virus, its virulence (as measured by the case fatality rate), and the fact that one-time exposure confers lifelong immunity. Any isolated community of less than a half million people may be struck by measles occasionally, but in a relatively short time the virus will die out. Why? Because it has consumed its opportunities among susceptible hosts. The adults and older children in the population are nearly all immune, having been previously exposed, and the number of babies born each year is insufficient to allow the virus a permanent circulating presence. When the population exceeds five hundred thousand, on the other hand, there will be a sufficient and continuing supply of vulnerable newborns.

Another crucial aspect of measles is that the virus is not zoonotic. If it were—if it circulated also in animals living near or among human communities—then the question of critical community size would be moot. There wouldn't *be* any necessary minimum size of the human population, because the virus could always remain present, nearby, in that other source. But bear in mind that measles, though it doesn't circulate in nonhuman animal populations, is closely related to viruses that do. Measles belongs to the genus *Morbillivirus,* which includes canine distemper and rinderpest; its family, *Paramyxoviridae*, encompasses also Hendra and Nipah. Although measles doesn't often pass between humans and other animals, its evolutionary lineage speaks of such passage sometime in the past.

Whooping cough, to take another example, has a critical community size that differs slightly from the measles number because it's a different disease, caused by a microbe with different characteristics: different transmission efficiency, different virulence, different period of infectivity, et cetera. For whooping cough, the CCS seems to be more like two hundred thousand people. Such considerations have become grist for a lot of fancy ecological mathematics.

Daniel Bernoulli, a Dutch-born mathematician from a family of mathematicians, was arguably the first person to apply mathematical analysis to disease dynamics, long before the germ theories of disease (there was a gaggle, not just one) became widely accepted. In 1760, while holding a professorship at the University of Basel in Switzerland, Bernoulli produced a paper on smallpox, exploring the costs versus the benefits of universal immunization against that disease. His career was long and eclectic, encompassing mathematical work on a wide range of topics in physics, astronomy, and political economy, from the movement of fluids and the oscillation of strings to the measurement of risk and ideas about insurance. The smallpox study seems almost anomalous amid Bernoulli's other interests, except that it also entailed the notion of calculating risk. What he showed was that inoculating all citizens with small doses of smallpox material (it wasn't known to be a virus then, just some sort of infectious stuff) had both risks and benefits, but that the benefits outweighed the risks. On the risk side, there was the fact that artificial inoculation sometimes—though rarely—led to a fatal case of the disease. More usually, inoculation led to immunity. That was an individual benefit from a single action. To gauge the collective benefits from collective action, Bernoulli figured the number of lives that would be saved annually if smallpox were entirely eradicated. His equations revealed that the net result of mass inoculation would be three years and two months of increased lifespan for the average person.

Life expectancy at birth wasn't high in the late eighteenth century, and those three years and two months represented a sizable increment. But because the real effects of smallpox are not averaged between the people who catch it and the people who don't, Bernoulli also expressed his results in a more stark and personal way. Among a cohort of 1,300 newborns, he projected, using life-table statistics for all causes of death as available to him at the time, 644 of those babies would survive at least to age twenty-five, if they lived in a society without smallpox. But if smallpox were endemic, only 565 of the same group would reach a twenty-fifth birthday. Health officials and ordinary citizens, imagining themselves among the seventy-nine preventable fatalities, could appreciate the force of Bernoulli's numerical argument.

Bernoulli's work, applying mathematics to understand disease, pioneered an approach but didn't create an immediate trend. Time passed. Almost a century later, the physician John Snow used statistical charts as well as maps to demonstrate which water sources (notably, the infamous Broad Street pump) were infecting the most people during London's cholera outbreak of 1854. Snow, like Bernoulli, lacked the advantage of knowing what sort of substance or creature (in this case it was *Vibrio cholerae*, a bacterium) caused the disease he was trying to comprehend and control. His results were remarkable anyway.

Then, in 1906, after Louis Pasteur and Robert Koch and Joseph

Lister and others had persuasively established the involvement of microbes in infectious disease, an English doctor named W. H. Hamer made some interesting points about "smouldering" epidemics in a series of lectures to the Royal College of Physicians in London.

Hamer was especially interested in why diseases such as influenza, diphtheria, and measles seem to mount into major outbreaks in a cyclical pattern—rising to a high case count, fading away, rising again after a certain interval of time. What seemed curious was that the interval between outbreaks remained, for a given disease, so constant. The cycle that Hamer plotted for measles in the city of London (population at that time: 5 million) was about eighteen months. Every year and a half came a big measles wave. The logic of such cycles, Hamer suspected, was that an outbreak declined whenever there weren't enough susceptible (nonimmune) people left in the population to fuel it, and that another outbreak began as soon as new births had supplied a sufficient number of new victims. Furthermore, it wasn't the sheer number of susceptible individuals that was crucial, but the density of susceptibles multiplied by the density of infectious people. In other words, contact between those two groups is what mattered. Never mind the recovered and immune members of the population; they just represented padding and interference so far as disease propagation was concerned. Continuation of the outbreak depended on the likelihood of encounters between people who were infectious and people who *could be* infected. This idea became known as the "mass action principle." It was all about math.

The same year, 1906, a Scottish physician named John Brownlee proposed an alternate view, contrary to Hamer's. Brownlee worked as a clinician and hospital administrator in Glasgow. For a paper delivered to the Royal Society of Edinburgh, he plotted sharp up-and-down graphs of case numbers, week by week or month by month, from the empirical records of several disease outbreaks—plague in London (1665), measles in Glasgow (1808), cholera in London (1832), scarlet fever in Halifax (1880), influenza in London (1891), and others—and then matched them with smooth rollercoaster curves derived from a certain mathematical equation. The equation expressed Brownlee's suppositions about what caused the outbreaks to rise and decline, and the good fits against empirical data proved (to him, anyway) that his suppositions were correct. Each epidemic had arisen, he argued, with "the acquisition by an organism of a high grade of infectivity," a sudden increase of the pathogen's catchiness or potency, which thereafter decreased again at a high rate. The epidemic's decline, which was generally not quite as abrupt as its start, resulted from this "loss of infectivity" by the disease-causing organism. The plague bacterium had shot its wad. The measles virus had slowed or weakened. Influenza had turned tame. Malign power had deserted each of them like air going out of a balloon. Don't waste your time worrying about the number or the density of susceptible people, Brownlee advised. It was "the condition of the germ," not the character of the human population, that determined the course of the epidemic.

One problem with Brownlee's nifty schema was that other scientists weren't quite sure what he meant by "infectivity." Was that synonymous with transmission efficiency, as measured by the number of transmissions per case? Or synonymous with virulence? Or a combination of both? Another problem was that, whatever he meant by infectivity, Brownlee was wrong to think that its inherent decline accounted for the endings of epidemics.

So said the great malaria man, Ronald Ross, in a 1916 paper presenting his own mathematical approach to epidemics. Ross by that time had received his Nobel Prize, and a knighthood, and had published a magnum opus, *The Prevention of Malaria,* which in fact dealt with understanding the disease in scientific and historical depth as well as preventing it. Ross recognized that, because of the complexity of the parasite and the tenaciousness of the vectors, malaria probably couldn't be "extirpated once and forever"—at least not until civilization reached "a much higher state." Malaria *reduction,* therefore, would need to be a permanent part of public health campaigns. Ross meanwhile had turned increasingly to his mathematical interests, which included a theory of diseases that was more general than his work on malaria, and a "theory of happenings" that was more general than his theory of diseases. By "happenings" he seems to have meant events of any sort that pass through a population, like gossip, or fear, or microbial infections, affecting individuals sequentially.

He began the 1916 paper by professing surprise that "so little mathematical work should have been done on the subject of epidemics," and noted without false modesty (or any other kind) that he himself had been the first person to apply a priori mathematical thinking (that is, starting with invented equations, not real-world statistics) to epidemiology. He nodded politely to John Brownlee's "excellent" work and then proceeded to dismiss it, rejecting Brownlee's idea about loss of infectivity and offering instead his own theory, supported by his own mathematical analysis. Ross's theory was that epidemics decline when, and because, the density of susceptible individuals in the population has fallen below a certain threshold. Look and see, he said, how nicely my differential equations fit the same sets of epidemic data that Dr. Brownlee adduced. Brownlee's hypothetical "loss of infectivity" was unnecessary for explaining the precipitous decline of an epidemic, whether the disease was cholera or plague or influenza or something else. All that was necessary was the depletion of susceptibles to a critical point—and then, *shazam*, the case rate fell drastically and the worst was over.

Ross's a priori approach may have been perilous, at such an early stage of malaria studies, and his attitude a little arrogant, but he produced useful results. His insight about susceptibles has met the test of time, coming down through the decades of theoretical work on infectious diseases to inform modern mathematical modeling. He was right about something else, too: the difficulty of extirpating malaria "once and forever." Although the control measures he advocated were effective toward reducing malaria in certain locales (Panama, Mauritius), in other places they failed to do much good (Sierra Leone, India) or the results were transitory. For all his honors, for all his mathematical skills, for all his combative ambition and obsessive hard work, Ronald Ross couldn't conquer malaria, nor even provide a strategy by which such an absolute victory would eventually be won. He may have understood why: because it's such an intricate disease, deeply entangled with human social and economic considerations as well as ecological ones, and therefore a problem more complicated than even differential calculus can express.

25

When I first wrote about zoonotic diseases, for *National Geographic* in 2007, I was given to understand that malaria was not one. No, I was told, you'll want to leave it off your list. Malaria is a *vector-borne* disease, yes, in that insects carry it from one host to another. But vectors are not hosts; they belong to a different ecological category from, say, reservoirs; and they experience the presence of the pathogen in a different way. Transmission of malarial parasites from a mosquito to a human is not spillover. It's something far more purposive and routine. Vectors seek hosts, because they need their resources (meaning, in most cases, their blood). Reservoirs do not seek spillover; it happens accidentally and it gains them nothing. Therefore malaria is not zoonotic, because the four kinds of malarial parasite that infect humans infect *only* humans. Monkeys have their own various kinds of malaria. Birds have their own. Human malaria is exclusively human. So I was told, and it seemed to be true at the time.

The four kinds of malaria to which these statements applied are caused by protists of the species *Plasmodium vivax, Plasmodium falciparum, Plasmodium ovale,* and *Plasmodium malariae*, all of them belonging to the same diverse genus, *Plasmodium*, which encompasses about two hundred species. Most of the others infect birds, reptiles, or nonhuman mammals. The four known for targeting humans are transmitted from person to person by *Anopheles* mosquitoes. These four parasites possess wondrously complicated life histories, encompassing multiple metamorphoses and different forms in series: an asexual stage known as the *sporozoite*, which enters the human skin during a mosquito bite and migrates to the human liver; another asexual stage known as the *merozoite*, which emerges from the liver and reproduces in red blood cells; a stage known as the *trophozoite*, feeding and growing inside the blood cells, each of which fattens as a *schizont* and then bursts, releasing more merozoites to further multiply in the blood, and causing a spike of fever; a sexual stage known as the *gametocyte*, differentiated into male and female versions, which emerge from a later round of infected red blood cells, enter the bloodstream en masse, and are taken up within a blood meal by the next mosquito; a fertilized sexual stage known as the *ookinete*, which lodges in the gut lining of the mosquito, each ookinete ripening into a sort of egg sac filled with sporozoites; and then come the sporozoites again, bursting out of the egg sac and migrating to the mosquito's salivary glands, where they lurk, ready to surge down the mosquito's proboscis into another host. If you've followed all that, at a quick reading, you have a future in biology.

This elaborate concatenation of life-forms and sequential strategies is highly adaptive and, so far as mosquitoes and hosts are concerned, difficult to resist. It shows evolution's power, over great lengths of time, to produce structures, tactics, and transformations of majestic intricacy. Alternatively, anyone who favors Intelligent Design in lieu of evolution might pause to wonder why God devoted so much of His intelligence to designing malarial parasites.

Plasmodium falciparum is the worst of the four in terms of its impact on human health, accounting for roughly 85 percent of reported malaria cases around the world—and for an even larger proportion of the fatalities. This form of the disease, known as falciparum malaria or malignant malaria, kills more than a half million people annually, most of them children in sub-Saharan Africa. Some scientists have suggested that the high virulence of *P. falciparum* reflects the fact that it's relatively new to humans, having shifted to us within the recent past from another animal host. That suggestion has led researchers to investigate its ancestral history.

Of course, everything comes from somewhere, and because we humans ourselves are a relatively new primate, it was always logical to assume that our oldest infectious diseases had come to us—transmogrified at least slightly by evolution—from other animal hosts. It was always sensible to recognize that the distinction between zoonotic diseases and nonzoonotic diseases is slightly artificial, involving a dimension of time. By a strict definition, zoonotic pathogens (accounting for about 60 percent of our infectious diseases, as I've mentioned) are those that *presently and repeatedly* pass between humans and other animals, whereas the other group of infections (40 percent, including smallpox, cholera, measles, and polio) are caused by pathogens descended from forms that must have made the leap to human ancestors sometime in the past. It might be going too far to say that *all* our diseases are ultimately zoonotic, but zoonoses do stand as evidence of the infernal, aboriginal connectedness between us and other kinds of host.

Malaria exemplifies this. Within the *Plasmodium* family tree, as revealed by molecular phylogenetics over the last two decades, the four human-afflicting kinds don't cluster on a single branch. They are each more closely related to other kinds of *Plasmodium*, infecting nonhuman hosts, than to one another. In the lingo of taxonomists, they are *polyphyletic*. What that suggests, besides the diversity of their genus, is that each of them must have made the leap to humans independently. Among the questions that continue to occupy malaria researchers are: Which other animals did they leap from, and when?

Falciparum malaria, because its global impact in death and misery is so high, has received particular attention. Early molecular research suggested that *P. falciparum* shares a close common ancestor with two different kinds of avian plasmodia, and that the parasite must therefore have crossed into humans from birds. A corollary to that idea, based on sensible deduction but not much evidence, is that the transfer probably happened just five or six thousand years ago, coincident with the invention of agriculture, which allowed for sedentary settlement—crop fields and villages—constituting the first sizable and dense aggregations of humans.

Such gatherings of people would have been necessary to sustain the new infection, because malaria (like measles, but for different reasons) has a critical community size and tends to die out locally if the hosts are too few. Simple irrigation works, such as ditches and impoundments, may have increased the likelihood of transfer by offering good breeding habitat for *Anopheles* mosquitoes. Domestication of the chicken, about eight thousand years ago in Southeast Asia, may have been another contributing factor, since one of the two forms of bird plasmodia in question is *Plasmodium gallinaceum*, known for infecting poultry.

That view of falciparum malaria's avian origins was propounded in 1991, a relatively long time ago in this field, and lately it doesn't look so persuasive. A more recent study suggested that the closest known relative of *P. falciparum* is *P. reichenowi*, a malarial parasite that infects chimpanzees.

Plasmodium reichenowi has been found in wild and (wild-born) captive chimps in both Cameroon and Côte d'Ivoire, suggesting that it's widespread across chimpanzee habitat in Central and West Africa. It contains a fair degree of genetic variation—more than *P. falciparum* worldwide—suggesting that it may be an old organism, or anyway older than *P. falciparum*. Furthermore, all known variants of *P. falciparum* seem to be twigs within the *P. reichenowi* branch of the *Plasmodium* family tree. These insights emerge from data gathered by a team of researchers led by Stephen M. Rich, of the University of Massachusetts, who proposed that *P. falciparum* has descended from *P. reichenowi* after spilling over from chimps into humans. According to Rich and his group, the spillover probably occurred just once, as early as 3 million years ago or as recently as ten thousand years ago. Some mosquito bit a chimpanzee (the insect becoming thereby infected with *P. reichenowi* gametocytes) and then also bit a human (delivering sporozoites). The transplanted strain of *P. reichenowi,* despite finding itself in an unfamiliar sort of host, managed to survive and proliferate. It passed from sporozoites into merozoites into gametocytes again, filled the bloodstream of that first human victim, and then caught itself another mosquito ride. From that insect it traveled onward, further vector-borne, to other humans as they foraged in the forest. Along the way it was changed by mutation and adaptation: *P. reichenowi* became *P. falciparum.*

This scenario implies that largish agricultural settlements *weren't* necessary for the disease to take hold among humans, since no such settlements existed in those areas of Africa ten thousand (let alone 3 million) years ago. Rich's group evidently considered the agricultural factor unnecessary. The genetic evidence they offered was compelling. Among Rich's coauthors were a handful of luminaries in the fields of anthropology, evolution, and disease. Their paper appeared in 2009. But it wasn't the last word.

Another group, led by a French anthropologist named Sabrina Krief and the malaria geneticist Ananias A. Escalante, published an alternative view in 2010. Yes, they agreed, *P. falciparum* may be more closely related to *P. reichenowi* than to any other known plasmodium. And yes, it seems to have spilled into humans within the relatively recent past. But look here, they said, we've located another host of *P. falciparum* itself—a host in which that parasite seems to have evolved *before* spilling into humans: the bonobo.

The bonobo (*Pan paniscus*) is sometimes known as the pygmy chimpanzee. It's an elusive beast, limited in numbers and distribution, not often displayed in Western zoos, and (though much prized, alas, as an item of cuisine by the Mongo people of the southern Congo basin) very closely related to humans. Its native range is along the left bank of the Congo River, in the forests of the Democratic Republic of the Congo, whereas the common chimpanzee (*Pan troglodytes*), more burly and familiar, lives only on the right bank of the big river. Screening blood samples from forty-two bonobos resident at a sanctuary on the outskirts of Kinshasa, the Krief group found four animals carrying parasites genetically indistinguishable from *P. falciparum*. The most plausible explanation, Krief's group wrote, is that falciparum malaria spilled over originally from bonobos into people, probably sometime within the last 1.3 million years. (An alternative explanation, offered by other researchers in a critical comment on the Krief paper, is that the bonobos in their small sanctuary, so near Kinshasa, had been infected by mosquitoes carrying *P. falciparum* from humans—sometime within recent years or decades.) The bonobos testing positive for *P. falciparum* had shown no overt signs of illness and low levels of parasites in their blood, which seemed consistent with an ancient association. To these descriptive and data-based results, Krief's team added a hypothesis and a caveat.

Their hypothesis: If bonobos carry a form of *P. falciparum* that is so similar to what humans carry, those parasites may still be passing back and forth between bonobos and us. In other words, falciparum malaria may be zoonotic—in the strict sense of the word, not just the loose sense. Humans in the forests of DRC might be infected on a regular basis with *P. falciparum* from the blood of bonobos, and vice versa.

Their caveat: If that's so, the great dream of malaria eradication becomes even less attainable. Krief and company didn't press the point but you might read that to mean: We can't hope to kill off the last parasite until we kill off (or cure) the last bonobo.

But wait! Still another study of *P. falciparum* origins, published in late 2010, pointed to still another candidate as its prehuman host: the western gorilla. This work appeared as a cover story in *Nature,* with Weimin Liu as first author and major contributions from the laboratory of Beatrice H. Hahn, then at the University of Alabama at Birmingham. Hahn is well known in AIDS-research circles for her role in tracing the origins of HIV-1 among chimpanzees, and for developing "noninvasive" techniques of sampling for virus in primates without having to capture the animals. Simply put: You don't need a syringe full of blood if a little poo will do. Fecal samples can sometimes yield the necessary genetic evidence, not just for a virus but also for a protist. Applying those techniques to the search for plasmodium DNA, Liu, Hahn and their colleagues were able to gather far more data than were previous researchers. Whereas the Krief group had looked at blood samples from forty-nine chimpanzees and forty-two bonobos, most of which were captive or confined within a sanctuary, Liu's group examined fecal samples from almost three thousand wild apes, including gorillas, bonobos, and chimps.

They found that western gorillas carry a high prevalence of plasmodium (about 37 percent of the population is infected) and that some of those gorilla parasites are nearly identical to *P. falciparum.* "This indicates," they wrote confidently, "that human *P. falciparum* is of gorilla origin, and not of chimpanzee, bonobo or ancient human origin."

Furthermore, they added, the entire genetic range of *P. falciparum* in humans forms "a monophyletic lineage within the gorilla *P. falciparum* radiation." In plain talk: The human version is one twig within a gorilla branch, suggesting that it came from a single spillover. That's one mosquito biting one infected gorilla, becoming a carrier, and then biting one human. By delivering the parasite into a new host, that second bite was enough to account for a zoonosis that still kills more than a half million people each year.

26

Mathematics to me is like a language I don't speak though I admire its literature in translation. It's Dostoyevsky's Russian, or the German of Kafka, Musil, and Mann. Having studied calculus hard in school, as I did Latin, I found that the deep knack wasn't in me, and the secret music of differential equations fell wasted on my deaf ears, just like the secret music of *The Aeneid.* So I'm an ignoramus, an outsider. That's why you should trust me when I say that two other bits of mathematical disease theory, derived from early twentieth century concerns over epidemic malaria and other outbreaks, are not only important but intriguing, their essence quite capable of comprehension by the likes of you and me. One came out of Edinburgh. The other had its roots in Ceylon.

The first bit was embedded in a 1927 paper titled "A Contribution to the Mathematical Theory of Epidemics," by W. O. Kermack and

A. G. McKendrick. Of these two partners, William Ogilvy Kermack has the more memorable story. He was a Scotsman, like Ross and Brownlee, educated in mathematics and chemistry before he began his career doing statistical analyses of milk yields from dairy cows. Every poet hears his first nightingale somewhere. Kermack went from milk yields into the Royal Air Force, emerged after brief service to do industrial chemistry as a civilian, and then around 1921 joined the Royal College of Physicians Laboratory in Edinburgh, where he worked on chemical projects until a lab experiment blew up in his face. I mean that literally. He was blinded by caustic alkali. Twenty-six years old. But instead of becoming an invalid and a mope, he became a theoretician. Gathering back resolve, he continued his scientific work with the help of students who read aloud to him and colleagues who complemented his extraordinary capacity for doing math in his head. Chemistry led Kermack into the search for new antimalarial drugs. Mathematics engaged him on the subject of epidemics.

In the meantime Anderson G. McKendrick, a medical doctor who had served in the Indian Medical Service (again like Ross), became superintendent of the Laboratory of the Royal College of Physicians and therefore in some sense Kermack's boss. On a level transcending hierarchy, they meshed. Sightless yet unquenchably curious, Kermack later worked on various subjects, such as comparative death rates in rural and urban Britain, and fertility rates among Scottish women, but the 1927 paper with McKendrick was his most influential contribution to science.

It contributed two things. First, Kermack and McKendrick described the interplay among three factors during an archetypal epidemic: the rate of infection, the rate of recovery, and the rate of death. They assumed that recovery from an attack conferred lifelong immunity (as it does, say, with measles) and outlined the dynamics in efficient English prose:

> One (or more) infected person is introduced into a community of individuals, more or less susceptible to the disease in question. The disease spreads from the affected to the unaffected by con-

> tact infection. Each infected person runs through the course of his sickness, and finally is removed from the number of those who are sick, by recovery or by death. The chances of recovery or death vary from day to day during the course of his illness. The chances that the affected may convey infection to the unaffected are likewise dependent upon the stage of the sickness. As the epidemic spreads, the number of unaffected members of the community becomes reduced.

This sounds like calculus cloaked in words; and it is. Amid a dense flurry of mathematical manipulations, they derived a set of three differential equations describing the three classes of living individuals: the susceptible, the infected, and the recovered. During an epidemic, one class flows into another in a simple schema, $S \rightarrow I \rightarrow R$, with mortalities falling out of the picture because they no longer belong to the population dynamic. As susceptible individuals become exposed to the disease and infected, as infected individuals either recover (now with immunity) or disappear, the numerical size of each class changes at each moment in time. That's why Kermack and McKendrick used differential calculus. Although I should have paid better attention to the stuff in high school, even I can understand (and so can you) that $dR/dt = \gamma I$ merely means that the number of recovered individuals in the population, at a given moment, reflects the number of infected individuals times the average recovery rate. So much for R, the "recovered" class. The equations for S ("susceptibles") and I ("infected") are likewise opaque but sensible. All this became known as an *SIR* model. It was a handy tool for thinking about infectious outbreaks, still widely used by disease theorists.

Eventually the epidemic ends. *Why* does it end? asked Kermack and McKendrick.

> One of the most important problems in epidemiology is to ascertain whether this termination occurs only when no susceptible individuals are left, or whether the interplay of the various factors of infectivity, recovery and mortality, may result in termination,

> whilst many susceptible individuals are still present in the unaffected population.

They were leading their readers toward the second of those two possibilities: that an epidemic might cease because some subtle interplay among infectivity, mortality, and recovery (with immunity) has stifled it.

Their other major contribution was recognizing the existence of a fourth factor, a "threshold density" of the population of susceptible individuals. This threshold is the number of concentrated individuals such that, given certain rates of infectivity, recovery, and death, an epidemic can happen. So you have density, infectivity, mortality, and recovery—four factors interrelated as fundamentally as heat, tinder, spark, and fuel. Brought together in the critical measure of each, the critical balance, they produce fire: epidemic. Kermack and McKendrick's equations calibrated the circumstances in which such a fire would ignite, would continue to burn, and would eventually smolder out.

One notable implication of their work was stated near the end: "Small increases of the infectivity rate may lead to large epidemics." This quiet warning has echoed loudly ever since. It's a cardinal truth, over which public health officials obsess each year during influenza season. Another implication was that epidemics don't end because *all* the susceptible individuals are either dead or recovered. They end because susceptible individuals are no longer sufficiently dense within the population. W. H. Hamer had said so in 1906, remember? Ross had made the same point in 1916. But the paper by Kermack and McKendrick turned it into a working principle of mathematical epidemiology.

27

The second bit of landmark disease theory came from George MacDonald. He was another malaria researcher of mathematical bent (is it inevitable that so many of them be Scottish?), who worked in the tropics for years and eventually became director of the Ross Institute of Tropical Hygiene, in London, which had been founded decades earlier for Ronald Ross himself. MacDonald got some of his field experience in Ceylon (now Sri Lanka) during the late 1930s, just after a calamitous malaria epidemic there in 1934–1935, which sickened a third of the Ceylonese populace and killed eighty thousand. The severity of the Ceylon epidemic had been surprising because the disease was familiar, at least in parts of the island, recurring as modest annual outbreaks that mostly affected young children. What happened differently in 1934–1935 was that, after a handful of years with little malaria at all, a drought increased breeding habitat for mosquitoes (standing pools in the rivers, instead of flowing current), whose population then multiplied hugely, carrying malaria into areas where it had been long absent and where most people—especially the young children—possessed no acquired immunity. Back in London, fifteen and twenty years later, George MacDonald tried to understand how and why malaria exploded in occasional epidemics, using math as his method and Ceylon as a case in point.

That was just about the time, in the mid-1950s, when the World Health Organization began formulating a campaign to eradicate malaria globally, rather than just controlling or reducing it in one country and another. WHO's vaunting ambition—total victory, no compromise—was partly inspired by the existence of a new weapon, the pesticide DDT, which seemed capable of exterminating mosquito populations and (unlike other insect poisons, which didn't linger as lethal residue) keeping them dead. The other crucial element of WHO's strategy was to eliminate malarial parasites from human hosts, also thoroughly, in order to break the human-mosquito-human cycle of infection. This would be achieved by treating every human case with malaria medicine, maintaining careful surveillance to detect any new or relapsing cases, and then treating those too, until the last parasite had been poisoned out of the last human bloodstream. That was the idea, anyway. George MacDonald's writings were meant to clarify and assist the effort. One of them, published in WHO's own *Bulletin* in 1956, was titled "Theory of the Eradication of Malaria."

In an earlier paper, MacDonald had made the point that "very small changes in the essential transmission factors" of malaria in any given place could trigger an epidemic. This affirmed Kermack and McKendrick's point about small increases in "infectivity" leading to large epidemics. But MacDonald was more specific. What were those essential transmission factors? He identified a whole list, including the density of mosquitoes relative to human density, the biting rate of the mosquitoes, the longevity of the mosquitoes, the number of days required for malarial parasites to complete a life cycle, and the number of days during which any infected human remains infectious to a mosquito. Some of these factors were known constants (a life cycle for *P. falciparum* takes about thirty-six days, a human case can remain infectious for about eighty days) and some were variable, dependent on circumstances such as which kind of *Anopheles* mosquito was serving as vector and whether pigs were present nearby to distract thirsty mosquitoes away from humans. MacDonald created equations reflecting his reasonable suppositions about how all those factors might interact. Testing his equations against what was known about the Ceylon epidemic, he found that they fit nicely.

That tended to confirm the accuracy of his suppositions. He concluded that a fivefold increase in the density of *Anopheles* mosquitoes in relatively disease-free areas of Ceylon, combined with conditions allowing each mosquito relative longevity (sufficient time to bite, become infected, and bite again), had been enough to launch the epidemic. One variable among many, increased by five—and the conflagration was lit.

The ultimate product of MacDonald's equations was a single number, which he called the basic reproduction rate. That rate represented, in his words, "the number of infections distributed in a community as the direct result of the presence in it of a single primary non-immune case." More precisely, it was the average number of secondary infections produced, at the beginning of an outbreak, when one infected individual enters a population where all individuals are nonimmune and therefore susceptible. MacDonald had identified a crucial index—fateful, determinative. If the basic reproduction rate was less than 1, the disease fizzled away. If it was greater than 1 (greater than 1.0, to be more precise), the outbreak grew. And if it was considerably greater than 1.0, then *kaboom*: an epidemic. The rate in Ceylon, he deduced from available data, had probably been about 10. That's very high, as disease parameters go. Plenty high enough to yield a severe epidemic. But it was the lower side of the range for circumstances such as those in Ceylon. On the upper side, MacDonald imagined this: that a single infected person, left untreated and remaining infectious for eighty days, exposed to ten mosquitoes each day, if those mosquitoes enjoyed reasonable longevity and reasonable opportunities to bite, could infect 540 other people. Basic reproduction rate: 540.

WHO's eradication campaign failed. In fact, by the judgment of one historian: "It all but destroyed malariology. It turned a subtle and vital science dedicated to understanding and managing a complicated natural system—mosquitoes, malarial parasites and people—into a spraygun war." After years of applying pesticides and treating cases, the healthocrats watched malaria resurge ferociously in those parts of the world, such as India, Sri Lanka (as then known), and Southeast Asia, where so much money and effort had been spent. Apart from the problem (which proved large) of acquired resistance to DDT among *Anopheles* mosquitoes, the planners and health engineers of WHO probably gave insufficient respect to another consideration—the consideration of small changes and large effects. Humans have an enormous capacity to infect mosquitoes with malaria. Miss one infected person in the surveillance-and-treatment program to eliminate malarial parasites from human hosts, and let that person be bitten by one uninfected mosquito—it all starts again. The infection spreads and, when its basic reproduction rate is greater than 1.0, it spreads quickly.

If you read the recent scientific literature of disease ecology, which is highly mathematical, and which I do not recommend unless you are deeply interested or troubled with insomnia, you find the basic reproduction rate everywhere. It's the alpha and omega of the field, the point where infectious disease analysis starts and ends. In the equations, this variable appears as R_0, pronounced aloud by the cognoscenti as "R-naught." (It's a little confusing, I concede, that they use R_0 as the symbol for basic *reproduction* rate and plain R as the symbol for *recovered* in an *SIR* model. That's just a clumsy coincidence, reflecting the fact that both words begin with the letter R.) R_0 explains and, to some limited degree, it predicts. It defines the boundary between a small cluster of weird infections in a tropical village somewhere, flaring up, burning out, and a global pandemic. It came from George MacDonald.

28

Plasmodium falciparum isn't the only malarial parasite of global concern. Outside of sub-Saharan Africa, most human cases are caused by *Plasmodium vivax*, the second-worst of the four kinds adapted particularly to infecting people. (The other two, *P. ovale* and *P. malariae*, are far more rare and not nearly so virulent, causing infections that usually pass without medical treatment.) *P. vivax* is less lethal than *P. falciparum* but it does create a lot of misery, lost productivity, and inconvenience, accounting for about 80 million cases of mostly nonfatal malaria each year. Its origins have lately been elucidated, again using molecular phylogenetics, and again one of the researchers involved is Ananias A. Escalante, formerly of the CDC, now at Arizona State University. Escalante and his partners have shown that, rather than emerging from Africa along with the earliest humans, as *P. falciparum* seems to have done, *P. vivax* may have been waiting for our ancestors when they arrived to colonize Southeast Asia. The evidence suggests that its closest relatives are plasmodia infecting Asian macaques.

I'm not going to summarize this body of work, because we're in deep enough already; but I want to alert you to one small aspect that leads off irresistibly on a peculiar tangent. Escalante's team reported in 2005 that *P. vivax* shares a recent ancestry with three kinds of macaque malaria. One of those is *Plasmodium knowlesi*, a parasite known from Borneo and Peninsular Malaysia, where it sometimes infects at least two native primates, the long-tailed macaque and the pig-tailed macaque. *P. knowlesi* occupies a strange place in medical annals, involving the treatment of neurosyphilis (syphilis of the central nervous system), which for a time in the early twentieth century was done using induced malarial fevers.

The story goes like this. Dr. Robert Knowles was a lieutenant colonel in the Indian Medical Service, assigned to Calcutta in the 1930s and doing malaria research. In July 1931 he came into possession of an unfamiliar new strain of malarial parasite, derived from an imported monkey. It was a plasmodium, he could see, but not any he recognized. Knowles and a junior colleague, an assistant surgeon named Das Gupta, decided to study it. They injected the bug into several other kinds of monkey and followed the progress of infection. This mystery strain proved devastating to rhesus macaques, causing high fevers and high loads of parasites in the blood, killing the animals quickly. In bonnet macaques, though, it had little effect. Knowles and Gupta also injected it into three human volunteers (that is to say, "volunteers," their freedom to decline having been a dubious matter), one of whom was a local man who had come to the hospital for treatment of a rat bite on his foot. This poor guy got very sick—not from the rat bite but from the injected malaria. In those experimental subjects (monkey and human) who suffered intermittent fevers, Knowles and Gupta noticed that the period of the fever cycle was one day, as distinct from the two-day or three-day cycles known for human malarias. Knowles and Gupta published a paper on the unusual parasite but didn't give it a name. Soon afterward another set of scientists did, labeling it *Plasmodium knowlesi* in honor of its senior discoverer.

Shift of scene: to Eastern Europe. Reading the literature, a well-connected malaria researcher in Romania named Mihai Ciuca got interested in the properties and potential uses of *Plasmodium knowlesi* and wrote to one of Knowles's colleagues in India, asking for a sample. When the monkey blood arrived, Professor Ciuca started injecting doses of *P. knowlesi* into patients with neurological syphilis. This was not nearly as crazy as it sounds, though even for Romania perhaps a little edgy, since the range of effects of *P. knowlesi* in humans was so little known. Still, Ciuca was merely following a line of therapy that had not only proven effective but had been scientifically canonized. Back in 1917 a Viennese neurologist named Julius Wagner-Juaregg had begun inoculating advanced syphilis patients with other strains of malaria, and not only had he escaped malpractice prosecution and accusations of criminal goofiness but he had also received a Nobel Prize in medicine. Wagner-Juaregg was a man of unsavory eminence in the old style, a bilious anti-Semite who advocated "racial hygiene," favored forced sterilization for the mentally ill, and wore a Nietzschean mustache, but his "pyrotherapy" using malaria seems to have helped many neurosyphilis patients, who otherwise would have suffered out their last days in asylums. There was cold logic—revise that, *hot* logic—to Wagner-Juaregg's mode of treatment. It worked because the syphilis bug is so sensitive to temperature.

Syphilis is caused by a spiral bacterium (aka a spirochete) known as *Treponema pallidum.* The bacterium is usually acquired during sexual contact, whereupon it corkscrews its way across mucous membranes, multiplies in the blood and lymph nodes, and, if a patient is especially unlucky, gets into the central nervous system, including the brain, causing personality change, psychosis, depression, dementia, and death. That's in the absence of antibiotic treatment, anyway; modern antibiotics cure syphilis easily. But there were no modern antibiotics in 1917, and the early chemical treatment known as Salvarsan (containing arsenic) didn't work well against late-stage syphilis in the nervous system. Wagner-Juaregg solved that problem after noting that *Treponema pallidum* didn't survive in a test tube at temperatures much above 98.6 degrees Fahrenheit. Raise the blood temperature of the infected person a few degrees, he realized, and you might cook the bacterium to death. So he began inoculating patients with *Plasmodium vivax.*

He would allow them to cycle through three or four spikes of fever, delivering potent if not terminal setbacks to the *Treponema*, and then dose them with quinine, bringing the plasmodium under control. "The effect was remarkable; the downward progression of late-stage syphilis was stopped," by one account, from the late Robert S. Desowitz, who was a prominent parasitologist himself as well as a lively writer. "Institutions for malaria therapy rapidly proliferated throughout Europe and the technique was taken up in several centers in the United States. In this way, tens of thousands of syphilitics were saved from a sure and agonizing death"—saved by malaria.

One of those European institutions was in Bucharest, with Professor Ciuca its vice-director. Romania had a long history of struggles against malaria, and presumably its share of syphilis too, but Ciuca evidently felt that *Plasmodium knowlesi* might be a better weapon against neurosyphilis than other kinds of the parasite. He inoculated several hundred patients and, in 1937, reported fairly good success. His program of treatments continued until, almost twenty years later, a problem arose. Repeatedly passaging *P. knowlesi* through a series of human hosts (injecting infected blood, allowing the merozoites to multiply, and then extracting infected blood) had made Ciuca's strain increasingly virulent—too virulent for comfort. After 170 such passages, he and his colleagues became concerned with its growing ferocity and stopped using it. That was a first cautionary signal, but still just a laboratory effect. (Passaging was necessary for replenishing a supply of the parasite, since it couldn't be cultured in a dish or a tube; but passaging it directly through humans liberated the parasite from whatever different evolutionary pressures had been entailed in completing its life cycle within mosquitoes. It became like the protist equivalent of a designated hitter—very capable of batting, and freed from the responsibility to play outfield.) Other evidence would eventually show that *P. knowlesi* could be dangerous enough to humans in its wild form.

In March 1965, a thirty-seven-year-old American surveyor employed by the US Army Map Service spent a month in Malaysia, including five days in a forested area northeast of the capital, Kuala Lumpur. For reasons of medical privacy (and possibly other reasons too), the surveyor's name has been occluded from the scientific literature, but his initials were BW. According to one report, BW did his work by night and slept during daylight. Hmm, stop to think: How odd for a surveyor. This wasn't the Sahara, where daytime heat was forbidding, nighttime cool, and moonshine more convenient for activity. It was tropical forest. Why the surveyor had arranged his labors that way, or what he could have been surveying (luminescent caterpillars? bat populations? natural resources? radio waves?) has never been explained, though there's some speculation that he was a spy. Malaysia at that time was struggling through its early years of independence, under pressure from the Communist-supported Sukarno government of nearby Indonesia, which must have made it a focus of US strategic concern; or maybe (as per one rumor) he was monitoring signals traffic from China. Anyway, for whatever political or cadastral reasons, this lone surveyor spent nights enough in the jungle to be bitten by more than a few *Anopheles* mosquitoes. He arrived back at Travis Air Force Base, in California, feeling sick—chills, fever, the sweats. What a surprise! Within three days, BW was admitted to the Clinical Center of the National Institutes of Health, in Bethesda, Maryland, and put into treatment for malaria. The NIH doctors diagnosed *Plasmodium malariae*, based on the look of the parasites in his blood smears under a microscope. But that identification was contradicted by the evidence of his fever cycle, just one day long. Then came the real surprise: Further testing revealed that he was infected with *P. knowlesi*, the monkey malaria. It wasn't supposed to be possible. "This occurrence," wrote a quartet of the doctors involved, "constitutes the first proof that simian malaria is a true zoonosis."

It was sometimes a human infection, in other words, as well as a disease of macaques.

But the case of BW was considered anomalous, just a one-time situation resulting from quirky circumstances. Many people spend nights out in the Malaysian jungle—local villagers while hunting, for instance—but few of them are American visitors, surveying or spying or whatever, and able later to get good medical diagnoses of their feverish ailments. That's roughly where things stood with *Plasmodium knowlesi* for thirty-five years, until two microbiologists in Malaysian Borneo, a married couple named Balbir Singh and Janet Cox-Singh, began looking into some peculiar patterns of malaria occurrence around a certain community in the Bornean interior.

29

Singh and Cox-Singh had arrived in Borneo by roundabout routes. He was born in Peninsular Malaysia, into a Sikh family with roots in the Punjab, and went to England for a university education. Eventually he got his PhD in Liverpool. Janet Cox came from Belfast to Liverpool, also to do a doctorate. They met at the Liverpool School of Tropical Medicine, in 1984, and found themselves sharing an interest in malaria, among other things. (The Liverpool School of Tropical Medicine, old and august, was a logical place to nurture such interest; Ronald Ross himself, after leaving the Indian Medical Service and before the Ross Institute was founded in London, had been a professor there.) Some years later, now married and with two young daughters, Singh and Cox-Singh moved back (for him) to the East: specifically, to Kelantan, on the east coast of Peninsular Malaysia. Then in 1999, offered a chance to do research under the auspices of a new medical school, they relocated to Sarawak, one of Malaysia's two Borneo states, establishing their lab within the University of Malaysia Sarawak, in Kuching, an exotic old city on the Sarawak River. Rajah Brooke had a palace there in the mid-nineteenth century. Alfred Russel Wallace passed through. It's a charming place if you want little backstreet hotels and riverboat commerce and Bornean jungle out your back door. Kuching means "cat," hence the nickname "Cat

City," and at the gateway to its Chinatown sits a huge concrete feline. Singh and Cox-Singh, though, didn't choose it for local color. They were tracking malaria. Soon after settling, they heard about some strange data coming from Kapit, a community along an upper tributary of the Rajang River in Sarawak.

Kapit town is the seat of Kapit Division, an area populated mainly by Iban people who live in traditional longhouses, travel the river by dugout, hunt in the forest, and raise rice and corn in gardens along the forest edges. *Plasmodium vivax* and *P. falciparum* are the most commonly reported malarial organisms in Sarawak, with *P. malariae* third in order, accounting for a small fraction. The blood-borne stages of those three can be distinguished under a microscope, rather quickly and easily, in a smear of blood on a slide—which was how malaria had been diagnosed for decades. But the reported statistics seemed skewed; a large portion of all the *P. malariae* cases in Sarawak, Singh and Cox-Singh learned, were coming from Kapit. Why? The division had a remarkably high incidence, it seemed, of this particular malaria. Furthermore, most of the Kapit cases were severe enough to require hospital treatment—rather than being mild or scarcely noticeable, as typical for *P. malariae*. Again, why? And the Kapit victims were mainly adults, who should have been immune because of prior exposure—rather than children, who as nonimmunes were the usual victims of *P. malariae*. What was going on?

Balbir Singh traveled by boat up to Kapit and took samples from eight patients, pricking the finger of each person and blotting the drop of blood onto a piece of filter paper. Back in Kuching, he and a young research assistant named Anand Radhakrishnan ran the samples through a molecular test using PCR, which was the new standard in malaria diagnostics, as in so many other areas, and a far more precise method of identification than peering at infected blood cells through a microscope.

PCR amplification of DNA fragments, followed by sequencing (reading out the genetic spelling) of those fragments, plumbs far deeper than microscopy. It allows a researcher to see below the level of cellular structure to the letter-by-letter genetic code. That code is written in nucleotides, which are components of the DNA and

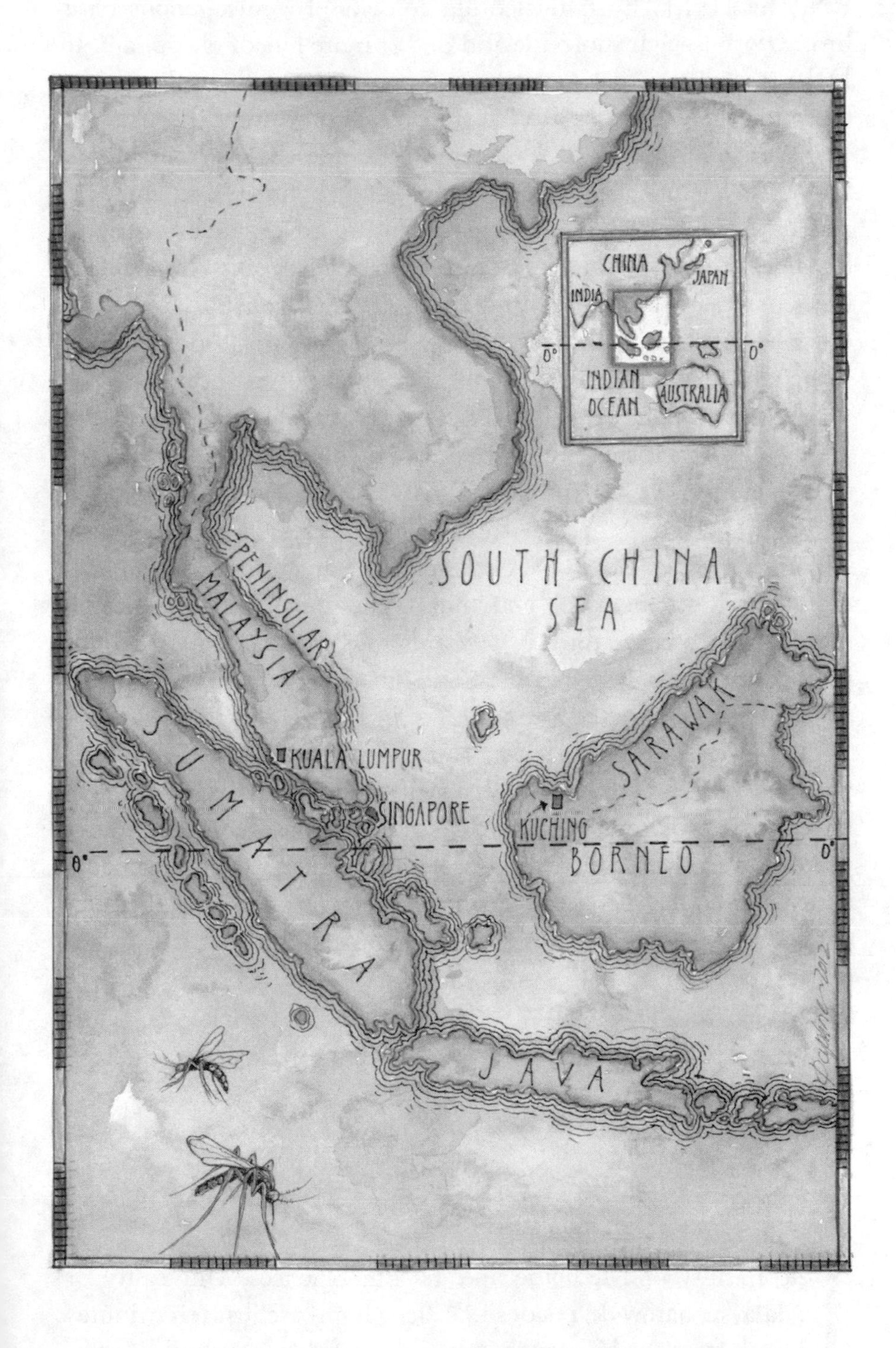
CHINA
JAPAN
INDIA
0°
0°
INDIAN
OCEAN
AUSTRALIA
SOUTH CHINA
SEA
PENINSULAR
MALAYSIA
SARAWAK
KUALA LUMPUR
SUMATRA
SINGAPORE
KUCHING
BORNEO
0°
0°
JAVA

RNA molecules. Each nucleotide consists of a nitrogenous base linked with a sugar molecule and one or more bits of phosphate. If DNA resembles a spiral staircase supported by two helical strands, those nitrogenous bases are the stair steps connecting the strands. There are four kinds of base in DNA—molecular components known as adenine, cytosine, guanine, and thymine, and abbreviated as A, C, G, and T, little pieces in the great game of genetic Scrabble. You've heard this before on the Discovery channel but it's elemental stuff that bears repeating, because genetic code is one crucial form of evidence by which disease scientists now recognize pathogens. In the RNA molecule, which serves for translating DNA into proteins (and has other roles, as we'll see), a different piece called uracil substitutes for thymine, and the Scrabble pieces are therefore A, C, G, and U.

Singh and Cox-Singh, with the help of Radhakrishnan, were looking for DNA and RNA fragments characteristic of *Plasmodium* parasites generally—and they found some. But these fragments hadn't come from *P. malariae*, nor from *P. vivax* nor *P. falciparum* either. They represented something new—or, anyway, something less expected and familiar.

Further testing and matching showed that five of the eight Kapit patients were infected with *Plasmodium knowlesi*. And there was no clustering of cases within a single longhouse, another unexpected clue. The absence of clustering meant that these people hadn't passed the parasite, via mosquitoes, to one another. Each patient seemed to have caught it from a mosquito that had bitten a macaque.

30

The Faculty of Medicine and Health Sciences, University of Malaysia Sarawak, resides in a sleek high-rise just ten minutes by cab from the big new hotels and the old market buildings of the Kuching riverfront. I found Balbir Singh there in his office on the eighth floor, a handsome and genial fiftyish man surrounded by books and papers and golf trophies. He wore a dark beard going gray, a purple-black turban, and a pair of reading glasses dangling around his neck. Despite the fact that he and his wife were leaving town the next day, for meetings with health officials elsewhere in Borneo, they had agreed to give me some time. Their discovery of *P. knowlesi* among the people of Kapit was still rather fresh, with implications for malaria treatment throughout Malaysia and beyond, and they were glad to talk.

From the high-rise, Balbir Singh and I walked across the street to a very modest South India café, his favorite, where he bought me a biryani lunch and told me about his Punjabi Sikh grandfather who had emigrated to Malaysia and his own circuit through Liverpool. I heard about *P. knowlesi* living successfully, asymptomatically, in long-tailed macaques amid the forest canopy. I heard about some surveyor, a spy, out in the Malaysian forest somewhere, but the information was flying and the food was good and I could hardly make sense of that part until later. Back in his office, Singh recounted with great élan the story of Julius Wagner-Juaregg and malaria pyrotherapy for syphilitics, Professor Ciuca's adaptation of *Plasmodium knowlesi* for that purpose in Romania, and again the mysterious American surveyor who got infected with this monkey disease in the wild. Singh showed me photos, on his computer screen, of Iban longhouses along the upper Rajang River. Eight different ethnic groups, but mostly Ibans, he said. Here's a longhouse, accommodating anywhere from five families to fifty. Great for doing blood surveys—you don't have to travel from house to house. Here's another typical scene: You see that greenery, you think it's grass, right? But it's not grass, it's hill paddy. Rice. They also grow corn. At harvest time the people stay out at night in huts by their fields, trying to haze off the macaques that come to raid the crops. They don't shoot the animals, because bullets are too expensive and a long-tailed macaque offers very little meat. Also, in some of the longhouses there's a taboo: Kill a monkey and its spirit will visit the womb of your pregnant wife, with dreadful effects on the baby. The monkeys are bold and persistent, and they've got to be kept off the paddy rice—evidently a matter of arm waving, shouting, clanging of pots. Two nights, three nights in a row the people stay out there. Of course they get bitten by nocturnal forest mosquitoes, including *Anopheles latens,* the main insect transmitting *P. knowlesi* hereabouts.

"So control is a problem," he said. "How are you going to control this?" Both men and women are infected. Their livelihood depends on going into the forest, where the macaques are abundant and so are the mosquitoes.

He showed me blown-up images from microscope slides full of malaria-infected human cells. Circles and dots, to me. Trophozoites, schizonts, gametocytes, to him. He was talking fast. Yes, easy to mistake *P. knowlesi* for *P. malariae* if that's what you're looking at, I agreed. No wonder the methods of molecular genetics have opened new vistas of discrimination. No wonder this zoonotic malaria was misdiagnosed for so long. Then we went downstairs to visit his wife in the lab.

Janet Cox-Singh is a small woman with short auburn-black hair and fine features, her speech reflecting almost no trace of her Belfast origins. She sat at a lab bench, not far from the PCR machine, before her own large computer monitor, and beneath shelves on which rested boxes filled with filter-paper samples of blood, dried and packed away, a precious archive of raw material from which she and her husband had extracted much of their data. Think of it as DNA jerky. "We developed this PCR method so we could take blood spots on filter paper and do very nice malaria epidemiology from very remote places," Cox-Singh told me. Kapit Division, Sarawak, is indeed a remote place if anywhere is.

Nearby on the floor rested several large liquid-nitrogen storage tanks for transport of frozen specimens, a more cumbersome method of bringing blood to the laboratory, not quite obsolete but now circumvented, for their purposes, by the filter-paper technique. After the first trip upriver, during which Singh had pricked eight fingers and blotted up eight samples, yielding the first signal of *P. knowlesi*, he and Cox-Singh continued their data gathering with visits to the Kapit hospital and nearby longhouses. They also expanded their reach by delegating the filter-paper technique. They sent kits of such papers to other parts of Sarawak, in the hands of trained helpers, and got back blood spots, dried but valuable. Using an old-fashioned paper punch (carefully sterilized to avoid contamination), they punched two small dark dots out of each paper and processed those dots through the PCR machine. Two crusty dots held about twenty microliters of blood, just enough for extracting DNA. Then the DNA had to be selectively amplified so they could work with it. Cox-Singh began describing to me the particular method they used, known as "nested PCR," diagramming it roughly on the back side of a journal paper as she spoke. Small subunits, fifteen hundred nucleotides, ribosomal RNA. I stared at the squiggles. Once they possessed amplified product, they sent that off to a mainland lab for genetic sequencing. The sequenced results were a longish series of letters, a passage written in genetic code as though to spell a choking expletive (*ACCGCAGGAGCGCT . . . !*), which could be entered into a vast online database for matching against known referents. That's how they had identified *P. knowlesi* in those first samples, she said, and in many more since.

Her husband pulled down a box and opened it. "This is our collection of blood spots," he said with quiet pride. Borneo is off the beaten path and, I suppose, not many science journalists visit. Inside the box was a neat file of plastic envelopes, each one containing a piece of porous paper no bigger than a business card; on each card was a rusty black spot. Near the center of the dark spot, on the card I inspected closely, was a perfectly round little hole. The punched dot, missing there, had already surrendered its secrets to science. DNA confetti.

During their first two years of work on the Kapit population, using filter-paper dots and PCR, the Singh–Cox-Singh team (like all scientists, they have helpers and colleagues) found 120 cases of *P. knowlesi*. Under earlier diagnostic assumptions and methods, most or all of those people would have been judged to have *P. malariae*, the benign form, and therefore received little or no medical care. They would have suffered, or worse. Properly diagnosed, and treated aggressively with drugs such as chloroquine, they had recovered. The paper describing those results appeared in an august British journal, *The Lancet*, delivering solid proof of what the strange case of BW the Surveyor had suggested: that *P. knowlesi* malaria is a zoonotic disease.

Expanding their search between 2001 and 2006, the team identified hundreds more cases of *P. knowlesi*, including 266 from Sarawak, 41 from Sabah (the other Malaysian state on the island of Borneo), and 5 from an area of Peninsular Malaysia just northeast of Kuala Lumpur—not far, probably, from where BW caught his case in 1965. They also found *P. knowlesi* in most of the long-tailed macaques from which they were able to take blood, confirming that those monkeys are a reservoir.

More dramatically, the team detected four human fatalities—four malaria patients, each of whom had gone to a hospital, been misdiagnosed with *P. malariae* (based on microscopy, the old way), developed severe symptoms, and died. Retrospective analysis of their blood samples by PCR showed that all four had suffered from *P. knowlesi*. These revelations suggested something more than that *P. knowlesi* is a zoonotic disease; they suggested that people were dying because doctors and microscopists were unaware of that fact. The paper in which Cox-Singh, Singh, and their colleagues presented the four-fatalities work, she told me, was initially rejected for publication. "Because we were saying that this was—"

Her husband completed the sentence: "—causing deaths."

"It was causing deaths," she concurred. "And they didn't like that." By "they" she meant anonymous manuscript reviewers for *The Lancet*. The editors of that journal, who had favored their first paper, declined this one on advice from such reviewers, in part because there was no absolute proof as to the cause of death in the four cases. There was no absolute proof, of course, because Cox-Singh and Singh had been working from archived blood samples, and reconstructing stories from medical files, to understand the illnesses of four people whose bodies were long since unavailable for postmortem. "So we ran into trouble with that one." But eventually the paper was accepted by another good journal and, published there in early 2008, caused a sizable stir. Its title stated the essence, which was that, far from being rare and innocuous, "*Plasmodium knowlesi* Malaria in Humans Is Widely Distributed and Potentially Life Threatening."

Science is a process performed in laboratories and in the field, but it's also a conversation conducted through the journals. Being part of this conversation is especially important, even in the age of email, if a scientist is separated by distance from most of his or her peers. Within that context, Singh and Cox-Singh had followed the second paper with an article in still another journal, summarizing their discoveries, reviewing previous knowledge, and offering some concrete recommendations. It was labeled "Opinion," a cautious editorial disclaimer, but it was really much more than that: a deeply informative overview, a thoughtful essay, and a warning. There was no list of coauthors; Cox-Singh and Singh spoke together, alone. The piece appeared in print not long before I met them, and I was carrying a copy.

Plasmodium knowlesi malaria, they wrote, is not a new emergent infection of humans. It has been getting into people for some while but it was overlooked. Three kinds of Asian primate serve as its reservoir hosts: the long-tailed macaque, the pig-tailed macaque, and the banded leaf monkey. Other monkeys, still unidentified, might be harboring the parasite too. Transmission from monkey to monkey (and from monkey to human) occurs by way of mosquitoes belonging to one group of closely related species, *Anopheles leucosphyrus* and its cousins, including *Anopheles latens* in Borneo. *Anopheles latens* is a forest-dwelling mosquito accustomed to biting macaques, but it will bite humans too, if presented with the necessity and the opportunity. As humans have increasingly entered the Bornean forests—killing and displacing macaques, cutting timber, setting fires, creating massive oil-palm plantations and small family farm plots, presenting themselves as an alternative host—both the necessity and the opportunity have increased. (Borneo has been deforested at a high rate within recent decades, to the point that its forest coverage is now less than 50 percent; meanwhile the island's human population has grown to about 16 million. Cox-Singh and Singh didn't cite these facts but clearly had them in mind.) Given such circumstances, Cox-Singh and Singh wrote, "it is possible that we are setting the stage for a switch of host for *P. knowlesi*, similar to the one postulated for *P. vivax*." A host switch, they meant, from macaques to humans.

They expressed the same concern to me. "Have we created this nice opening for *knowlesi* to come into?" It was Cox-Singh voicing the question. By "opening" she meant an ecological opportunity. "What's a mosquito going to do? If we start taking so much of the habitat, will the mosquito adapt then to being in a less-forest environment?"

She let that thought trickle off, paused, and then started again. "I honestly believe we're at a sort of critical point. And we should be watching. We should be watching the situation very, very carefully," she said. "And hopefully nothing will happen." But of course, as she well knew, something always does happen. It's just a question of what and when.

31

Months and years after my conversation with Balbir Singh and Janet Cox-Singh, I was still wondering about *Plasmodium knowlesi*. I remembered a curious point the two scientists had made: that, unlike other malaria parasites, *P. knowlesi* is capable of reproducing in several kinds of primate. Its tastes in warm-blooded hosts are eclectic. It infects long-tailed macaques and pig-tailed macaques and banded leaf monkeys without distressing them much. It infects humans, sometimes, causing malaria that can be severe. It infects rhesus macaques—as laboratory experiments have shown—killing them quickly and surely. Further experimental work has revealed that it can infect a wide range of primates, including marmosets from South America, African baboons, and other kinds of Asian macaque. So with regard to hosts for the asexual phase of its life cycle—the sporozoite-to-gametocyte phase, occurring in mammalian blood and livers—it is a generalist. Generalists tend to do well in changing ecological circumstances.

I remembered also a vivid illustration from their overview article. It was a sketched map of the region, showing India, Southeast Asia, and the island realm of which Borneo sits at the center. The map showed, at a glance, how widely *Anopheles leucosphyrus* mosquitoes and long-tailed macaques are distributed. A solid line demarcated the native range of the mosquitoes, encircling southwestern India and Sri Lanka in a small loop, separate to themselves, and then a much larger, irregular loop sprawling over the map like a monstrous continental amoeba. The larger loop encompassed Bhutan and Myanmar and half of Bangladesh; the northeastern Indian states, including Assam; southern China, including Yunnan and Hainan and Taiwan; Thailand and Cambodia and Vietnam and Laos; all of Malaysia, all of the Philippines; and most of Indonesia, stretching eastward beyond Bali and Sulawesi. The area within that line, by my rough calculations, contains about 818 million people—that is, roughly one-eighth of the world's human population, living within the greater ambit of *Anopheles leucosphyrus* mosquitoes. The distributional range of the long-tailed macaque was also traced on the map: a line of dashes, encircling almost the same area as the mosquitoes' range, though not quite so large.

Would it be excessive to say that those 818 million people are all at risk of *P. knowlesi* malaria? Yes, it would. For one thing, long-tailed macaques are only patchily present within that vast area; they live mainly in edge habitats, where human-modified landscape meets forest. For another thing, the level of human jeopardy depends on other factors besides the geographical ranges of the mosquitoes and the monkeys. It depends on whether those mosquitoes come out of the forest to bite humans, and whether people go into the forest to be bitten. It depends on whether sizable expanses of forest are left standing within that region and, if not, how the mosquitoes react. As deforestation proceeds, do the forest mosquitoes go extinct, or do they adapt? It depends on whether the parasite becomes so well established within human populations that monkey hosts are no longer necessary. It depends on whether the parasite colonizes a new vector, achieving transmission via some other kind of mosquito—members of a species more willing to seek out humans in their longhouses, their villages, their cities. In other words, it depends on chance and ecology and evolution.

Awareness of *P. knowlesi* malaria, thanks in large part to Singh and Cox-Singh, has begun to spread. What's harder to know is whether the parasite itself is spreading. Reports have appeared in the journals, documenting a few cases throughout the wider region. There was a Bangkok man who spent several weeks in a forested area of southern Thailand and got bitten by mosquitoes at dawn and dusk. There was a young soldier in Singapore who had trained in a forest full of mosquitoes and macaques. There were five cases from Palawan, a heavily forested island in the Philippines. There was an Australian man who worked in Kalimantan (Indonesian Borneo), near a forested area, and later sought treatment at a hospital in Sydney. There was a Finnish tourist who spent a month in Peninsular Malaysia, including five days in the jungle without a bed net, and then turned up sick in Helsinki. There have been cases from China and Myanmar. They all tested positive for *P. knowlesi*. No one knows how many more cases have gone unreported or unrecognized.

We are a relatively young kind of primate, we humans, and therefore our diseases are young too. We have borrowed our troubles from other creatures. Some of those infections, such as Hendra and Ebola, visit us only occasionally and, when it happens, arrive soon at dead ends. Others do as the influenzas and the HIVs have done—take hold, spread from person to person, and achieve vast, far-flung, enduring success within the universe of habitat that is us. *Plasmodium falciparum* and *Plasmodium vivax*, from their origins in nonhuman primates, have done that also.

Plasmodium knowlesi may be at a transitional stage—or anyway, a straddling stage—and we can't know its future plans. It's a protist, after all; it doesn't *have* plans. It will simply react to circumstances. Possibly it will adapt to the changing trend among primate hosts—fewer monkeys, more humans—as its plasmodium cousins have adapted over the epochs. Meanwhile it serves as a nice reminder of what's crucial about any zoonosis: not just where the thing comes from but how far it goes.

IV

DINNER AT THE RAT FARM

32

In late February 2003, SARS got on a plane in Hong Kong and went to Toronto.

Its arrival in Canada was unheralded but then, within days, it began to make itself felt. It killed the seventy-eight-year-old grandmother who had carried it into the country, killed her grown son a week later, and spread through the hospital where the son had received treatment. Rather quickly it infected several hundred other Toronto residents, of whom thirty-one eventually died. One of the infected was a forty-six-year-old Filipino woman, working in Ontario as a nursing attendant, who flew home to the Philippines for an Easter visit, started feeling sick the day after arrival (but remained active, shopping and visiting relatives), and began a new chain of infections on the island of Luzon. So SARS had gone halfway around the world and back, in two airline leaps, over the course of six weeks. If circumstances had been different—less delay on the ground in Toronto, an earlier visitor headed from there to Luzon or Singapore or Sydney—the disease could have completed its global circuit far more quickly.

To say that "SARS got on a plane," of course, is to commit metonymy and personification, both of which are forbidden to the authors of scientific journal articles but permissible to the likes of me. And you know what I mean: that what actually boarded an airplane in each of those cases was an unfortunate woman carrying some sort of infectious agent. The seventy-eight-year-old Toronto grandmother and the younger nursing attendant remain anonymous in the official reports, identified only by age, gender, profession, and initials (like BW the malarious surveyor), for reasons of medical privacy. As for the agent—it wasn't identified and named until weeks after the outbreak began. No one could be sure, at that early stage, whether it was a virus, a bacterium, or something else.

In the meantime it had also arrived in Singapore, Vietnam, Thailand, Taiwan, and Beijing. Singapore became another epicenter. In Hanoi, a Chinese American businessman who brought his infection from Hong Kong became ill enough to merit examination by Dr. Carlo Urbani, an Italian parasitologist and communicable-diseases expert stationed there for the World Health Organization. Within ten days the businessman was dead; within a month, Dr. Urbani was too. Urbani died at a hospital in Bangkok, having flown over for a parasitology conference in which he was never able to take part. His death, because of his much-admired work within WHO, became a signal instance of what emerged as a larger pattern: high rates of infection, and high lethality, among medical professionals exposed to this new disease, which seemed to flourish in hospitals and leap through the sky.

It reached Beijing by at least two modes of transport, one of which was China Airlines flight 112, from Hong Kong, on March 15. (The other route into Beijing was by car, when a sick woman drove up from Shanxi province seeking better treatment in the national capital; how she had become infected, and whom she infected in turn, is a different branch of the story.) Flight CA112 took off from Hong Kong that day carrying 120 people, including a feverish man with a worsening cough. By the time it landed in Beijing, three hours later, twenty-two other passengers and two crewmembers had received infectious doses of the coughing man's germs. From them it spread through more than seventy hospitals just in Beijing—yes, *seventy*—infecting almost four hundred health-care workers as well as other patients and their visitors.

Around the same time, officials at WHO headquarters in Geneva issued a global alert about these cases of unusual pulmonary illness in Vietnam and China. (Canada and the Philippines weren't mentioned because this was just before their involvement was recognized.) In Vietnam, said the statement, an outbreak had begun with a single patient (the one Carlo Urbani examined) who was "hospitalized for treatment of severe, acute respiratory syndrome of unknown origin." The little comma after "severe" reflects the fact that those three adjectives and one noun hadn't yet been codified into a name. Several days later, as the pattern of hopscotching outbreaks continued to unfold, WHO issued another public statement of alarm. This one, framed as an emergency travel advisory, marked the transformation of a descriptive phrase into a label. "During the past week," it said, "WHO has received reports of more than 150 new suspected cases of severe acute respiratory syndrome (SARS), an atypical pneumonia for which cause has not yet been determined." The advisory quoted WHO's director-general at the time, Dr. Gro Harlem Brundtland, speaking starkly: "This syndrome, SARS, is now a worldwide health threat." We had all better work together, Brundtland added (and do so quickly, she implied), to find the causal agent and stop its spread.

Two aspects of what made SARS so threatening were its degree of infectiousness—especially within contexts of medical care—and its lethality, which was much higher than in familiar forms of pneumonia. Another ominous trait was that the new bug, whatever it might be, seemed so very good at riding airplanes.

33

Hong Kong wasn't the origin of SARS, merely the gateway for its international dispersal . . . and very *close* to its origin. The whole phenomenon had begun quietly, several months earlier, in the southernmost province of mainland China, Guangdong, a place of thriving commerce and distinctive culinary practices, to which Hong Kong is attached like a barnacle to the belly of a whale.

Once a British colony, Hong Kong in 1997 was subsumed into the People's Republic of China—but subsumed on a special basis, retaining its own legal system, its capitalist economy, and a degree of political autonomy. The Hong Kong Special Administrative Region, which includes Kowloon and other mainland districts as well as Hong Kong Island and several other islands, shares a border with Guangdong and a fluid exchange of visitors and trade. More than a quarter million people cross that border by land travel every day. Despite the easy commercial relations and visiting privileges, though, there's not much direct contact between Hong Kong officialdom and Guangdong's provincial capital, Guangzhou, a city of 9 million people that sits about two hours by road from the crossing. Political communications are filtered through the national government in Beijing. That constraint applies also, and unfortunately, to the scientific and medical institutions in both places—such as Hong Kong University, with its excellent medical school, and the Guangzhou Institute of Respiratory Diseases. Lack of basic communication, let alone resistance to collaborative work and sharing of clinical samples, caused problems and delays in responding to SARS. The problems were eventually solved but the delays were consequential. When the infection first crossed the border, from Guangdong to Hong Kong, very little information crossed with it.

Guangdong is drained by the Zhu (Pearl) River, and the whole coastal area encompassing Hong Kong, Macau, Guangzhou, and a new border metropolis called Shenzhen, as well as Foshan, Zhongshan, and other surrounding cities, is known in English as the Pearl River Delta. On November 16, 2002, a forty-six-year-old man in Foshan came down with fever and respiratory distress. He was the first case of this new thing, so far as epidemiological sleuthing can determine. No samples of his blood or mucus were later available for laboratory screening, but the fact that he triggered a chain of other cases (his wife, an aunt who visited him in the hospital, the aunt's husband and daughter) strongly suggests that SARS was what he had. His name too goes unmentioned, and he has been described simply as a "local government official." The only salient aspect of his profile, in retrospect, is that he had helped prepare some meals, of which the ingredients included chicken, domestic cat, and snake. Snake on the menu wasn't unusual in Guangdong. It's a province of ravenous, unsqueamish carnivores, where the list of animals considered delectable could be mistaken for the inventory of a pet store or a zoo.

Three weeks later, in early December, a restaurant chef in Shenzhen fell ill with similar symptoms. This fellow worked as a stir-fry cook, and though his tasks didn't include killing or gutting wild animals, he would have handled their chopped and diced pieces. Feeling sick in Shenzhen, he commuted home to another city, Heyuan, and sought medical treatment there at the Heyuan City People's Hospital, where he infected at least six health-care workers before being transferred to a hospital in Guangzhou, about 130 miles to the southwest. One young doctor who rode to Guangzhou in the ambulance with him also became infected.

Not long afterward, during late December and January, other such illnesses started occurring in Zhongshan, sixty miles south of Guangzhou and just west across the Pearl River Delta from Hong Kong. Within the next several weeks, twenty-eight cases were recognized there. Symptoms included headache, high fever, chills, body aches, severe and persistent coughing, coughing up bloody phlegm, and progressive destruction of the lungs, which tended to stiffen and fill with fluid, causing oxygen deprivation that in some cases led to organ failure and death. Thirteen of the Zhongshan patients were health-care workers and at least one was another chef, whose bill of fare included snakes, foxes, civets (smallish mammals, distantly related to mongooses), and rats.

Authorities at Guangdong's provincial health bureau noticed the Zhongshan cluster and sent teams of "experts" to help with treatment and prevention, but nobody was really an expert, not yet, on this mystifying, unidentified disease. One of those teams prepared an advisory document on the new ailment, labeling it "atypical pneumonia" (*feidian* in Cantonese). That was the phrase, a common though vague formulation, used weeks later by WHO in its global alert. An atypical pneumonia can be any sort of lung infection not attributable to one of the familiar agents, such as the bacterium *Streptococcus pneumoniae*. Applying that familiar label tended to minimize, not accentuate, the uniqueness and potential severity of what was occurring in Zhongshan. This "pneumonia" was not just atypical; it was anomalous, fierce, and scary.

The advisory document, which went to health offices and hospitals throughout the province (but was otherwise kept secret), also supplied a list of telltale symptoms and recommended measures for controlling against wider spread. Those recommendations were too little and too late. At the end of the month, a seafood wholesaler who had recently visited Zhongshan checked into a Guangzhou hospital and triggered the chain of infections that would circle the world.

This seafood merchant was a man named Zhou Zuofeng. He holds the distinction of being the first "superspreader" of the SARS epidemic. A superspreader is a patient who, for one reason or another, directly infects far more people than does the typical infected patient. While R_0 (that important variable introduced to disease mathematics by George MacDonald) represents the average number of secondary infections caused by each primary infection at the start of an outbreak, a superspreader is someone who dramatically exceeds the average. The presence of a superspreader in the mix, therefore, is a crucial factor in practical terms that might be overlooked by the usual math. "Population estimates of R_0 can obscure considerable individual variation in infectiousness," according to J. O. Lloyd-Smith and several colleagues, writing in the journal *Nature*, "as highlighted during the global emergence of severe acute respiratory syndrome (SARS) by numerous 'superspreading events' in which certain individuals infected unusually large numbers of secondary cases." Typhoid Mary was a legendary superspreader. The significance of the concept, Lloyd-Smith and his coauthors noted, is that if superspreaders exist and can be identified during a disease outbreak, then control measures should be targeted at isolating those individuals, rather than applied more broadly and diffusely across an entire population. Conversely, if you quarantine forty-nine infectious patients but miss one, and that one is a superspreader, your control efforts have failed and you face an epidemic. But this useful advice was offered from hindsight, in 2005, too late for application to the fishmonger Zhou Zuofeng in early 2003.

No one seems to know where Mr. Zhou picked up his infection, though presumably it wasn't from seafood. Fish and marine crustaceans have never been implicated among the possible reservoirs for the pathogen causing SARS. Zhou ran a shop in a major fish market, and possibly his sphere of activities intersected with other live markets, including those that offered domestic and wild birds and mammals. Whatever its source, the infection took hold, went to his lungs, caused coughing and fever, and drove him to seek help at a Guangzhou hospital on January 30, 2003. He remained at that hospital only two days, during which he infected at least thirty health-care workers. His condition worsening, he was transferred to a second hospital, a place that specialized in handling cases of atypical pneumonia. Two more doctors, two nurses, and another ambulance driver were infected during his transfer, as Zhou gasped for breath, vomited, and spattered phlegm around the ambulance. At the second hospital he was intubated to save him from suffocation. That is, a flexible tube was inserted deep into his mouth, past his glottis, and down his windpipe into his lungs, to help with breathing. This event represents another important clue toward explaining how SARS spread so effectively through hospitals around the world.

Intubation is a simple procedure, at least in theory, but it can be difficult to execute amid the gag reflexes, sputters, and expectorations of the patient. The task was especially hard with Zhou, a portly man, sedated and feverish, and though his disease hadn't yet been identified, the attending doctors and nurses seem to have had some sense of the danger to which they were being exposed. They knew by then that this atypical pneumonia, this whatever, was more transmissible and more lethal than pneumonias of the common sort. "Each time they began to insert the tube," according to an account by Thomas Abraham, a veteran foreign correspondent based in Hong Kong, there was "an eruption" of bloody mucus. Abraham continues:

> It splashed on to the floor, the equipment and the faces and gowns of the medical staff. They knew the mucous [*sic*] was highly infectious, and in the normal course of things, they would have cleaned themselves up as quickly as possible. But with a critically ill patient kicking and heaving around, a tube half-inserted into his windpipe and mucous and blood spurting out, there was no way any of them could leave.

At that hospital, twenty-three doctors and nurses became infected from Zhou, plus eighteen other patients and their relatives. Nineteen members of his own family also got sick. Zhou himself would eventually become known among medical staff in Guangzhou as the Poison King. He survived the illness, though many people who caught it from him—directly, or indirectly down a long chain of contacts—did not.

One of those secondary cases was a sixty-four-year-old physician named Liu Jianlun, a professor of nephrology at the teaching hospital where Zhou had first been treated. Professor Liu began feeling flulike symptoms on February 15, two weeks after his exposure to Zhou, and then seemed to get better—well enough, he thought, to follow through on plans to attend his nephew's wedding in Hong Kong. He and his wife took the three-hour bus ride from Guangzhou on February 21, crossed the border, spent an evening with family, and then checked into a large, midrange hotel called the Metropole, favored by businessmen and tourists, in the Kowloon district of Hong Kong. They were given room 911, across from the elevators in the middle of a long corridor, a fact that became central to later epidemiological investigations.

Two fateful things happened that night at the Metropole Hotel. The professor's condition worsened; and at some point he seems to have sneezed, coughed, or (depending on which account you believe) vomited in the ninth-floor corridor. In any case, he shed a sizable dose of the pathogen that was making him sick—enough to infect at least sixteen other guests and a visitor to the hotel. Professor Liu thereby became the second known superspreader of the epidemic.

Among the hotel guests sharing floor nine was a seventy-eight-year-old grandmother from Canada. I mentioned her earlier. She had come to visit family and then spent several nights at the Metropole, along with her husband, as part of an airline-hotel package. Her room was 904, just across the corridor and a few steps down from Professor Liu's. Her stay overlapped with his presence for only one night—the night of February 21, 2003. Maybe they shared a ride on the elevator. Maybe they passed in the hallway. Maybe they never laid eyes on each other. No one knows, not even the epidemiologists. What's known is that, the next day, the professor awoke feeling too sick to attend any wedding and instead checked himself into the nearest hospital. He would die on March 4.

One day after Professor Liu left the Metropole, the Canadian grandmother left too, having finished her Hong Kong visit. Infected but not yet symptomatic, and presumably feeling fine, she boarded her flight home to Toronto, taking SARS global.

34

Another route of international dispersal from the Metropole Hotel led to Singapore, when a young woman named Esther Mok returned from a shopping vacation in Hong Kong, feeling feverish. That was February 25. For the previous four nights, she and a female friend had shared room 938 at the Metropole, about twenty steps from Professor Liu's room.

Back home in Singapore, Mok's fever lingered and she developed a cough. On March 1 she consulted doctors at Tan Tock Seng Hospital, a large public facility housed in gleaming new buildings just north of the city center. After a chest X-ray showed white patches on her right lung, Mok was admitted under a diagnosis of atypical pneumonia. One of the doctors who saw her was Brenda Ang, a senior consultant for infectious diseases, who happened also to be in charge of infection control at Tan Tock Seng. There was no particular alarm about infection control, though, when Esther Mok brought her condition to the hospital. "At that time," Brenda Ang told me later, "we didn't know what it was."

Ang agreed to take me through the story from memory, half a dozen years after the events, and though she warned that her recollections might be patchy, on many points they seemed rather precise. We met in a conference room within a small, detached structure on the landscaped grounds of Tan Tock Seng; it was a room that served intermittently for staff meetings and as a classroom for medical students on rounds, but we had it for an hour. Ang was a tiny, forthright woman in a lilac print dress. Observing medical discretion, she didn't use Esther Mok's name but spoke instead of "a young lady" who had been "the first index case." In her role as infectious disease consultant, Dr. Ang had seen the first index case herself. She was assisted by her registrar (a younger doctor in specialty training), who took a mucus sample from Mok for culturing. The registrar wasn't wearing a mask, Ang told me. No one at Tan Tock Seng was masked against this infection at the start, but unlike Ang herself, the registrar got sick.

His case, with some dramatic complications, unfolded later. In the meantime, Ang and her colleagues dealt with Esther Mok's worsening pneumonia, unaware that the young woman was becoming another superspreader of this disease that had not yet been identified or named.

At first Mok was placed in an open ward, with closely spaced beds, in proximity to other patients and staff members coming and going. After a few days, now gasping for air, she was transferred to the Intensive Care Unit. It seemed unusual, Ang told me, for such a young person to be struck by pneumonia so severely—unusual enough that, on the Friday of that week, when doctors from the other Singapore hospitals visited Tan Tock Seng for weekly grand rounds, Ang and her colleagues presented the atypical pneumonia case for discussion. Having heard the symptoms and the history, one doctor from Singapore General Hospital spoke up, saying, That's odd, we have an atypical pneumonia case too, another young woman, and she too has recently returned from Hong Kong. With a little checking, they learned that the Singapore General case was Esther Mok's friend, who had shared room 938 at the Metropole. This brought a moment of chill recognition.

In coming days, more atypical pneumonias arrived at Tan Tock Seng, most or all of them with connections to Esther Mok. First was her mother. Three days later, the pastor of her church, who had visited Esther at the hospital to pray, came back as a patient. Then her father showed up, suffering a cough with blood-streaked sputum. Then her maternal grandmother, then her uncle. By mid-month they were all patients at Tan Tock Seng. And as the Mok family cluster began to generate alarm, another bit of ominous news reached Brenda Ang. It was Thursday, March 13, when an administrative assistant informed her that four nurses from Mok's original ward had called in sick. Four nurses out sick on one day—that wasn't anywhere near within the boundaries of normal. "Defining moment for me," Ang said dryly, as I sat before her scribbling notes. "Everything was accelerating."

And related events were accelerating worldwide, not just at Tan Tock Seng—though Ang and her colleagues didn't yet know it. In Geneva, at almost precisely the same time, WHO issued its global alert about a "severe, acute respiratory syndrome of unknown origin." Officials at Singapore's Ministry of Health were soon in the loop, made aware that three cases of atypical pneumonia (Esther Mok and her friend, plus another) had turned up at once, all traceable to Hong Kong's Metropole Hotel. That put Mok's case into a much larger picture. Someone from the ministry seems to have called the CEO of Tan Tock Seng, whereupon a meeting of senior hospital staff was convened. The CEO, the chairman of the medical board, the nursing director, Ang herself as head of infection control, and others—they all came to this room, Ang said, to discuss what was happening.

"Came to *this* room?" I asked.

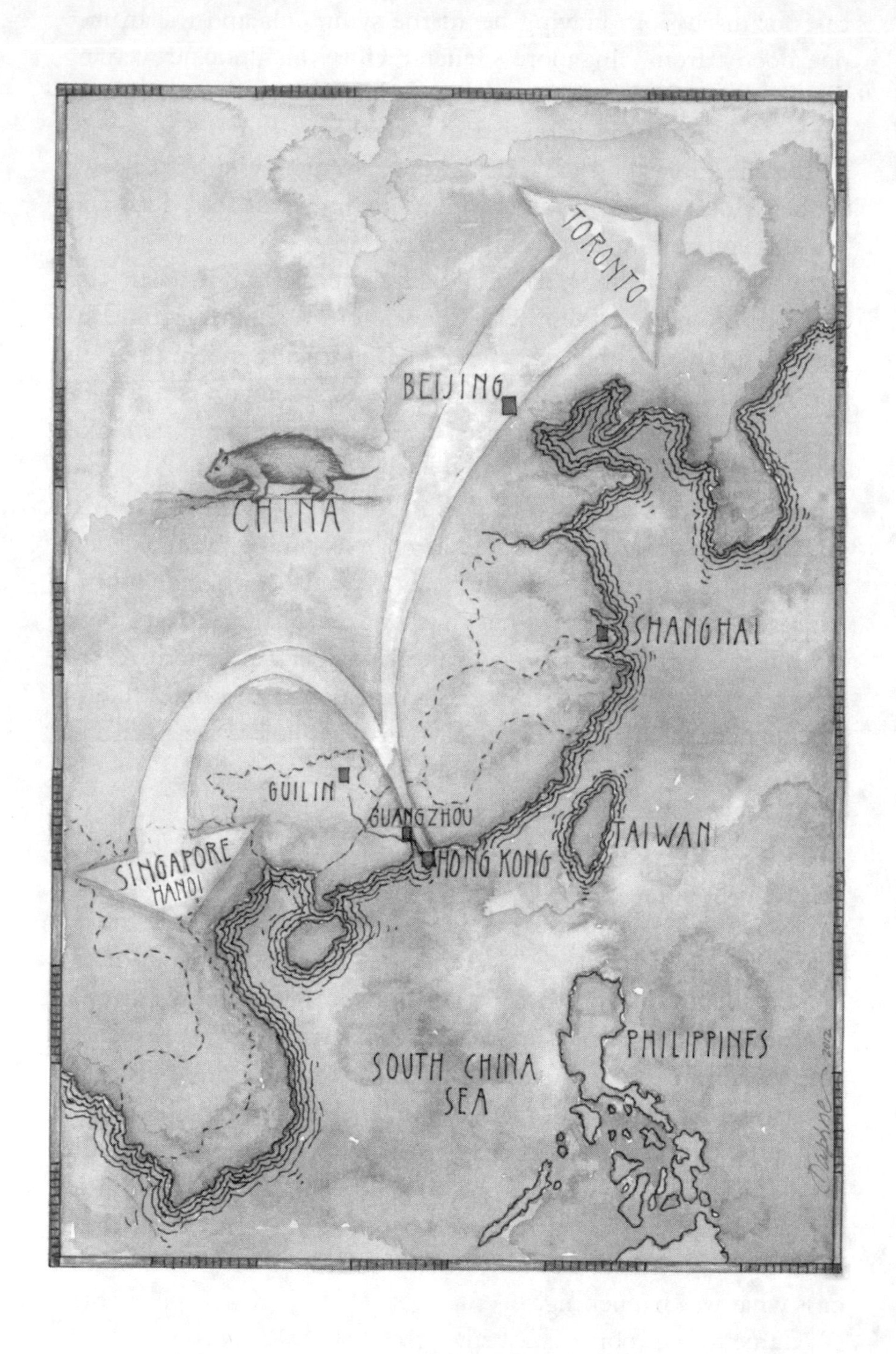
TORONTO
BEIJING
CHINA
SHANGHAI
GUILIN
GUANGZHOU
HONG KONG
TAIWAN
SINGAPORE
HANOI
SOUTH CHINA SEA
PHILIPPINES

"This room," she said. "Same room." That's when the CEO told them: "I think we've got an outbreak on our hands. And we need to organize."

A doctor named Leo Yee Sin, with previous experience of handling a Nipah outbreak, was charged with mobilizing special measures of response. The Ministry of Health advised Tan Tock Seng's leadership: Get ready to accept cases, because we're starting to see more—friends and relatives of the first group, now showing symptoms. Leo Yee Sin got people moving. They set up a tent outside one ward, for screening patients, and brought down an X-ray machine to check possible cases for lung involvement. Most of the patients were admitted to general wards, but the sicker ones went to Intensive Care. As the first Intensive Care Unit filled up, two others were converted into SARS ICUs, exclusively for handling additional cases. Isolation and barrier nursing were important control measures, though Ang and her colleagues still didn't know what they were isolating. "Remember," she told me, "all this time there are no diagnostic tests." No tests, she meant, that detected presence or absence of the culpable infectious agent—because no one had yet identified that agent. "We are going purely based on epidemiology—whether there is contact with some of the source patients." It was blind man's bluff.

On Friday of that week, March 14, the hospital's annual dinner and dance, long planned and anticipated, would occur at the Westin Hotel. It went ahead as scheduled, more or less, although Brenda Ang and some colleagues sat at half-empty tables wondering, Where's Leo Yee Sin, where's this colleague, where's that one? Well, they were absent in extremis—back at the hospital, shifting beds and other furniture to put the place on an emergency footing. Ang herself rejoined the scramble on Saturday morning.

In her capacity as head of infection control, Ang started getting all staff members into gowns, gloves, and high-filtration N95 masks, the kind that fit more snugly than mere surgical masks. But she faced a shortage of those supplies, and then also black-market inflation; N95 masks in Singapore went from $2 to $8 apiece. Still, they were doing the best that could be done. On March 23, by which point the disease had an internationally recognized name, Tan Tock Seng became the designated SARS hospital for Singapore, with all patients to be transferred there from other hospitals. Visitation was restricted. Staff members were masked, gloved, and gowned.

Before the isolation and protection measures were fully implemented, though, another superspreader event occurred, this one in the hospital's Coronary Care Unit. A middle-aged woman with multiple health problems, including diabetes and heart disease, had been admitted to one of the open wards; she was infected there by a health-care worker, who had in turn been infected by Esther Mok. Then the older woman suffered a heart attack and was moved to the CCU. Her atypical pneumonia symptoms hadn't yet manifested—not enough, anyway, to be weighed against her coronary crisis. In the CCU she was intubated by the attendant cardiologist, with assistance from a cardiology resident. Again, as with the Poison King in Guangzhou, intubation seems to have been an occasion for transmission. Eventually twenty-seven people became infected in the CCU, including five doctors, thirteen nurses, one ultrasound technician, two cardiac technicians, one attendant, and five visitors. I found that tally in a later report. Brenda Ang's account was more personal. She recollected that the cardiologist, a pregnant woman, had worn a mask while performing the intubation, and though that doctor got ill afterward, she recovered. The resident, standing nearby, had worn no mask. "It was a guy. He was sick for a while and brought it home. His mother," Ang said. "His own mother nursed him and *she* became sick."

"Did they survive?"

"No."

"Neither one of them," I said.

"It was one of the most painful things. Because he was a young, twenty-seven-year-old doctor. And his mother also died."

Another young doctor who faced similar exposure was Brenda Ang's registrar—remember him?—who had taken a throat swab from Esther Mok. His story reflects the dawning awareness that this syndrome was caused by some highly infectious bug, maybe a bacterium, maybe a virus, which spread readily through face-to-face contact, especially in crowded or intimate circumstances. Days after assisting Ang with her examination of Mok, the registrar boarded a plane. He flew to an infectious-disease conference in New York, twenty hours' worth of air travel from Singapore, and was there when he began feeling sick. Before embarking to come home, via Frankfurt, he phoned a colleague in Singapore and mentioned that he was ill. The colleague alerted Singaporean authorities, who alerted WHO, who alerted German officials, who met the plane when it landed in Frankfurt and took the doctor away into quarantine. He spent almost three weeks in a Frankfurt hospital, along with his wife and his mother-in-law, who by then were sick too. One crewmember from the plane, but only one, had also been infected. Unlike the cardiology resident who assisted the intubation, though, these patients in Frankfurt all survived.

Back in Singapore, health officials and government authorities cooperated to stanch further transmission. They enacted firm measures that reached far beyond the hospitals—such as enforced quarantine of possible cases, jail time and fines for quarantine breakers, closure of a large public market, school closures, daily temperature checks for cab drivers—and the outbreak was brought to an end. Singapore is an atypical city, firmly governed and orderly (that's putting it politely), therefore especially capable of dealing with an atypical pneumonia, even one so menacing as this. On May 20, 2003, eleven people were taken to court and fined $300 each for spitting.

By the middle of July, when the last SARS patient left Tan Tock Seng Hospital, more than two hundred cases had been recognized. Thirty-three of those people died, among whom were Esther Mok's father, her pastor, her mother, and her uncle, in that order of demise. Esther herself survived.

35

Dead or recovered, they had all been infected—but infected with *what*?

As the disease spread internationally, scientists on three continents worked in their laboratories with samples of tissue, blood, mucus, feces, and other vital, unsavory materials taken from one patient or another, trying to isolate and identify a causal agent. The very name coined during that early period, SARS, reflects the fact that this thing was known only by its effects, its impacts, like the footprints of a large, invisible beast. Ebola is a virus. Hendra is a virus. Nipah is a virus. SARS is a syndrome.

The search for the SARS pathogen proceeded urgently in those laboratories, but it was hampered by some confusing signals and false leads. For starters, the symptoms looked a little bit too much like influenza—or, more precisely, like influenza at its worst. One form of influenza at its worst is the so-called bird flu, caused by a virus designated as H5N1, with which Hong Kong in particular had had fearful experience just half a dozen years earlier, when eighteen people became infected by spillovers from domestic poultry. Eighteen doesn't sound like a large number of patients; the fearful aspect was that six of those eighteen died. Health authorities had responded quickly, ordering the closure of live poultry markets and the destruction of every chicken in Hong Kong—amounting to 1.5 million doomed, squawking birds—followed by a seven-week hiatus for decontamination. This draconian response, combined with the fact that H5N1 didn't transmit well from human to human, only from bird to human, had succeeded in ending the 1997 Hong Kong outbreak. But in February 2003, just when alarming rumors about "a strange contagious disease" began to emerge by email and text message from Guangdong, avian flu struck again in Hong Kong. It was entirely distinct from the SARS outbreak, but that couldn't easily be seen at the time.

The flu killed a thirty-three-year-old man and sickened (but didn't kill) his eight-year-old son. It probably also killed the man's seven-year-old daughter, who had died two weeks earlier of a pneumonia-like illness during a family visit to Fujian, the Chinese province just northeast of Guangdong. Possibly the little girl had consorted too closely with Chinese chickens; her brother had definitely done that, according to his own later testimony. Samples of nasal mucus from both the father and the son showed positive for H5N1, which seemed to suggest that the wider flurry of case reports from Guangdong might likewise pertain to avian flu. So the scientists tested their SARS samples for H5N1. But that was a false lead.

Another wrong notion was that SARS might be caused by some form of chlamydia, a diverse group of bacteria that includes two kinds associated with respiratory disease in humans (as well as another, more famous among teenagers, that's sexually transmitted). One of the respiratory chlamydias is zoonotic, leaping from birds (notably, pet parrots) into humans. During late February, a very senior Chinese microbiologist found what looked like chlamydia in some SARS specimens and, based on his tenuous evidence—also, his august standing in the respectful milieu of Chinese science—the chlamydia hypothesis was embraced overconfidently by high health officials in Beijing. At least one other eminent Chinese researcher dissented, arguing that, if a chlamydia was the cause, SARS cases should have responded to treatment with antibiotics—which they did not. But that fellow was down in Guangdong, at the Institute for Respiratory Diseases, and Beijing didn't want to hear him.

The laboratory scientists meanwhile explored other possibilities too, quite a list: plague, spotted fever, Legionnaires' disease, typhus, several kinds of bacterial pneumonia, seasonal influenza, *E. coli* in the blood, Old and New World hantaviruses, and more. Part of what made the task difficult was that, in pursuing the SARS agent, they didn't know whether they were looking for something familiar, something newish but closely resembling something familiar, or something entirely new.

And there was one other possible category: something familiar to veterinarians but entirely new as an infection of humans. In other words, an emerging zoonosis.

The sorts of lab methodology I've described earlier, involving

PCR to screen for recognizable fragments of DNA or RNA, combined with molecular assays to detect antibodies or antigens, are useful only in searching for what's familiar—or, at least, for what closely resembles something familiar. Such tests essentially give you a positive, negative, or approximated answer in response to a specific question: Is it *this*? Finding an entirely new pathogen is more difficult. You can't detect a microbe by its molecular signature until you know roughly what that signature is. So the lab scientist must resort to a slightly older, less automated approach: growing the microbe in a cell culture and then looking at it through a microscope.

At the University of Hong Kong, high on the side of a peak overlooking the downtown neighborhoods, a team led by Malik Peiris took this approach to its fruitful conclusion. Peiris is an Oxford-educated microbiologist, born and raised in Sri Lanka, soft-spoken and judicious, with fine dark hair that hugs his skull roundly. He is known primarily as an influenza researcher and, having come to Hong Kong in 1995, just before the big bird-flu scare, he had reason to consider avian influenza as a leading hypothesis for what was now coming out of Guangdong. "The first thing going through our minds was that the H5N1 virus had possibly acquired the ability to transmit from human to human," he told a reporter in 2003. But after testing their SARS samples for H5N1, as well as for a roster of the usual suspects, and finding no evidence of any, his team moved toward the idea that they were dealing with a new virus.

They focused then on trying to culture it. This meant, first of all, giving the mystery creature an environment of living cells in which it was able to replicate, until it grew abundant enough in the culture, and caused enough damage to the cells, that its presence could be seen. The living cells of the culture had to be one or another "immortalized" lineage (such as the famous HeLa cells from an unfortunate woman named Henrietta Lacks), so that they would continue replicating indefinitely until something killed them. Peiris's team began by offering the new bug five different cell lines that had variously proven hospitable to familiar respiratory pathogens: cells from a dog's kidney, cells from a rat's tumor, cells from the lung of an aborted human fetus, and others. No luck. There was no sign of cell damage and therefore no evidence of viral growth. Then they tried another line, derived from kidney cells of a fetal rhesus monkey. *Yes* luck. By the middle of March, they could see "cytopathic effect" in their monkey-cell culture, meaning that something had begun to replicate within those cells and destroy them, spilling from one cell to another and creating a visible zone of devastation. Within a few more days, the team had electron microscope images of round viral particles, each particle encircled by a corona of knobs. This was so unexpected that one microscopist on the team had recourse to what amounted to a field guide; he browsed through a book of viral micrographs, looking for a match, as you or I might do for a new bird or a wildflower. He found his match among a group known as the coronaviruses, characterized by a corona of knobby proteins rimming each viral particle.

So the culturing work had established that an unknown coronavirus was present in SARS patients—some of them, anyway—but that didn't necessarily mean it had caused the disease. To establish causality, Peiris's team tested blood serum from SARS patients (because it would contain antibodies) against the newfound virus in culture. This was like splashing holy water at a witch. The antibodies recognized the virus and reacted strongly. Less than a month later, based on that evidence plus other confirming tests, Malik Peiris and his colleagues published a paper cautiously announcing this new coronavirus as "a possible cause" of SARS.

They were right, and the virus became known as SARS coronavirus, inelegantly abbreviated as SARS-CoV. It was the first coronavirus ever found to inflict serious illness upon humans. (Several other coronaviruses are among the many viral strains responsible for common colds. Still others cause hepatitis in mice, gastroenteritis in pigs, and respiratory infection in turkeys.) SARS-CoV has no ominous ring. In older days, the new agent would have received a more vivid, geographical moniker such as Foshan virus or Guangzhou virus, and people would have run around saying: *Watch out, he's got Guangzhou!* But by 2003 everyone recognized that such labeling would be invidious, unwelcome, and bad for tourism.

Several other teams, working independently to isolate a SARS causal agent, had gotten the same answer at about the same time. In the United States, it was a group based at the CDC in Atlanta, with a long list of international partners. In Europe, it was a set of collaborators spread among research institutions in Germany, France, and the Netherlands. In China, it was a small squad of earnest, adept, and deferential researchers who had isolated a coronavirus and photographed it weeks before Peiris's group did the same. These unfortunate Chinese scientists, based at the Academy of Military Medical Sciences, let themselves be cowed by the chlamydia theory and its august promoter in Beijing, passing up their opportunity to announce the real discovery first. "We were too cautious," one of them said later. "We waited too long."

The next logical step for Malik Peiris and his gang, after having identified the virus, sequenced a portion of its genome, and placed it within a family tree of other coronaviruses, was to wonder about its origin. The thing hadn't come out of nowhere. But what was its usual habitat, its life history, its natural host? One scientist involved in the work, a young biologist named Leo Poon, touched on that during a conversation with me in Hong Kong.

"The data that we found in human samples," said Poon, "suggested that this virus is novel to humans. What I mean is that humans had not been infected by this virus before. So it must have been coming from some kinds of animals."

But which animals, and how did they happen to transmit the infection to people? Those questions could only be answered by going into the forests, the streets, the markets, the restaurants of southern China to gather evidence. Nudging him toward that subject, I wondered: "Were you part of the fieldwork?"

"No, I'm a *molecular* scientist," he said. It had been like asking Jackson Pollock if he painted houses, I suppose, but Leo Poon didn't take my question amiss. He was happy to give credit where due. No, another of their colleagues, a wildcat fellow named Guan Yi, with the instincts of an epidemiologist and the balls of a brass macaque, had crossed into China and, with cooperation from some local officials, taken swabs from the throats, the anuses, and the cloacae of animals on sale in the biggest live market in Shenzhen. Those samples were what first led Leo Poon (who did the molecular analysis), Malik Peiris, Guan Yi himself—and, eventually, scientists and health officials all over the world—to cast their suspicious attentions upon a mammal called the civet cat.

36

In a crowded country with 1.3 billion hungry citizens, it should be no surprise that people eat snake. It should be no surprise that there are Cantonese recipes for dog. Stir-fried cat, in such a context, seems sadly inevitable rather than shocking. But the civet cat (*Paguma larvata*) is not really a cat. More accurately known as the masked palm civet, it's a member of the viverrid family, which includes the mongooses. The culinary trade in such unusual wild animals, especially within the Pearl River Delta, has less to do with limited resources, dire necessity, and ancient traditions than with booming commerce and relatively recent fashions in conspicuous consumption. Close observers of Chinese culture call it the Era of Wild Flavor.

One of those observers is Karl Taro Greenfeld, who served as editor of *Time Asia* in Hong Kong during 2003, oversaw the magazine's coverage of SARS, and soon afterward wrote a book about it, *China Syndrome*. Before his editing role, Greenfeld had covered "the new Asia" as a journalist for some years, giving him opportunity to see what people were putting in their stomachs. According to him:

> Southern Chinese have always noshed more widely through the animal kingdom than virtually any other peoples on earth. During the Era of Wild Flavor, the range, scope, and amount of wild animal cuisine consumed would increase to include virtually every species on land, sea, or air.

Wild Flavor (*yewei* in Mandarin) was considered a way of gaining "face," prosperity, and good luck. Eating wild, Greenfeld explained, was only one aspect of these new ostentations in upscale consumption, which might also involve patronizing a brothel where a thousand women stood on offer behind a glass wall. But the food vogue arose easily from earlier traditions in fancy cuisine, natural pharmaceuticals, and exotic aphrodisiacs (such as tiger penis), and went beyond them. One official told Greenfeld that two thousand Wild Flavor restaurants were now operating within the city of Guangzhou alone. Four more received licenses during the hour Greenfeld spent in the man's office.

These eateries drew their supplies from the "wet markets" of Guangdong province, vast bazaars filled with row after row of stalls purveying live animals for food, such as the Chatou Wildlife Market in Guangzhou and the Dongmen Market in Shenzhen. Chatou began operating in 1998 and within five years had become one of the largest wild-animal markets in China, especially for mammals, birds, frogs, turtles, and snakes. Between late 2000 and early 2003, a team of researchers based in Hong Kong conducted an ongoing survey of wild animals on sale at Chatou, Dongmen, and two other big Guangdong markets. Compared to an earlier survey done in 1993–1994, the team found some changes and new trends.

First, the sheer volume of the wild-animal trade seemed to have increased. Second, there was more cross-border commerce, legal or covert, drawing wildlife from other Southeast Asian countries into southern China. Meaty but precious individuals of endangered species, such as the Bornean river turtle and the Burmese star tortoise, were turning up. Third, greater numbers of captive-bred animals had become available from commercial breeders. Certain kinds of frogs and turtles were being farmed. Snakes, according to rumor, were being farmed. Small-scale civet farms, operating in central Guangdong and southern Jiangxi (an adjacent province), helped supply the demand for that animal. In fact, much of the trade in three popular wild mammals—the Chinese ferret badger and the hog badger in addition to the masked palm civet—seemed to come from farm breeding and rearing. Evidence for this supposition, made by the survey team, was that the animals appeared relatively well fed, uninjured, and tame. Caught from the wild, they would more likely show trap wounds and other signs of desperation and abuse.

But even if they arrived healthy and robust from the farm, conditions in the markets weren't salubrious. "The animals are packed in tiny spaces and often in close contact with other wild and/or domesticated animals such as dogs and cats," the survey team wrote. "Many are either sick or with open wounds and without basic care. Animals are often slaughtered inside the markets in several stalls specialising in this." Open wire cages, stacked vertically, allowed wastes from one animal to rain down onto another. It was zoological bedlam. "The markets also provide a conducive environment," the team noted, almost passingly, "for animal diseases to jump hosts and spread to humans."

Guan Yi, the intrepid microbiologist from Hong Kong University, waded into these conditions at Dongmen Market, in Shenzhen, and persuaded sellers to let him take swab samples and blood from some of their animals. Exactly how he persuaded them is still mystifying—force of personality? eloquent arguments? clear explication of scientific urgency?—although holding a thick wad of Hong Kong dollars in his hand apparently helped. He anaesthetized twenty-five animals one by one, swabbed for mucus, swabbed for feces, drew blood, and then took the samples back to Hong Kong for analysis. The hog badgers were clean. The Chinese hares were clean. The Eurasian beavers were clean. The domestic cats were clean. Guan had also sampled six masked palm civets, which weren't clean; all six carried signs of a coronavirus resembling SARS-CoV. In addition, the fecal sample from one raccoon dog (a kind of wild canid, which looks like an overfed fox with raccoon markings), tested positive for the virus. But the data overall pointed most damningly at the civet.

This discovery, the first concrete indication that SARS is a zoonotic disease, was announced at a Hong Kong University press conference on May 23, 2003. One day later, the *South China Morning Post*, Hong Kong's leading English-language newspaper, ran a front-page story (amid all its other SARS coverage) on the announcement, headlined: SCIENTISTS LINK CIVET CATS TO SARS OUTBREAK. Residents of the city were quite aware, by then, that the SARS contagion traveled on human respiratory emissions from person to person, not just in the juices and flesh of wild meat. Earlier editions of the *Morning Post*, as well as other Hong Kong newspapers, had carried articles accompanied by vivid photos of people in surgical masks—a masked couple kissing, a hospital official demonstrating a mask and visor, a comely model at an auto show wearing a mask decorated with car advertising—as well as hospital staff and soldiers doing infection control in full hazmat suits. Hong Kong's governmental supplies department distributed 7.4 million masks to schools, medical personnel, and health officials on the front line of response, and demand was high too among the general public. Circle K, the convenience store chain, had sold almost a million masks; Sa Sa Cosmetics had moved 1.5 million. Prices per mask had quadrupled. Despite the widespread alarm over person-to-person transmission, though, there was still great interest in learning where this virus had its zoological source.

Using a press conference to break the news about civets, rather than publishing first in a scientific journal, was unorthodox but not unprecedented. Journal publication would have taken longer, because of editorial work, peer review, backlogs of articles, and lead times. Circumventing that process reflected haste, driven by civic concern and the urgency of the outbreak but also possibly by scientific competition. The CDC in Atlanta had shown its own haste just two months earlier in announcing, also by way of a press conference, that scientists there had identified a new coronavirus as the likely cause of SARS. The CDC announcement didn't mention that Malik Peiris and his team had found the same virus and confirmed its connection with SARS three days before. That act of claiming priority by the CDC, unnoticeable to the world at large, probably put the Hong Kong University scientists on edge against their competitors in Atlanta and elsewhere, and contributed to the decision to trumpet Guan Yi's discovery at the earliest reasonable chance.

One immediate consequence of Guan's findings was that the

Chinese government banned the sale of civets. In its uncertainty, the government also banned fifty-three other Wild Flavor animals from the markets. The ban inevitably caused economic losses, generating such foofaraw from animal farmers and traders that in late July, after an official review of the risks, it was rescinded. The rationale for reversal was that another group of researchers had screened masked palm civets and found no evidence whatsoever of a SARS-like virus. Under the revised policy, farm-raised civets could be legally traded again but the sale of wild-caught animals was prohibited.

Guan Yi showed some annoyance at the doubts about his findings. But he forged ahead through scientific channels, presenting a detailed explication and supporting data (tables, figures, genome sequences) in a paper published in *Science* the following October. Leo Poon and Malik Peiris, his HKU colleagues, were included in the long list of coauthors. Guan and company worded their conclusions judiciously, noting that infection of civets didn't necessarily mean that civets were the reservoir host of the virus. The civets might have becomc infected "from another, as yet unknown, animal source, which is in fact the true reservoir in nature." They might have functioned as amplifier hosts (like those Hendra-infected horses in Australia). The real point, according to Guan and his colleagues, was that the wet markets such as Dongmen and Chatou provided a venue for SARS-like coronaviruses "to amplify and to be transmitted to new hosts, including humans, and this is critically important from the point of view of public health."

By the time that paper appeared, the SARS epidemic of 2003 had been stopped, with the final toll at 8,098 people infected, of whom 774 died. The last case was detected and isolated in Taiwan on June 15. Hong Kong had been declared "SARS-free." Singapore and Canada had been declared "SARS-free." The whole world was supposedly "SARS-free." What those declarations meant, more precisely, was that no SARS infections were currently raging in humans. But the virus hadn't been eradicated. This was a zoonosis, and no disease scientist could doubt that its causal agent still lurked within one or more reservoir hosts—the palm civet, the raccoon dog, or whatever—in Guangdong and maybe elsewhere too. People celebrated the end of the outbreak, but those best informed celebrated most guardedly. SARS-CoV wasn't gone, it was only hiding. It could return.

In late December, it did. Like an aftershock to a quake, a new case broke in Guangdong. Soon afterward, three more. One patient was a waitress who had been exposed to a civet. On January 5, 2004, the day the first case was confirmed, Guangdong authorities reversed policy again, ordering the death and disposal of every masked palm civet held at a farm or a market in the province. Wild civets were another question, left unanswered.

Eradication teams from the Forestry Department (which regulates the wild animal trade) and the Health Department went out to civet farms. During the days that followed, more than a thousand captive civets were suffocated, burned, boiled, electrocuted, and drowned. It was like a medieval pogrom against satanic cats. This campaign of extermination seemed to settle the matter and made people more comfortable. That sense of comfort remained for, oh, a year or more—until other scientists showed that the doubts about reservoir identification were well-founded, that the judicious language of Guan Yi was percipient, and that the story was just a little deeper and more complicated. Woops, civets aren't the reservoir of SARS. Never mind.

37

It was Leo Poon who told me about the wild civets of Hong Kong. We were sitting in a small meeting room by the elevator on an upper floor of the Medical Faculty building at Hong Kong University, on its hillside above the towering banks and other sleek skyscrapers rising like spikes of obsidian above Central district. Below and beyond, across Victoria Harbor, were the funky streets, market stalls, alleys, shops, noodle parlors, housing projects, and tourist destinations of Kowloon, including the Metropole Hotel, now sterilized and renamed, where I was staying. I hadn't imagined there was much of *anything* wild in such a hectic environment of people and vehicles and vertical concrete, but only because I'd been limited to a cityside view of Hong Kong. Wild civets, *oh yes*, out in the New Territories, Poon assured me. Those so-called New Territories (new to the colonial British when they leased them from China in 1898 for ninety-nine years) still encompass the less developed areas of the Hong Kong Special Administrative Region, from Boundary Street on the north edge of Kowloon to the Guangdong border, plus outlying islands, with forests and mountains and nature reserves that show green on a map. These are places where, even into the twenty-first century, masked palm civets might survive in the wild. "They're all over the countryside!" Poon said.

Just after the epidemic ended, his HKU team started trapping animals out there to look for evidence of coronavirus. They focused first on civets, capturing and sampling almost two dozen. From each animal they took a respiratory swab and a fecal swab—zip zap, thank you very much—and then released the civet back to the Hong Kong wilds. Each sample was screened by PCR methodology using what the technical lingo calls "consensus primers," meaning generalized molecular jump-starters that would amplify RNA fragments shared commonly among coronaviruses, not just those unique to the SARS-like coronavirus that Guan Yi had found in his civets. So how much coronavirus did Poon find? I asked. "None at all," he said. That absence suggested that the civet is not the reservoir for SARS coronavirus. "We were quite disappointed."

But disappointment, in science, is sometimes a gateway to insight. If not the civet, then what? "We hypothesized that, if this animal"—this unidentified creature—"is the natural reservoir for SARS, it must be quite widespread." So they trapped, in several sylvan locations, whatever wild and feral animals they could find. The eventual list was richly various, ranging from rhesus macaques to porcupines, from rat snakes to turtle doves, from wild boars to black rats, and including at least one Chinese cobra. Again the

PCR results were almost universally negative—almost. Only three kinds of animal out of forty-four showed any sign of infection with a coronavirus. All three were microchiropterans. To you and me: little bats.

Only one of those registered high prevalence as a group, with most of the sampled individuals testing positive, as measured by virus shed in their feces: a delicate thing called the small bent-winged bat.

Poon gave me a copy of the paper he published (sharing credit, again, with Guan and Peiris among its coauthors) in the *Journal of Virology* in 2005, about a year after the great civet slaughter. He wanted me to be clear about his findings. "This bat coronavirus is quite different from SARS," Poon said. That is, he didn't claim to have found the reservoir of SARS-CoV. "But this *is* the first coronavirus in a bat." That is, he had turned up a strong clue.

Soon afterward, an international team of Chinese, American, and Australian researchers published an even more revealing study, based on sample collections they made in Guangdong and three other Chinese sites. This team, led by a Chinese virologist named Wendong Li, also included Hume Field, the laconic Australian who had found the reservoir of Hendra, and two scientists from the Consortium for Conservation Medicine, based in New York. Unlike the Hong Kong sampling study, Li's focused specifically on bats. The team trapped animals from the wild, drew blood, took fecal and throat swabs, and then analyzed duplicate samples of the material independently at labs in China and Australia, creating a double-check on themselves that strengthened the certitude of their results. What they found was a coronavirus that, unlike Leo Poon's, closely resembled SARS-CoV as seen in human patients. They called it SARS-like coronavirus. Their sampling showed that this SARS-like virus was especially prevalent in several bats belonging to the genus *Rhinolophus*, known commonly as horseshoe bats. Horseshoe bats are delicate little creatures with large ears and flanged, opened-out noses that, homely but practical, seem to play a role in directing their ultrasonic squeaks. They roost mainly in caves, of which southern China has an abundance; they emerge at night to feed on moths and other insects. The genus is diverse, encompassing about seventy species. Li's study showed bats of three species in particular carrying SARS-like virus: the big-eared horseshoe bat, the least horseshoe bat, and Pearson's horseshoe bat. If you ever notice these animals on the menu of a restaurant in southern China, you might want to choose the noodles instead.

High prevalence of antibodies to the virus among horseshoe bats, compared with zero prevalence among wild civets, was an important discovery. But there was more. Li's team also sequenced fragments of viral genome extracted from fecal samples. Comparative analysis of those fragments showed that the SARS-like virus contained, from sample to sample, considerable genetic diversity—more diversity than among all the isolates of SARS-CoV as known from humans. This virus seemed to have been in the bat populations for some time, mutating, changing, diverging. In fact, the totality of diversity known in the human SARS virus nested *within* the diversity of the bat virus. That sort of nesting relationship can best be depicted as a family tree. Li and company drew one. It appeared as a figure in the paper they published in *Science*. Human SARS virus was a single branch, skinny and small, within a limb of branches representing what lives within horseshoe bats.

What did this mean? It meant that horseshoe bats are a reservoir, if not *the* reservoir, of SARS-CoV. It meant that civets must have been an amplifier host, not a reservoir host, during the 2003 outbreak. It meant that no one knew just what had happened in Guangdong that winter to trigger the outbreak, although Li and his colleagues could speculate. ("An infectious consignment of bats serendipitously juxtaposed with a susceptible amplifying species," they wrote, "could result in spillover and establishment of a market cycle while susceptible animals are available to maintain infection." Infection by association. Susceptible animals might include not just masked palm civets but also raccoon dogs, ferret badgers, who knows what. So many different candidates pass through the wildlife supply chain.) It meant that you could kill every civet in China and SARS would still be among you. It meant that this virus existed—facing its ecological limits and opportunities—within a culture where "an infectious consignment of bats" might arrive at a meat market as a matter of course. It meant, Let the diner beware. And it meant that further research was needed.

38

Aleksei Chmura is a young American researcher of mild demeanor, clean-cut appearance, diverse experience, and catholic tastes. He grew up in Connecticut, quit college, traveled, worked as a baker, trained as a chef, shifted to furniture restoration, and reentered academia after ten years to study environmental science. Employed, when I first encountered him, in an administrative capacity by the Consortium for Conservation Medicine (a program of Wildlife Trust, which has since been renamed Eco-Health Alliance), he was also gathering data toward a doctorate on the ecology of zoonotic diseases in South Asia, particularly SARS. For that he was collecting samples from bats. He invited me to come out and see some of the work. On the agreed date he met my flight in Guangzhou, and I suppose the durian should have been my first signal that he was a temerarious eater.

Just in from the airport, Chmura and I joined a group of his friends at Sun Yat-sen University and plunged into a snack of the world's stinkiest fruit. It's a large spiky thing, a durian, like a puffer fish that has swallowed a football; pried open, it yields individual gobbets of glutinous creamy pulp, maybe eight or ten gobbets per fruit, and an unwelcoming bouquet. The pulp tastes like vanilla custard and smells like the underwear of someone you don't want to know. We ate barehanded, slurping the goo between our fingers as it oozed and dripped. This was before dinner, in lieu of peanuts and beer. Then we went out to a restaurant where Chmura ordered us a dish featuring congealed pig's blood—in little hepatic cubes, like diced liver—with bean sprouts and hot red peppers. By late evening my shirt was soaked with sweat. Welcome to China. But I was keen to learn what Aleksei Chmura knew, to benefit from his voracious curiosity, and I would eat my way toward insight at his side, if necessary.

Next day we flew onward to the city of Guilin, northwest of Guangzhou, in a river valley famed for its karst-mountain vistas and its caves. The mountains rose abruptly, like croquettes on a plate, but they were forested in green and riddled with natural cavities, chutes, potholes, and nooks weathered out through the soluble limestone of the karst. It was a good place to be a tourist, if you wanted dramatic scenery, and a good place to be a bat, if you wanted to roost. We hadn't come for the scenery.

But before the bat work began, Aleksei took me out to a food market for a glimpse of what's presently available in Guilin's aboveground economy. Strolling the narrow corridors between stalls, I saw vegetables laid out in neat bundles. The fruits were carefully piled. The mushrooms were gnomic. The red meat was sold mainly in slabs, joints, and pieces by women at large plywood tables, wielding sharp cleavers. The catfish, the crabs, and the eels churned slowly in aerated tanks. The bullfrogs huddled darkly in scrums. It was grim to be reminded how we doom animals with our appetite for flesh, but this place seemed no more odd or morbid than a meat market anywhere. That was the point. This was the "after" condition in a "before/after" contrast revealing how SARS had put a damper on yewei. What had changed here in recent years, Aleksei told me, was the disappearance of the trade in wildlife. Things had been far different in 2003—and even in 2006, when he first started visiting wet markets in southern China.

At the Chatou market in Guangzhou, for instance, he had seen storks, seagulls, herons, cranes, deer, alligators, crocodiles, wild pigs, raccoon dogs, flying squirrels, many snakes and turtles, many frogs, as well as domestic dogs and cats, all on sale as food. There were no civets, not when he saw the place; they had already been demonized and purged. The list he recited was just a selection, from memory and from his own discreet inspections, of what food markets were offering then. You could also buy leopard cat, Chinese muntjac, Siberian weasel, Eurasian badger, Chinese bamboo rat, butterfly lizard, and Chinese toad, plus a long list of other reptiles, amphibians, and mammals, including two kinds of fruit bat. Quite an epicure's menu. And of course birds: cattle egrets, spoonbills, cormorants, magpies, a vast selection of ducks and geese and pheasants and doves, plovers, crakes, rails, moorhens, coots, sandpipers, jays, several flavors of crow. One fellow, a Chinese colleague of Aleksei's, told me that the bird-and-bat trade was covered by an adage: "People in south China will eat everything that flies in the sky, except an airplane." He was a northerner himself.

After the SARS outbreak and the civet publicity, local governments (presumably with some pressure from Beijing) had tightened down, enacting new restrictions against wildlife in the markets. The Era of Wild Flavor hadn't ended but it had been driven underground. "There's still a lot of people in China that believe eating fresh, wild animals is good for your respiratory system, it's good for sexual potency, whatever," Aleksei said. But tracing the traffic now, let alone measuring it, was difficult. Market sellers had gotten wary, and especially wary of obvious outsiders such as Aleksei, a westerner speaking hesitant Mandarin, who might come snooping around. Wild animals were still available, no doubt, but they would be under the counter, or going out the back door, or traded from a van that stopped on a certain street corner at 2 a.m. If you wanted to feast on a Burmese star tortoise or a muntjac nowadays, you would need to know somebody who knew somebody, pay premium rates, and make your arrangements beyond the sight of the crowds.

Aleksei himself, I discovered as we shared time and meals, harbored a robustly unusual attitude on the subject of carnivorism—unusual, anyway, for an American. He didn't judge yewei harshly. He didn't disapprove of eating an animal, virtually any animal, so long as it hadn't been illegally harvested, it didn't belong to a threatened species, and it wasn't contaminated with the sort of pernicious microbes he'd come to study. One evening as we sat together over a pot of delicate little fish and bamboo shoots, crunching the fish heads and backbones as we chewed, I tried to push him to articulate his scruples. I suppose my questions were obvious and simplistic. What animals *won't* you eat, Aleksei? Tell me what kinds are off limits. Primates? Would you dine on a monkey? Without a blink he said yes, with a proviso: that the monkey meat seemed appetizing. What about ape? If you were in Africa, would you eat gorilla or chimpanzee? "I can't draw the line there," he answered. "It's either eat meat, or don't eat meat. You'd have to test me by putting human flesh in front of me." This could have sounded ghoulish, provocative, or just silly, but it didn't, because he was earnestly trying to answer my hypothetical with candor and logic. Taxonomy simply wasn't among his guiding standards of diet. Back in New York, he had told me, he lives mainly on fruit.

We spent the following days, in and around Guilin, trapping bats. The karst mountains, with all their erosional hollows, offered plenty of roosting sites. The trick was to find which caves were presently in use. For scouting the good spots, and for help with the netting and processing, Aleksei was assisted by several Chinese students, including a young ecologist named Guangjian Zhu, from East China Normal University in Shanghai. With years of experience, Guangjian was an expert handler of bats, sure-fingered and steady with the delicate little animals as they tried to wriggle free from a mist net, bite him, and escape. He was small, lean, and strong, an agile climber, an unhesitating spelunker, traits that serve well for studying bats in the wild. Yang Jian, another student, knew the local terrain and led the way to the caves. Late on the third afternoon, we four took a taxi to the outskirts of Guilin and, armed with our nets and poles, began walking down a narrow village lane. Late afternoon is when a person goes trapping for cave-roosting bats, so that they can be caught as they emerge for a night's feeding.

Just outside the village, with the sun sinking blearily behind Guilin's smog, we tromped through a citrus grove, then a pea field, then a zone of high weeds, and ascended on a faint tunnel-like trail through the hillside vegetation, a thicket of thorns and vines and bamboo. After a brief traverse, we came to a hole in the slope, not much larger than an old cellar door. Guangjian and Jian climbed down into it and disappeared; Aleksei and I followed. Beyond the hole was a small foyer and, on the far side of that, a low slot, like a mountain's smirk, leading onward. We belly crawled through and came up dirty in a second small chamber. Not for the claustrophobic. We crossed that chamber and then butt skidded through another low gap, down another rabbit hole into a third chamber (this all felt a little like being swallowed through the multiple stomachs of a cow), which opened out wider and deeper. Here we found ourselves perched high above the floor, as though on the sill of a second-floor window. We could feel the flutter of little bats whirling through the air around our faces. Which of them carries this deadly virus? I wondered.

Bats everywhere, that was good—but would we, from our perch in a high corner, be able to catch any? I couldn't see how. Then again, I couldn't see much of anything. By the light of my headlamp, I found myself a small ledge of knobby limestone on the sloping wall of the room, settled my rear upon it, and waited for whatever would happen next. What happened, to my surprise, was that Aleksei and Guangjian spread a mist net across the hole we had just come through, sealing us inside the chamber. Now the bats were sealed in too. The air was cozily warm. Mmm, yum. The net immediately began stopping little creatures, scarcely audible as they hit and stuck, like flies in a spider's web. Exit blocked, they couldn't escape us. We were the spider.

Aleksei and Guangjian untangled the bats quickly, dropping each into a cloth bag and handing the bags to me. My assigned job was to hang the bags, like laundry, on a horizontal pole I had rigged into place between rocks. It seems that bats remain more calm and comfortable—even bats in cloth bags—when they dangle. Jian meanwhile stood at the bottom of the chamber, sweeping the air with a butterfly net to catch other bats in flight, and cursing at them mildly in English when he missed.

At this moment I became conscious of a dreary human concern: Though we were searching for SARS-like coronavirus in these animals, and sharing their air in a closely confined space, none of us was wearing a mask. Not even a surgical mask, let alone an N95. Um, why *is* that? I asked Aleksei. "I guess it's like not wearing a seat belt," he said. What he meant was that our exposure represented a calculated, acceptable risk. You fly to a strange country, you jump into a cab at the airport, you're in a hurry, you don't speak the language—and usually there's no seat belt, right? Do you jump out and look for another cab? No, you proceed. You've got things to do. You might be killed on the way into town, true, but probably you won't. Accepting that increment of risk is part of functioning within exigent circumstances. Likewise in a Chinese bat cave. If you were absolutely concerned to shield yourself against the virus, you'd need not just a mask but a full Tyvek coverall, and gloves, and goggles—or maybe even a bubble hood and visor, your whole suit positive-pressurized with filtered air drawn in by a battery-powered fan. "That's not very practical," Aleksei said.

Oh, I said, and continued handling the bagged bats. I couldn't disagree. But what I thought was, Catching SARS—*that's* practical?

Back at the laboratory in Guilin, Aleksei divided the processing chores into a sort of assembly line, with Guangjian as chief handler, Jian assisting, Aleksei himself intervening at delicate moments; all three of them had pulled on blue latex gloves. Guangjian coaxed each bat out of its bag, gripping it gently but firmly. He weighed it, measured it, and identified it by species, while Jian recorded those data. *Rhinolophus pusillus,* least horseshoe bat. *Rhinolophus affinis,* intermediate horseshoe bat. *Hipposideros larvatus,* intermediate roundleaf bat. From each animal, Guangjian took mouth-swab and anal-swab samples, handing the swabs to Jian, who broke off the cotton tips and let them drop into tubes for preservation. Then Aleksei leaned in with a needlelike tool to puncture a certain small vein near the bat's tail—just a light prick, yielding one or two drops of blood. You can't take five milliliters by syringe from such a small animal, he had explained, as you might from a monkey or a civet; you'd suck the poor bat dry. Two drops were enough for two samples, duplicates, each of which could be screened independently for virus. Jian drew the blood away with a delicate pipette, drop by drop, and released it into a tube of buffer. One complete set of blood samples and swabs would go to Shanghai, the other to New York.

The three men worked smoothly together, all tasks assigned and routinized. The routine reduced risk of jabbing one another, stressing a bat unnecessarily by clumsiness or delay, or losing data. After processing, the bats were released alive from the third-floor laboratory window—most of them, anyway. There were some unintended fatalities, as there often are in any capture and handling of wild animals. Tonight, among twenty bats caught, two died. One was a least horseshoe bat, tiny as a shrew, killed instantly in the cave by a blow from the rim of Jian's butterfly net. If he couldn't release it, Aleksei decided, he should at least dissect the dead bat, salvaging what data he could.

I watched over his shoulder as he worked with a small scissors, puncturing the skin and then zipping upward across the little bat's chest. He spread the pelt back with his fingers—a light pull was enough—to reveal huge breast muscles, reddish purple as sirloin. This animal was built like Mighty Mouse. Aleksei cut through those flight muscles and then through the bones beneath, too delicate to give much resistance to his scissors. With a pointy aliquot, he drew some blood directly from the heart. He snipped out the liver and spleen, dropping them into separate tubes. And for these tasks, I noticed, the seatbelt analogy didn't apply; in addition to his blue gloves, Aleksei donned an N95 mask. Still, it was very undramatic. Only later did I notice the connection between least horseshoe bats and what Wendong Li's group had discovered. The least horseshoe bat is one of the suspected reservoir hosts of the virus.

Once finished, with the blood and organs preserved, Aleksei dropped the carcass into a Ziploc bag. He added the other bat carcass, after dissection, to the same bag. Where do those go? I asked. He pointed to a biohazardous waste box, specially designed for accepting suspect materials.

"But if they were food," he added, "they'd go there," indicating an ordinary trash basket against the wall. It was a shrug back toward our dinner discussions and the tangled matter of categorical lines: edible animals versus sacrosanct animals, safe animals versus infected animals, dangerous offal versus garbage. His point again was that such lines of division, especially in southern China, are arbitrarily and imperfectly drawn.

39

Several days later we traveled down to the city of Lipu, about seventy miles south of Guilin, to visit a rat farm that interested Aleksei. The trip took two hours on a rather luxurious bus—one offering seat belts and bottled water. At the bus station in Lipu, while waiting for our local contact to arrive, I noticed a sign stipulating security restrictions. The sign was in traditional Chinese characters but I could tell from the illustrations what was disallowed on board Lipu–Guilin busses: no bombs, no fireworks, no gasoline, no alcohol, no knives, and no snakes. We weren't carrying any.

Mr. Wei Shangzheng eventually pulled up in a white van. He was a short, stocky, amiable man who laughed easily and often, especially after his own statements, not because he thought he was funny but from sheer joy at life's curious sweetness. That's the impression I took, anyway, as his words came translated by Guangjian and his attitude shone merrily through. We climbed into his van and rode six miles to a village northeast of Lipu, where Mr. Wei turned onto a narrow lane, then through a gate, above which was a line of calligraphy announcing: SMALL HOUSE IN THE FIELD BAMBOO RAT RAISING FARM. Beyond was a courtyard surrounded on three sides by cinderblock buildings. Two wings of the building were filled with low concrete pens. The pens contained silver-gray creatures, small-eyed and blunt-headed, that looked like gigantic guinea pigs: Chinese bamboo rats. Mr. Wei gave us a tour up and down the rows.

The pens were clean and well-drained, each furnished with a water dish and holding one to four animals. The Chinese bamboo rat is native to southern China and thereabouts, and the chewed-upon stalks of bamboo in some cages signaled that its diet is true to its name. The front teeth are beaverlike, well suited for gnawing those stalks, but in disposition a bamboo rat is more comparable to a pussycat. Mr. Wei lifted one by the scruff of its neck, turned it over, and gently poked at its sizable scrotum. Don't try that with a beaver. The animal barely wriggled. Up and down the line we could see adults, juveniles, one female nursing two mouse-size pups, a mounting in progress. They breed readily, Mr. Wei explained. He kept mostly females, plus a few good studs. Last month he sold two hundred rats, and now he was expanding his operation, building new sheds. Already he was the largest bamboo-rat farmer in southern China! he told us exuberantly. Southern China, yes, and maybe beyond! After the expansions, with capacity for five thousand animals, he might be the largest bamboo-rat farmer in all China! He stated this not to brag, it seemed, but in joyous amazement at the vagaries of fortune. Business was good. Life was good. He laughed—ha ha ha!—at the thought of life's goodness. He's famous! he told us. He had been featured on Chinese TV! We could Google him! His ventures in bamboo-rat husbandry began in 2001, when he lost his job at a factory and decided to try something new.

Enterprising and innovative, Mr. Wei now also had two pairs of large, rather menacing porcupines, which sulked in larger pens at the end of one room. He was diversifying. He had begun to breed them and, yes, their offspring too would be sold as food. A special product for special occasions, targeting the wealthier, more jaded epicure. A pair of porcupines was worth $1,000, Mr. Wei said. He did not lift one and poke its scrotum.

I noticed several hypodermic syringes lying ready along the edge of a pen. Was he concerned about the health of his bamboo rats? I asked. Yes, very, said Mr. Wei, especially regarding viruses. They're invisible. They're dangerous. And you can't run a bamboo-rat farm if the animals are sick. He showed us how he would inject an ailing rat on the inside of its calf. He didn't mention what sort of medicine he injected, and most likely it was an antibiotic (therefore useless against viruses), not a newly developed SARS vaccine already available at the level of bamboo-rat wholesaling. But at least Mr. Wei's animals might be free of common bacterial infections at time of sale. What they encountered thereafter—confined to their cages among tenements of other creatures, coughed upon, peed upon, shat upon by bats or civets or raccoon dogs in a warehouse or a wet market—that was a different matter.

After the tour, Mr. Wei insisted we stay for dinner. He had commanded his family to prepare a small banquet. We sat at a low table on tiny chairs with an electric burner amid us, atop which Mr. Wei's elderly mother assembled a formidable hotpot. Into the boiling broth she slid portions of chopped pork, chopped duck, some sort of potatolike tuber, enoki mushrooms, bean sprouts, bok choy, and greens from a plant related to morning glory. She stirred. She added dabs of salt. The ingredients cooked quickly, floated up, and combined to a savory stew, which we picked at with chopsticks and ladled into our rice bowls. Separately, on a cool platter, she offered us roasted gobbets of bamboo rat.

The rat meat was mild, subtle, faintly sweet. There were many small femurs and ribs. One eats bamboo-rat hocks with one's fingers, I learned, sucking clean the bones and piling them politely on the table beside one's bowl, or else dropping them on the floor (the preferred method of Mr. Wei's father, a shirtless old man seated to my left), where they would be scavenged by the skinny cat who slept under the table. The hotpot was scorching. Mr. Wei, an exemplary host, brought out some big bottles of Liquan beer, Guilin's finest brew, nicely chilled. After a few glasses, I got into the spirit of the meal and found myself turning back to the rat platter, browsing for choice morsels.

I had begun to see Aleksei's point: If you're a carnivore, you're a carnivore, so what's the merit of fine distinctions? And if you're going to eat bamboo rat, I figured, best to do it here, at the source—before the poor animals get shipped, stacked amid other animals, and sick. Wild Flavor doesn't need to be seasoned with virus.

40

Apart from the aftershock cases in early 2004, SARS hasn't recurred . . . so far. The known events of the 2003 outbreak are still being interpreted. Many bits aren't known. Many questions remain unanswered. Are bats the sole reservoir hosts of SARS-like coronavirus? If so, which kinds of bats? Is the coronavirus that was detected in least horseshoe bats the direct ancestor of SARS-CoV as found in humans? If so, how did the original spillover occur? Was it just a single transmission—from one bat into one civet—or several such happenings? And from civet into human—how many occurrences, how many independent spillovers? Did a cage full of infected civets, sold one by one in a market, send the disease off in multiple directions at once? What exactly happened on the ninth floor of the Metropole Hotel? Did Professor Liu vomit in the corridor, or did he merely sneeze, merely cough—merely exhale? How did the virus evolve during its passage through 8,098 humans? What role did the unique culinary culture of southern China play in bringing a dangerous pathogen out to Hong Kong and then to the world? Where do Mr. Wei's bamboo rats go after leaving the Small House in the Field Bamboo Rat Raising Farm? How are they handled, amid what other animals, what piles of cages, what flying excretions, before reaching the restaurants of Guilin, Guangzhou, and Shenzhen? Why are some people superspreaders, when infected with this virus, but not others? What is the numerical value of R_0 for SARS? When will the virus emerge again? Aleksei Chmura is just one researcher among many trying to add new data to the dossier in which these questions reside.

Much has been written about SARS in the scientific literature since spring of 2003. Most of those papers are narrowly technical, addressing the details of molecular evolution, reservoir relationships, or epidemiology, but some take a broader view, asking *What is it that makes this virus unusual?* and *What have we learned from the SARS experience?* One thought that turns up in the latter sort is that "humankind has had a lucky escape." The scenario could have been very much worse. SARS in 2003 was an outbreak, not a global pandemic. Eight thousand cases are relatively few, for such an explosive infection; 774 people died, not 7 million. Several factors contributed to limiting the scope and the impact of the outbreak, of which humanity's good luck was only one. Another was the speed and excellence of the laboratory diagnostics—finding the virus and identifying it—performed by Malik Peiris, Guan Yi, their partners in Hong Kong, and their colleagues and competitors in the United States, China, and Europe. Still another was the brisk efficiency with which cases were isolated, contacts were traced, and quarantine measures were instituted in southern China (after some early confusion and denial), Hong Kong, Singapore, Hanoi, and Toronto; and the rigor of infection-control efforts within hospitals, such as those overseen by Brenda Ang at Tan Tock Seng. If the virus had arrived in a different sort of big city—more loosely governed, full of poor people, lacking first-rate medical institutions—it might have escaped containment and burned through a much larger segment of humanity.

One further factor, possibly the most crucial, was inherent to the way SARS-CoV affects the human body: Symptoms tend to appear in a person before, rather than after, that person becomes highly infectious. The headache, the fever, and the chills—maybe even the cough—precede the major discharge of virus toward other people. Even among some of the superspreaders, in 2003, this seems to have been true. That order of events allowed many SARS cases to be recognized, hospitalized, and placed in isolation before they hit their peak of infectivity. The downside was that hospital staff took the first big blasts of secondary infection; the upside was that those blasts generally weren't emitted by people still feeling healthy enough to ride a bus or a subway to work. This was an enormously consequential factor in the SARS episode—not just lucky but salvational. With influenza and many other diseases the order is reversed, high infectivity preceding symptoms by a matter of days. A perverse pattern: the danger, then the warning. That probably helped account for the scale of worldwide misery and death during the 1918–1919 influenza: high infectivity among cases before they experienced the most obvious and debilitating stages of illness. The bug traveled ahead of the sense of alarm. And that infamous global pandemic, remember, occurred in the era *before* globalization. Everything nowadays moves around the planet faster, including viruses. If SARS had conformed to the perverse pattern of presymptomatic infectivity, its 2003 emergence wouldn't be a case history in good luck and effective outbreak response. It would be a much darker story.

The much darker story remains to be told, probably not about this virus but about another. When the Next Big One comes, we can guess, it will likely conform to the same perverse pattern, high infectivity preceding notable symptoms. That will help it to move through cities and airports like an angel of death.

Two days after our dinner at the rat farm, I rose early in Guilin, said my farewell to Aleksei Chmura, and caught a plane back to Guangzhou. I killed some hours in the airport there, paying more yuans for a ham sandwich and two lattes than I'd spent on a week's meals in the cafés and noodle parlors of Guilin. Then I boarded my onward flight. In the row beside me were two young Japanese tourists, a couple, possibly returning from a romantic vacation amid the hotels, parks, malls, markets, restaurants, and crowded streets of Guangzhou or other cities of southern China. They took their seats unobtrusively and settled in for the short ride to Hong Kong. Maybe they felt a bit cowed by their own adventurousness and relieved to be headed home to a tidier nation; maybe they remembered the news stories about SARS. I didn't intrude on them with questions. I wouldn't have noticed them at all, except they were both wearing surgical masks.

Yes, I thought, if only it were that simple.

V

THE DEER, THE PARROT, AND THE KID NEXT DOOR

41

Although the drumbeat has quickened in recent decades, the emergence of new zoonotic diseases isn't unique to our era. Three stories exemplify that point.

Q fever. Sixty years before Hendra, sixty years before Vic Rail's horses started dying in that suburb of Brisbane, a very different sort of pathogen made its first recognized spillover in almost the same locale. It wasn't a virus, though in some measure it behaved like one. It was a bacterium, but unlike most other bacteria. (An ordinary bacterium differs from a virus in several obvious ways: It's a cellular organism, not a subcellular particle; it's much larger than a virus; it reproduces by fission, not by invading a cell and commandeering the cell's machinery of genetic copying; and it can usually be killed by antibiotics.) This new bug caused an illness that resembled influenza or maybe typhus. The earliest cases, occurring in 1933, were among abattoir workers in Brisbane, whose jobs involved slaughtering cattle and sheep. The affliction they suffered, known initially as "abattoir fever" among the doctors who treated them, acquired a more opaque name that stuck: Q fever. Never mind, for the moment, the origin of that name. The most notable thing about Q fever is that, even now in the age of antibiotics, for reasons related to its anomalous biology, it's still capable of causing serious devilment.

Psittacosis. Around the same time as Q fever emerged, in the 1930s, another peculiar bacterial zoonosis hit the news. This one also had links to Australia, but its scope was global, and it seems to have first reached the United States by way of a shipment of diseased parrots from South America. That was in late 1929, just in time for the Christmas season of parrot-giving. One unlucky recipient was Lillian Martin, of Annapolis, Maryland, whose husband bought her a parrot from a pet store in Baltimore. The bird keeled over dead on Christmas Day, a bad omen, and Mrs. Martin started feeling ill about five days later. Psittacosis is the medical name for the ailment she contracted; it passes from birds (especially those of the order Psittaciformes, meaning parrots and their kin) to humans, causing fever, aches, chills, pneumonia, and sometimes death. "Parrot fever" was the label under which it raised alarm in the United States during early 1930, when people exposed to those unhealthy imported birds started getting sick, especially in Maryland. PARROT FEVER HITS TRIO AT ANNAPOLIS was a typical headline, bruiting a story that ran in *The Washington Post*, on January 8, about Lillian Martin and two of her close relatives. Three days later, also in *The Post*: BALTIMORE WOMAN'S DEATH BLAMED ON PARROT DISEASE. Over the next several months psittacosis would become a national concern, causing enough reaction or overreaction that one commentator called the whole thing an example of "public hysteria," commensurate with flagellation zeal and St. John's fire in the Middle Ages.

And then there's Lyme disease. This seems to be a more recent version of the spooky-new-bacteria phenomenon. In the mid-1970s, two alert mothers in Lyme, Connecticut, near Long Island Sound, noticed that not only their children but a high incidence of other youngsters nearby had been diagnosed with juvenile rheumatoid arthritis. The odds were against any such concentration of cases occurring by chance. Once the Connecticut Department of Health and the Yale University School of Medicine had been alerted, researchers noticed that these arthritis diagnoses coincided with a particular pattern of skin rash—a red ring, spreading outward from a point—known to occur sometimes around tick bites. Ticks of the genus *Ixodes,* commonly called "deer ticks," were abundant in the forests of eastern Connecticut and surrounding areas. In the early 1980s, a microbiologist named Willy Burgdorfer found a new bacterium in the guts of some *Ixodes* ticks, a likely suspect as the causative agent. It was a spirochete, a long spiral form, closely resembling other spirochetes of the genus *Borrelia.* After further research confirmed its role in the arthritis-like syndrome, that bacterium was named *Borrelia burgdorferi* in honor of its principal discoverer. Lyme disease is now the most common tick-borne disease in North America and one of the fastest-increasing infectious diseases of any sort, especially in New England, the mid-Atlantic states, and Wisconsin. Part of what makes it problematic is that the life history of *Borrelia burgdorferi* is very complex, involving much more than ticks and people.

Lyme disease, psittacosis, Q fever: These three differ wildly in their particulars but share two traits in common. They are all zoonotic and they are all bacterial. They stand as reminders that not every bad, stubborn, new bug is a virus.

42

Parrot fever was recognized as far back as 1880, when a Swiss physician named Ritter described a household outbreak, of something resembling typhus, in which seven people got sick and three died. Because the illness showed certain pneumonia-like aspects, suggesting airborne transmission, Dr. Ritter called it "pneumotyphus," but he was groping. Although he couldn't identify what caused it, he did manage to pinpoint the site of common exposure: the house's study. The only thing remarkable about that room was that it happened to contain a dozen caged birds, including finches and parrots.

A larger outbreak occurred in Paris in 1892, after two animal dealers received a shipment of five hundred parrots imported from Buenos Aires. The dealers became infected, several of their customers became infected, and then so did relatives, friends, and one doctor in attendance. Sixteen people died. Soon the disease had cropped up also in Germany, in New York, and at a department store (which sold birds) in Wilkes-Barre, Pennsylvania. In 1898 it struck the annual exhibition of the Berlin Union of Canary Fanciers, demonstrating that parrots and their kin weren't the only birds capable of carrying this "parrot fever" microbe, whatever it was. (Canaries belong to the order Passeriformes, not to the Psittaciformes.) Half a dozen canary fanciers fell ill and, by an account in a Berlin newspaper, "three died in agony."

Then came a hiatus, if not in the incidence of parrot-borne infections at least in the attention they received. The Great War, followed immediately by the great influenza, gave people a surfeit of death and disease to engage their sorrows and fears. The 1920s were decidedly more cheerful and carefree, until they weren't. "The year 1929 marked a turning point in the revival of interest concerning the etiology of human psittacosis," according to one historical survey of the disease. Etiology, that was the crux. Outbreaks might come and go. What differed in 1929, besides the Crash and a general lowering of spirits, was a sufficiency of parrot-fever cases to make studying the *cause* not only more practical but also more urgent.

Lillian Martin of Annapolis had been among the first of this new wave, and though she eventually recovered, others weren't so lucky. *The Washington Post* continued to track the story, reporting parrot-fever fatalities in Maryland, Ohio, Pennsylvania, New York—and Hamburg, Germany. On January 13, the Surgeon General telegraphed health officials in nine states, asking for help in tracking the situation. Two weeks later, with cases now reported also from Minnesota, Florida, and California, President Hoover declared an embargo against imported parrots. The director of the Bureau of Bacteriology within Baltimore's health department, who had been doing necropsies on infected birds, got sick and died. A laboratory technician at the Hygienic Laboratory, which was part of the

US Public Health Service, got sick and died. That technician had been assisting a researcher, Charles Armstrong, with bird-to-bird transmission experiments in the laboratory basement. Their working conditions were less than ideal: two small basement rooms full of distressed parrots held in garbage cans, wire mesh over the tops, feathers and bird shit flying out, curtains soaked in disinfectant to contain the airborne drift. It wasn't BSL-4. Charles Armstrong got sick but did not die. Nine other personnel of the Hygienic Laboratory also became infected, none of whom had even entered the basement bird rooms. The laboratory director, realizing that his building was broadly contaminated with whatever wafting agent caused psittacosis, closed the place down. Then he descended to the basement himself, chloroformed all the remaining parrots, chloroformed the guinea pigs and pigeons and monkeys and rats involved in the same experimental work, and threw their dead bodies into the incinerator. This forthright man, this hands-on administrator, described in one source as "tall with a gnarled Lincolnian face," was Dr. George W. McCoy. For reasons explicable only in terms of the wonders of the immune system and the vagaries of fortune, Dr. McCoy didn't get sick.

The psittacosis epidemic of 1930 was winding down, and probably also, though more slowly, the psittacine panic. On March 19, the Acting Secretary of the Navy issued a general order for sailors on shipboard to get rid of their parrots. George McCoy reopened the Hygienic Laboratory, Charles Armstrong returned from convalescence, and the search for a cause of the disease continued.

43

Within a month, a culprit had been identified. It was a small bacterium with some unusual properties, seemingly similar to the agent that causes typhus (*Rickettsia prowazekii*) and therefore given the name *Rickettsia psittaci*. Where did it come from? Argentina had been implicated as a source of sick birds at the start of the 1930 outbreak; President Hoover's embargo would have stanched that source. But then latent psittacosis was detected in some commercial California aviaries, where parakeets for the domestic pet trade were produced—meaning that American breeders were harboring an endemic reservoir of the infection and distributing it by way of interstate commerce. So a proposal was made to destroy all those infected flocks and then reestablish the trade with healthy birds from Australia. This seemed to make sense on two counts. First, what we Americans call a "parakeet" is a native Australian bird, widespread and abundant in the wild, known to Australians as the budgerigar. Second, Australia itself (despite a high diversity of psittacine birds) was thought to be psittacosis-free. Starting over with wild birds might free the American bird trade of psittacosis. That was the idea, anyway.

A pair of American scientists got permission, despite the embargo, to import a consignment of two hundred Australian parakeets lately captured in the vicinity of Adelaide. They wanted to do an experiment. Their plan was to infect the imported birds, whose immune systems were assumed to be naïve, with American strains of psittacosis. But when one of the imports fell dead, not long after arrival, the scientists opened it up and found *Rickettsia psittaci*. They also noticed that some others of their birds, seemingly healthy, carried the bacterium as a latent infection, like the birds in those California aviaries. That raised fresh concern about what might be lurking in other aviaries, in zoos, and in pet shops around America, and strongly suggested that Australia might not be as clean as it seemed.

This is where Frank Macfarlane Burnet, a great figure in Australian science, enters the story. Burnet was a complicated, brilliant, crotchety man and a signal character in the study of infectious diseases. Eventually he would earn a knighthood, a Nobel Prize, and a number of other fancy honors, but long before those he had made a name for himself in zoonoses. Born in 1899, second child among an eventual seven, he was a solitary, opinionated schoolboy who read H. G. Wells, disapproved of his own father's shallow morality, preferred beetle collecting to more sociable activities, despised his roommates, read about Charles Darwin (who became one of his heroes) in an encyclopedia, forced himself (despite an inaptitude for sports) to achieve competence as a cricketer, and became an agnostic during his undergraduate years. Unfit for a career in the Church, ambivalent toward the law, he chose medicine. He trained as a doctor in Melbourne but then, recognizing his lack of empathy with patients, went to London for a PhD in virology. Declining a chair at the University of London, he returned to Australia to do research. He was a nationalist, stoutly Aussie. Much later in life, laden with honors and fame, Burnet kept his edge by publishing cranky pontifications on a wide range of subjects including euthanasia, infanticide for handicapped babies, Aboriginal land rights, population control, tobacco advertising, French nuclear testing in the Pacific, the futility of trying to cure cancer, and the merits (low, in his view) of molecular biology (as distinct from his discipline, microbiology). Burnet received his Nobel, in 1960, for helping illuminate the mechanisms of acquired immune tolerance. His role in understanding zoonotic diseases began much earlier. In 1934, as a young microbiologist based at the Walter and Eliza Hall Institute, back in Melbourne, he got interested in psittacosis.

Keying off the American study, Burnet ordered himself a crate of parrots and cockatoos from Adelaide. He found that a third of them were infected. He ordered another dozen from Melbourne. At least nine of those were probable carriers. Another two dozen from Melbourne yielded still more positives. So much for the myth of Australia as a prelapsarian psittacosis-free Eden.

But if the country's wild bird populations were riddled with this bacterium, how could the country's people—so many of whom doted upon their pet budgerigars and talking cockatoos—be entirely as unaffected as they seemed? The likely answer, Burnet guessed, was not some magical form of immunity but ignorance and underdiagnosis. Australian doctors didn't know psittacosis when it wheezed in their faces. To test that guess, Burnet started chasing down cases of human illness that *looked* like psittacosis but might have been diagnosed as influenza or typhoid. He and a co-investigator found seventeen people, sick with fever, cough, headache, pneumonia, et cetera, all of whom had been exposed to pet birds—either captive-bred budgerigars or parrots and cockatoos lately caught from the wild. His most interesting cluster was a group of twelve people infected from one batch of sulphur-crested cockatoos.

Those birds, all forty-nine of them, had been sold by the bird catcher to a Melbourne man, a laborer, who dabbled in bird dealing for a bit of seasonal income. Burnet called the man Mr. X, giving him the usual medical anonymity. Mr. X kept his avian merchandise in a small, dark, backyard shed. The first signal of disease in the birds, several weeks after their transfer to his "aviary," was that eight or nine of them died. But by then Mr. X, wasting no time, had sold seven others to people in the neighborhood and sent his twelve-year-old son off to the local market with twenty more. Mr. X's son got sick, and his daughter, and his wife, and his mother-in-law. Five neighbors and three other people, each of whom lived in a house with a cockatoo bought from Mr. X or his son, also fell ill, some of them severely. Nobody died. Mr. X himself didn't sicken, not on this occasion—possibly because there is no justice in the world, though more likely because exposure to *Rickettsia psittaci* during his earlier bird dealings had given him some acquired immunity.

Macfarlane Burnet, as a biologist as well as a physician, was interested in the birds and the bacterium, not just in the people. He knew that the sulphur-crested cockatoo nests in tree holes, producing two or three eggs in a clutch, and that bird catchers typically raided the nest holes just before fledging. He suspected that almost all the young became infected with the bacterium as hatchlings, before leaving (or being taken from) the nest. "If the young cockatoo, after capture, is kept under good conditions," he and his coauthor wrote, "it remains healthy and presents no danger to human beings." Likewise, the wild bird populations might carry a high prevalence of infection but suffer little impact in terms of damaged health or mortality. "When, on the other hand, birds are crowded into small spaces, with inadequate food and sunlight, their latent infection is lit up." The bacterium multiplies and "is excreted in large amounts." It floats out of the cages along with downy feathers, powdered dung, and dust. It rides the air like a Mosaic plague. People inhale it and become ill. Burnet acknowledged that no government in Australia was likely to prohibit the sale of cockatoos, not in those days, nor even to insist they be kept under decent conditions. But that's what is needed, he added gruffly. Then he turned to another disease.

44

The other disease was Q fever. Remember those abattoir workers in Brisbane, during the early 1930s, who suffered mysterious, feverish ailments resembling typhus? The job of investigating that cluster of cases fell first to a man named Edward H. Derrick, newly appointed as director of the microbiology laboratory at the Queensland Health Department. Using guinea pigs inoculated with patients' blood to start a sequence of infections and then infecting one guinea pig from another, Derrick established the presence of "a distinct clinical entity," a new sort of pathogen, not recognizable by any of the standard lab tests for typhus, undulant fever, or other familiar possibilities. But he couldn't see the new thing through a microscope, nor could he get it to grow in a dish. That led him to suspect it was a virus. So he sought help from Macfarlane Burnet.

In October 1936, Derrick sent Burnet a sample of guinea-pig liver, infected experimentally with whatever had been raging through the abattoir workers. From that sample, Burnet and a laboratory assistant continued the chain of infection in more guinea pigs, and also in a series of inoculated mice. Like Derrick, Burnet and his assistant checked for bacterial pathogens and found none. So they suspected "a filterable virus," meaning an agent so small it would pass through a fine filter designed to screen out bacteria.

They took a thin smear of puréed spleen from an infected mouse, stained it for microscopy, and looked through the scope. Thirty years later, Burnet recalled: "Most significant discoveries just grow on one over weeks or months. Recognition of Q fever as a rickettsiosis was, however, an exception datable to the minute." What he saw were tiny rod-shaped "inclusions" within some of the spleen cells. For a better view, he tried another slide of spleen using a different stain. This one showed an abundance of the rods, some within spleen cells and some floating free. "From that moment, there was no doubt in my mind about the nature of the agent responsible for Q fever." It was another new rickettsia, he concluded, not too unlike the one that caused parrot fever.

In his later recollection, characteristically blunt, Burnet told how the disease got its name:

> Problems of nomenclature arose. The local authorities objected to "abattoir's fever", which was the usual name amongst the doctors in the early period. In one of my annual reports I referred to "Queensland rickettsial fever", which seemed appropriate to me, but not to people concerned with the good name of Queensland. Derrick, more or less in desperation, since "X disease" was preoccupied by [*sic*, meaning "already applied to"] what is now Murray Valley encephalitis, then came out for "Q" fever (Q for "query"). For a long time, however, the world equated Q with Queensland, and it was only when the disease was found to be widespread around the world that "Q fever" came to stand firmly in its own right as the name of the disease.

For the scientific binomial, Derrick proposed *Rickettsia burnetii*, to honor Burnet's role in finding and identifying the bug. The genus name, *Rickettsia*, would eventually change due to a taxonomic revision, but Burnet's half stuck.

Meanwhile, nine thousand miles away, the same pathogen came under scrutiny by a much different route, when two bacteriologists at the Rocky Mountain Laboratory, in Hamilton, Montana, found it in ticks from a place called Nine Mile, a Civilian Conservation

Corps camp in the mountains northwest of Missoula. These two weren't looking for abattoir fever. Gordon Davis, the first on the hunt, had brought the ticks into his lab for research on the ecology of two other diseases, Rocky Mountain spotted fever and tularemia. Setting the ticks onto guinea pigs, he watched one guinea pig become sick with something he couldn't identify. For a while it was simply "the Nine Mile agent." Herald Cox, joining the laboratory a year later, helped Davis isolate it and recognize that it was probably a rickettsia. Then another man entered the fray, an infectious disease expert who was also a powerful administrator at the National Institutes of Health, with supervisory responsibility for Cox, Davis, and their colleagues at the Rocky Mountain Laboratory. His name was Dr. Rolla Dyer. Dr. Dyer seems to have been a bit of a bullhead, but not irredeemably so. Strongly skeptical of Cox's claim to have found that the Nine Mile agent was a rickettsia, he stormed out to Montana and into Cox's lab. Cox showed him evidence on a microscope slide. Dyer reversed himself, acknowledged the discovery, and stayed around in Hamilton just long enough, assisting Cox with the work, to catch a dose of Q fever himself. Ten days after returning to Washington, he felt "sharp pains in the eyeballs," followed by chills, followed by fever and night sweats for a week. Maybe there's some justice to zoonotic diseases after all. But probably not, just a high degree of infectiousness in Q fever, because by that time Macfarlane Burnet had caught it too. Both he and Rolla Dyer recovered.

As for Herald Cox, he was further vindicated when, in 1948, the pathogen was recognized as different enough from all other *Rickettsia* to deserve its own genus and was renamed *Coxiella burnetii*, honoring him as well as Macfarlane Burnet. That name remains today.

"There is no disease to match Q fever for queer stories," wrote Burnet, in the little memoir he published in 1967. First, he claimed, it was "a record-breaker" for producing laboratory infections, such as his own, Dyer's, and similar illnesses in two secretaries at the Hall Institute. (He may have wrongly ignored the laboratory-infection claims of psittacosis.) Second, he noted the high incidence of what had been called "Balkan grippe" during the Great War, especially among German troops in Greece and New Zealanders in Italy. Furthermore, a shipload of American soldiers had been assembled "for a night or two near Bari in southern Italy, prior to embarkation," more than half of whom took sick by the time their boat reached home. "Sooner or later, all these episodes were established as Q fever." After the war, research showed "the extraordinary versatility of *C. burnetii* as a parasite," infecting dairy cows in California, sheep in Greece, rodents in North Africa, and bandicoots back home in Queensland. It passed from one species to another in the form of minuscule airborne particles, often dispersed from the placenta or the dried milk of an infected female animal, inhaled, and then activated through the lungs, or taken directly into the bloodstream from the bite of a tick. As he said, it was versatile.

"One of the more bizarre episodes concerns an English class of art students," Burnet recounted with some enthusiasm. "Around 1950, a collection of casts from classic statuary was ordered from Italy. The crates arrived with the casts packed in straw, and everyone in the class lent a hand in unpacking. Most of them got Q fever, but no one knows how the straw was contaminated." All of this, Burnet wrote, "was the beginning of an ever-widening recognition of Q fever across the world." He was right. Though *Coxiella burnetii* is now known as a bacterium, not an anomalous form halfway between bacteria and viruses, its impact on human health didn't disappear with the development and mass production of antibiotics during the 1940s. As recently as 2007, Q fever caused serious trouble in a modern European country, far removed from both Queensland and Montana: the Netherlands.

45

Fifty miles southeast of Utrecht, amid the flat landscape and tangled roadways of the Dutch province of Noord-Brabant, lies a little back-road village called Herpen. It's a tidy place, largely assembled from red brick: redbrick farmhouses on the outskirts, redbrick cottages in town, cobbled sidewalks, and a handsome old redbrick church. The farmhouses, some shielded behind pruned hedges and prim gardens, command fields of hay and corn, grown for fodder to feed livestock that shelter in large, low, redbrick barns. Although it looks like a farm village, Herpen nowadays is a bedroom community for laborers and contractors in the building business. A few workhorses stand idle in pastures, kept company by a modest number of cows, sheep, and pigs. But the agricultural component of the local economy, insofar as it still exists, is committed more heavily to dairy goats. They seem to have been the source of the problem in 2007.

Nannies had given birth to their kids during the usual kidding season, which can stretch from January to as late as April. Mostly those births had gone well, though on certain farms of the province, including at least one in the Herpen area, many females aborted during the last month of pregnancy. Even full-term kids seemed a bit weak and puny, with a higher mortality rate than usual. Evidently something was troubling the goats, an infection of some sort, possibly new, and veterinarians took note, trying to forestall the abortions with antibiotics. That didn't help. The general public noticed this situation little or not at all.

Then came a balmy spring—far warmer and drier than normal. In April, by the recollection of one resident, "there was no drop of rain." Even before summer arrived, lands surrounding the village had gotten dusty. Breezes blew. In early May, people began to get sick.

A local physician named Rob Besselink, with an office in Herpen, saw an odd, flulike ailment in several of his patients: high fever, severe headache, muscle aches, shortness of breath, coughing. Was it a bacterial pneumonia? "We started treating them,"

Besselink said later, "and it turned out that they didn't react as we expected them to react on the antibiotics they were given." He discussed it with a colleague. "After that first week we said to each other, 'There's something strange going on,' because I had three or four people having the same symptoms, and he had also two or three." Within a couple weeks, the two doctors had seen about twenty patients fitting the profile, of whom almost a dozen, unresponsive to antibiotics, had to be hospitalized.

Around the same time, in another part of Noord-Brabant, a medical microbiologist named Ineke Weers, employed at a regional laboratory, heard murmurings about a similar cluster. Despite Weers's broad training and experience—she was an MD with a PhD in microbiology and twenty-one years of work in the diagnostics of infectious diseases—this turned out to be something new to her. An internist at one of the hospitals mentioned that doctors there had lately seen quite a few patients with an atypical, antibiotic-resistant pneumonia. Did Weers know what it might be? Had she read anything about such a syndrome? No, nothing, she answered. But she offered to call the Municipal Health Service in Den Bosch, a large city nearby, and ask whether those authorities could offer some glimmer of insight or advice. They could not; they had heard no other such reports.

Four days later, Rob Besselink called the same office of the MHS about his situation in Herpen. Two weeks after that, another general practitioner in Noord-Brabant made a similar report to the MHS. This aggregation of puzzling cases was enough to trigger the beginnings of a response. The physicians took blood samples, some of which went to a nearby laboratory, some to a more specialized lab, where the sera were tested for antibodies. After a bit of confusion about what sort of microbe might be causing such "atypical pneumonia," both labs eventually converged on an answer: It was *Coxiella burnetii*, the agent of Q fever.

Q fever wasn't unknown in the Netherlands but for fifty years it had been blessedly rare. Although the bacterium seemed to be endemic among livestock populations, based on occasional surveys, it had seldom caused noticeable disease in cows, in sheep, or in humans. Now the outbreak in Noord-Brabant caught the attention of the National Institute for Public Health and the Environment (commonly known by its Dutch initials, RIVM), up near Utrecht. Scientists there made an informed guess that maybe the high incidence of abortions on dairy goat farms, which had begun back in 2005, and which had been diagnostically linked to Q fever, might be a source of the human cases. *Coxiella burnetii* was known to be capable of airborne transmission. At this point, RIVM sent people south to the village of Herpen and surrounding areas to conduct a study. Someone had to learn what was happening downwind of the goats.

46

I drove down from Utrecht to Herpen myself, three years later, on a dreary day in February when the gray of the sky and the fog seemed to blend almost seamlessly, along a flat line of horizon, with the gray of the snow. Dr. Rob Besselink received me, just after working hours, in his little medical office on the village's main street. He was a thin man, in his late forties, with a wide smile that pinched creases into his narrow face. Wearing a black sport coat, a blue paisley shirt, and faded jeans, he looked more like lead guitarist in a rock band than what you'd expect of a rural Dutch physician. Among the first things he mentioned, when I asked about the character of Herpen as a community, was the big change that had come in local farming practices within the past decade: the increase in goats.

This change had actually started back in 1984, when the European Community established quotas on cow milk that pushed Dutch farmers away from dairy cattle. Many continued as dairymen but started milking goats. The dairy-goat trend grew stronger after 1997 and 1998, when outbreaks of classical swine fever (caused by a virus, but not zoonotic) led to mass cullings of pigs, and many pig farmers, hard hit financially and scared about a recurrence, sought an alternative line of husbandry. "So they started keeping goats, in quite some amounts," Besselink told me. It was true in Noord-Brabant and true across the country. From a low of about 7,000 animals in 1983, the total Dutch goat population had increased to 374,000 by 2009, of which 230,000 were dairy goats. Most of those lived indoors—stabled year-round inside buildings such as the large, redbrick sheds I had seen on the outskirts of Herpen. You might think that keeping the goats within four walls and a roof should minimize chances of their releasing an infection. But circumstances in the nature of Dutch goat husbandry, as I learned from Besselink and others, conspired to bring *C. burnetii* out of those sheds in great quantity and launch it on the wind.

Coxiella burnetii is an assertive bug. It not only causes abortion in goats but also concentrates massively in the placental material expelled during those abortive deliveries. A single gram of placenta from an aborting goat can contain as many as 1 billion bacterial particles. It is also excreted in milk, urine, feces, and during normal deliveries of kids carried to term. Assuming those deliveries and abortions occur within the kidding shed, how does the stuff escape? Very simply, Besselink explained: Goat feces and dirty bedding straw are shoveled up and carried outside by the farmers to fertilize their fields. From there the bacteria can waft into a nearby village as easily as the pleasant, autumnal smell of smoke from a pile of leaves.

Two goat farms in the Herpen vicinity attracted attention. One was a sizable commercial operation with almost four thousand goats, which had suffered a storm of abortions in April. The other was a "hobby farm" with less than ten animals. When the study team came down from RIVM to look for the source of the outbreak, they visited both places, taking samples of urine, milk, manure, and straw from the stable floors; insects from a light trap; and water from drinking buckets. The hobby farm seemed to be clean. From the commercial farm, every category of sample included evidence of *Coxiella burnetii* except the milk, the urine, and the water. "There were a lot of *Coxiella* bacteria in the farm," Besselink recalled. It was only a kilometer south of the village—virtually right next door. That farmer and his family endured some obloquy during the following year. "He has a wife, he has kids, the kids go to school here, so they were having a hard time because they were having the blame, of course, of what was going on," Besselink said. The goat farmer hadn't done anything illegal, merely been unlucky and maybe a little careless, but he suffered lost revenue, sapped energy, sleepless nights. A village doctor comes to know these things. The farmer's children were stigmatized and his kids—that is, his nanny goats' kids—were suspect also, having been born under circumstances that included a plume of virulent microbes.

Arnout de Bruin, a molecular biologist with a background in evolutionary studies, was part of the RIVM team that went to Herpen. When I met him at the institute's headquarters, a fenced complex in a suburb of Utrecht, he wore a light stubble of beard and a brown T-shirt reading VARSITY TEAM—NORTH DAKOTA. He was a bright young man with a dark sense of whimsy. The funny thing about his involvement with the outbreak down there, de Bruin told me cheerily, was that it only happened because he'd been studying Q fever as a possible bioterrorist threat. (The bacterium had a history of attracting dark interest; biological warfare researchers in the United States had worked on it during the 1950s, so had the Soviets, and four decades later the Japanese cult Aum Shinrikyo seems to have considered it, before using sarin gas for their 1995 attack on the Tokyo subway.) De Bruin's group on that project, a "biological calamities" team, had developed PCR primers for detecting *Coxiella burnetii* in a sample. So when the cases started piling up in Noord-Brabant, both among goats and among people, and the health authorities wanted urgently to trace the source, they asked de Bruin's team for help. Okay, yeah, sure. He and his partners jumped at the chance for a field test of their new molecular tools. On the advice of veterinary officials, who knew of the abortion wave on the big commercial farm, they went to that place.

"And the farmer said, 'This is the secure area, and *this* is the nonsecure area, because here the goats have been standing which had aborted,'" de Bruin told me. "So we took all kinds of samples. Surface area swabs, water from the drinking buckets, vaginal swabs from the goats. What did we take more? Oh yeah, for instance, insects, from the insect lamp. Dust particles, hay, manure." He laughed grimly. "We found it everywhere."

What sort of protection were you wearing? I asked. Masks, respirators? None, he said, laughing again, at his own foolishness and the laxity of supervisory vigilance. "But nobody got sick." Maybe he and his colleagues were lucky. Anyway, the farmer was wrong about which parts of his property should be scrutinized. "We found it everywhere," de Bruin repeated. "There was no secure/unsecure area because the whole farm was infested."

On the basis of this field sampling and the lab results, he told me, some health officials became overly eager, inclined toward concluding too much. "They said immediately, 'Oh, that's the source!' And we said, 'Well, it is *a* source.'" But no one had checked the other farms in the neighborhood, any of which might also have been leaking *Coxiella burnetii* into the air. You should test those too, de Bruin advised. Meanwhile his team worked on other aspects of the outbreak-response study.

They gathered blood samples from 443 people in the Herpen area and, in 73 of those individuals, found evidence of recent infection with *C. burnetii*; another 38 had been infected sometime in the past. From questionnaire information, the study team matched positives against different forms of potential exposure. The most revealing result from this analysis was that direct contact with animals was *not* a significant risk factor for infection. Nor was drinking raw milk. Some of the cases—but only a minority, less than 40 percent—involved contact with agricultural products such as hay, straw, and manure. From these data, the team narrowed it down to "windborne transmission" as the most likely source of Q fever in the area. The high incidence of infection among goats, the cascade of abortions, the practice of fertilizing fields with manure from the kidding sheds, the nature of the bacterium itself (more on this below), the dry April weather, and the easterly winds had combined to becloud the village of Herpen with *Coxiella burnetii.*

De Bruin himself, having helped gather and analyze these data, was acutely aware how well the bacterium went airborne. Later, as the epidemic continued into 2008 and 2009, he grew more wary about field sampling. "I said, 'Hey, we're not going anymore without protection—because we're lab people, we're not immune.'" If you're a farmer, he said, you may have developed immunity from prior exposure to Q fever at a level that never caused overt illness. That turns out to be quite common among Dutch farmers and veterinarians—but not among molecular biologists. "So we went with masks." Still, it's hard to work in a mask—your breathing constrained, your glasses or goggles fogging up—and you find that you don't want to wear such gear a minute longer than necessary. De Bruin saw more dark amusement in the absurdity of drawing a line between what was impracticable and what was safe. He recalled driving down to another major outbreak site in the south. "I came to that farm, and the only place I could park my car was in front of the stable. So I opened my car, and there was a big wind blowing through the stable." He got out. He breathed the wind. He thought, "And *now* I'm going to put on my mask?" This time we both laughed.

The outbreak continued, growing worse in 2008, worse still in 2009. By the end of that year, 3,525 human cases had been recorded since the first alerts in May 2007, most of those still in Noord-Brabant. The infection generally made itself manifest as fever, pneumonia, and in some cases hepatitis. At least twelve people died—not a high lethality compared to some of the grisly viruses, but fairly severe when you remember that this is a *bacterial* infection, supposedly treatable with antibiotics.

One cluster of cases, in 2008, occurred at a psychiatric care institution in the town of Nijmegen. After three of the psychiatric patients came down with atypical pneumonia and were hospitalized, the Municipal Health Service screened patients, employees, and visitors, finding twenty-eight cases of *C. burnetii* infection. What was the source? A goat farm near Nijmegen had suffered a storm of abortions, and Q fever was confirmed from vaginal swabs. The bacteria could have traveled downwind from those aborted kids. But in this instance, there was also a more immediate possibility. The psychiatric institution maintained a small flock of sheep on a meadow within the premises. During that year's lambing season, one lamb had been abandoned by its mother—and was then adopted by a patient, who took it into her bedroom and bottle-fed it six times a day. The pet lamb was also cuddled consolingly by several other patients. This seems to have been somebody's idea of therapy, until the lamb tested positive for Q fever.

On the day after my conversation with Arnout de Bruin, I drove north to the Central Veterinary Institute, a university-affiliated facility near the city of Lelystad, with an annex devoted partly to research on dangerous zoonotic agents. Whatever was happening in the Netherlands to account for these sequential outbreaks, it was clearly a veterinary concern as well as a matter of human health. The CVI annex, tucked among trees off a secondary road, was so discreet that I had to circle the neighborhood twice to find it. There I was welcomed by Hendrik-Jan Roest, a slim veterinary scientist in rimless glasses and a casual blue sweater, tall enough to play forward on the Dutch national basketball team, who led me back outside immediately so we could peer in the window of a BSL-3 lab where he and his technician were growing *C. burnetii*. Through the little window I could see incubators and a negative-airflow hood, like the fan hood above a stove, meant to suck away ambient bacteria as his technician worked at her bench. In this building, Roest told me, we work also on West Nile virus, Rift Valley fever, and foot-and-mouth disease, among other things. Rift Valley fever, I said, you have *that* in the Netherlands? Not yet, he said.

Back in his office, Roest sketched a verbal portrait of *Coxiella burnetii*, listing the traits that make it so unusual and problematic. First of all, it's an intracellular bacterium, meaning that it reproduces within cells of its host—as does a virus, though by dissimilar mechanisms—not out in the bloodstream or the gut, where it could be more easily targeted by immune response. Furthermore, it exists in two forms of bacterial particle, one large and one small, each with different characteristics suited to different phases of its life history. The large form replicates prolifically inside host cells and then transmogrifies to the small form, which is tougher and more stable. The small form, almost like a spore, is packaged for survival in the external environment. (The smallness of this small form may account for why Macfarlane Burnet and some others mistook it for "a filterable virus," a microbe so tiny it passed through filters designed to scoop away ordinary bacteria.) It is resistant to desiccation, resistant to acids, resistant to high and low temperatures, and resistant to ultraviolet light. It can live in salt water for more than six months. No wonder it travels so well, not just from host to host but from place to place—even from continent to continent.

"Does anyone know where it came from?"

"I think it was always there," Roest said.

Always where? Always *everywhere*? In Montana, where Herald Cox found it, and in Australia, where Macfarlane Burnet found it, and in the Netherlands, where you're finding it now? No, not quite everywhere, he said. There is no record of *Coxiella burnetii* in New Zealand. So far.

Then why had the disease just lately—since 2007—become so troublesome in Noord-Brabant? When I asked him about the increase in dairy goats, he brushed that idea aside as too simplistic and began showing me photos and charts on his computer. One image revealed a vast building, like a train depot, filled with white goats.

"This is the *way* they are goat farming."

"Wow."

"They are huge, huge barns."

"Big barns," I agreed.

Another shot gave a clearer view of what he called a "deep litter shed," the standard arrangement for housing hundreds or thousands of dairy goats. The shed had a concrete floor, recessed below ground level so that it could contain weeks' or months' worth of bedding straw, goat shit, and urine, a savory mulch of organic waste that grew ever deeper and, warmed by decay, offered a lovely culture medium for microbes. New straw was added regularly, as long as possible, to stiffen and mitigate the mess. "Very slowly the package of manure and straw is getting thicker and thicker," Roest explained, "and so the level where the animals live on is coming up." Shin-deep in their own ordure, the nannies milled around, converting their feed to milk. As the manure rose, composting gently, it harbored uncountable abundances of *C. burnetii*, "alive and kicking, down deep in the litter." By the time such a shed had filled to its brim, any single infected goat could have passed its infection to many or most of the others. Then the goats were moved out, machinery came in, shoveling began, the valuable manure was transferred to crop fields and pastures—and billions more particles of the bacterium, in its small and resistant form, were launched on the breezes.

High-density dairy-goat husbandry, Dutch style—that's one factor among several to account for the recent outbreaks, Roest said. Factor two was concomitant to factor one: proximity of humans. The Netherlands is a crowded country, containing 16 million people within an area half the size of Indiana, and many of those high-density goat farms are sited near towns and cities. Factor three was the weather: Yes, very dry springtime conditions, during each year since 2007, had doubtless exacerbated the airborne spread of the bacterium. And Roest suspected a fourth factor: It might be, he said, that the nature of the bug itself had changed. An evolutionary twitch could have enabled an ecological leap.

His molecular data showed that one particular genetic strain of the bacterium—one among fifteen that his team identified—had come to predominate. "On all farms in the high-risk area," by which he meant Noord-Brabant and some adjacent zones, "and on the two dairy farms outside," which also tested positive, "there is one genotype present in 90 percent of all samples. And that is what we call the CbNL-01." CbNL-01 seems a fancy cryptogram but it connotes simply "*Coxiella burnetii*, Netherlands, genotype #1." Such disproportionate representation suggested that a mutation in that strain might have made it especially aggressive, efficient, transmissible, and fierce.

Dutch officials tried to cope with this crisis by means of some forceful, if inconsistent, regulatory measures. In June 2008, shortly after the outbreak among patients at the psychiatric facility in

Nijmegen, Q fever became a "notifiable" disease for dairy goats and dairy sheep, meaning that veterinarians were required to notify the government about any abortion storms. (It had been a notifiable disease with regard to human cases since 1975.) Another regulation, issued the same day, prohibited farmers from removing manure from an infected stable or deep litter shed for three months following notification of an outbreak. Almost a year later, in April 2009, as the pattern of outbreaks continued on dairy-goat farms and the number of human cases ascended faster than ever, a program of mandatory vaccination against Q fever went into effect. This order applied to all dairy goats and sheep on farms with more than fifty animals, and to zoos and "care farms" such as the one at Nijmegen, where the general public might come into close contact with infected animals. By November 2009, more than a quarter million goats and sheep had been vaccinated, at government expense—but the human case count for the year was alarmingly high, and concern had spread widely through the Dutch media. So in early December 2009, a ban was decreed on the breeding of goats: no more pregnant nannies allowed until further notice. Closer consideration revealed that was too little, too late. Many females had already been bred. One week later, on advice from an expert panel, the government announced that all currently pregnant goats and sheep (including those recently vaccinated) on affected dairy farms would be culled.

Veterinary teams went out to do the deeds. One dairyman, awaiting the cullers, told a reporter that his animals would be less agitated if he remained with them, but "I just don't know if I can watch it." The eventual toll included about fifty thousand dead goats and scores of angry, frustrated farmers, who were compensated for the value of each animal but not for lost revenue as they faced rebuilding their herds, nor for emotional stress. "It was also distressing for the veterinarians," Hendrik-Jan Roest told me—and distressing too, as he could say from experience, "for the veterinary advisors."

Despite all these measures, and the disappearance of pregnant goats from the Dutch landscape, Q fever did not disappear—not entirely, not at once. The bacterium was still out there in some abundance. In its small, sturdy form, it could survive in the fetid wastes on infected farms for as long as five months. In its large form, it could replicate in a variety of animals. Highly robust but not too specialized, it was capable of invading a wide range of hosts, and had been found not just in goats and sheep but also in cattle, rodents, birds, amoebae, and ticks. An enterprising organism and, as Macfarlane Burnet had noted, quite versatile.

In time the regulatory measures had some effect, and another spring passed, this one without many newborn or aborted kids. The rate of new human cases declined from its 2009 peak. By the middle of July 2010, only 420 more Netherlanders had been diagnosed with Q fever. The ministry officials could feel guardedly optimistic that their public health crisis had been brought under control. The doctors could relax slightly. The dairy farmers could lament their losses. But the scientists knew that *Coxiella burnetii* wasn't gone. It had waited for ideal conditions before, and it could wait again.

47

Back in Australia, around the time of his work on Q fever and psittacosis, smart and curmudgeonly Macfarlane Burnet began thinking more broadly about infectious diseases, not so much from the medical perspective as from the viewpoint of a biologist. During the late 1930s he drafted a book on the subject, in the opening pages of which he paid tribute to the great nineteenth-century founders of bacteriology, especially Pasteur and Koch, who had finally provided a rational basis for concerns over clean drinking water, decent sewage disposal, food untainted with rot, and antiseptic surgical techniques. It was a qualified tribute, concluding on page two, after which Burnet got to his real point.

Those men and their colleagues, he wrote, "were on the whole too busy to think of anything but the diseases for which bacteria were responsible, and how these might be prevented." They gave little consideration to the microbes as beings in their own right, or to "how their nature and activities fitted into the scheme of living things." Most bacteriologists were trained as medical men—Burnet himself had been, before going into bacteriological research—and "their interest in general biological problems was very limited." They cared about curing and preventing diseases, which was well and good; less so about pondering infection as a biological phenomenon, a relationship between creatures, equal in fundamental importance to such other relationships as predation, competition, and decomposition. Burnet's purpose in the book was to rectify that slight. He published *Biological Aspects of Infectious Disease* in 1940, a landmark along the route to modern understandings of zoonoses on a crowded, changing planet.

Burnet didn't claim that the broader perspective was uniquely his own. He recognized it as a salubrious trend. Biochemists had begun applying their methods to disease-related questions, with considerable success, and there was also new interest at the level of organisms (even single-celled organisms) as highly adapted creatures with their own life histories in the wild. He wrote:

> Other workers with an appreciation of modern developments in biology are finding that infectious disease can be thought of with profit *along ecological lines* as a *struggle for existence* between man and micro-organisms of the same general quality as many other types of competition between species in nature.

The italics are mine. Thinking "along ecological lines," and about the "struggle for existence" (a phrase that came straight from Darwin), was what Burnet specially offered: a book on the ecology and evolution of pathogens.

He preferred the term "parasites," used in its looser sense. "The parasitic mode of life is essentially similar to that of the predatory carnivores. It is just another method of obtaining food from the tissues of living animals," though with parasites the consumption tends to be slower and more internalized within the prey. Small creatures eat bigger ones, generally from the inside out. This is just what I was getting at, back at the start, when I mentioned lions and wildebeests, owls and mice. The main problem facing a parasite over the long term, Burnet noted, is the issue of transmission: how to spread its offspring from one individual host to another. Various methods and traits have developed toward that simple end, ranging from massive replication, airborne dispersal, environmentally resistant life-history stages (like the small form of *C. burnetii*), direct transfer in blood and other bodily fluids, behavioral influence on the host (as exerted by the rabies virus, for instance, causing infected animals to bite), passage through intermediate or amplifier hosts, and the use of insect and arachnid vectors as means of transportation and injection. "It will be clear, however," Burnet wrote, "that no matter by what method a parasite passes from host to host, an increased density of the susceptible population will facilitate its spread from infected to uninfected individuals." Increased density: Crowded hosts allow pathogens to thrive. Macfarlane Burnet may or may not have been influenced by those early mathematical works on infectious disease—the differential equations of Ronald Ross, the 1927 paper by Kermack and McKendrick—but he was putting some of the same points into plain English prose in a book that was both authoritative and accessible.

Biological Aspects of Infectious Disease was later revised and reissued, in 1972, as *Natural History of Infectious Disease.* Though even its revised version seems antiquated today (new diseases have emerged, as well as new insights and new methods), the book was a valuable contribution in its time. It offered no erudite mathematical models but it spoke plainly on the subject of what disease scientists do, and what they should do. What they should do, by his lights, was to think about infectious pathogens in ecological and evolutionary terms as well as medical ones.

Parrot fever was one of his exemplary cases. It had the attractions of an Australian connection (for him, a local bug) plus global reach, and it illustrated a favorite point. "Like many other infectious diseases, psittacosis was first recognized as a serious epidemic disease of human beings, but as its nature became gradually understood, it grew clear that the epidemic phase was only an accidental and relatively unusual happening." The bacterium had its own life to lead, that is, of which infecting humans was just one part—and arguably a digression.

Burnet retold the tale of the California-bred parakeets, the wild Australian cockatoos, the infection of working-class Melbourne bird fanciers by animals sold out of Mr. X's dismal backyard shed. Psittacosis, Burnet noted, is not normally very infectious. It exists endemically among wild bird populations, causing little trouble. One could reasonably suppose that "those cockatoos, left to a natural life in the wild, would never have shown any symptoms." But the bird catcher, and then Mr. X as middleman, had disrupted their natural life. "In captivity, crowded, filthy and without exercise or sunlight, a flare-up of any latent infection was only to be expected." The stressful conditions had allowed *Chlamydophila psittaci* (as *Rickettsia psittaci* later became known, after another of those taxonomic revisions) to replicate and erupt.

This case and similar ones, Burnet wrote, embodied a general truth about infectious disease. "It is a conflict between man and his parasites which, in a constant environment, would tend to result in a virtual equilibrium, a climax state, in which both species would survive indefinitely. Man, however, lives in an environment constantly being changed by his own activities, and few of his diseases have attained such an equilibrium." Burnet was right on the big ideas, including that one: environmental disruption by humans as a releaser of epidemics. Still, he couldn't foresee the particulars of what would come. Publishing in 1940, he focused on several infectious diseases in addition to psittacosis: diphtheria, influenza, tuberculosis, plague, cholera, malaria, yellow fever. These were the old, familiar, infamous scourges, fairly easy to recognize though not well enough understood. Our modern age of emerging viruses was just beyond the reach of his headlights.

48

Burnet didn't mention Lyme disease but, because it shares one important characteristic with Q fever and psittacosis, I will. The most fundamental thing about this newly emergent or re-emergent infection is that it's not caused by a virus. The Lyme agent, like *Coxiella burnetii* and *Chlamydophila psittaci*, is an anomalous, crafty bacterium.

Lyme disease is hotly controversial, though, in a way that Q fever and psittacosis are not. Segments of the scientific and medical communities, plus victims or supposed victims, can't even agree (*especially* they can't agree) on who has the disease and who doesn't. Roughly thirty thousand cases of Lyme disease were reported in the United States during a recent year, and more than twenty thousand per year as a ten-year average. You probably know someone who has had it; you may well have had it yourself. By any standard, it's the most commonly reported vector-borne disease in the United States. But do those thirty thousand cases in one year represent the true total of affected Americans or only a small fraction of the real number of cases, most of which go undiagnosed? Is there such a thing as "chronic Lyme disease," which eludes detection by conventional diagnostics, persists despite prescribed treatment with antibiotics, and causes gruesome suffering among people who can't persuade their doctors or their insurance companies that they are genuinely infected? Does *Borrelia burgdorferi* hide in the body and somehow later recrudesce?

Disagreements on such points have stretched all the way from the examining room to the courtroom, making Lyme not just the most common infection of its kind but also the most confusingly politicized. For instance, in 2006 the Infectious Diseases Society of America suggested in its guidelines for treatment that "chronic Lyme disease" is an illusion. More precisely, the IDSA wrote: "No convincing biologic evidence exists for symptomatic chronic *B. burgdorferi* infection in patients after recommended treatment regimens for Lyme disease." The recommended treatment regimens, involving two to four weeks on an antibiotic (such as doxycycline or amoxicillin), should cure the disease itself. What the IDSA carefully labeled "post-Lyme disease syndrome" was another matter. Implication: These people are head cases. That dismissiveness about the possibility of lingering Lyme infection infuriated many mysteriously tortured patients, who believed that they had it and who (counseled by certain private physicians, contra the IDSA) felt they should be treated with high doses of antibiotics given intravenously over a much longer term—months or years. Such treatments, by the conventional view, might actually be harmful to a patient's health. They are also a matter of consequence to insurance companies that don't want to pay for them.

In late 2006, the attorney general of Connecticut (Richard Blumenthal, later a US senator) began an antitrust investigation into the IDSA and the way it formulated its Lyme treatment guidelines. Had there been conflicts of interest? Blumenthal thought so. The IDSA's Lyme disease guideline panel undercut its own credibility, he said, "by allowing individuals with financial interests—in drug companies, Lyme disease diagnostic tests, patents and consulting arrangements with insurance companies—to exclude divergent medical evidence and opinion." He emphasized, though, that his scrutiny was directed at the guideline formulation process, not at the science itself. Two years later the IDSA and Blumenthal agreed on a compromise settlement, whereby the guidelines would be reviewed by a new, independent panel. In 2010, the independent panel unanimously reaffirmed the original guidelines. They too saw "no convincing evidence for the existence of chronic Lyme infection." Furthermore, they warned, long-term intravenous treatment with antibiotics was worse than useless; it could lead to deadly blood infections, severe drug reactions, disruption of normal gut flora (the beneficent bacteria that help people digest), consequent diarrhea as other bacteria took hold, and the creation of antibiotic-resistant "superbugs," menacing not just to patients under such treatment but to all the rest of us also.

Another complication of the whole story is that, though Lyme disease seems like a new problem, unnoticed before 1975, it has probably been around for a long time, not just in the United States but also in Europe and Asia. For decades it was detected marginally and piecemeal, by some of its symptoms, but not recognized as a single syndrome with a single cause. Only in retrospect were the pieces assembled into a pattern with a name.

This prehistory period began in 1909, when a Swedish dermatologist named Arvid Afzelius reported the case of a woman, bitten by a sheep tick, who suffered a rosy rash that spread like concentric ripples. Afzelius called the condition *erythema migrans* ("spreading redness") and wrote about it for a German journal devoted largely to syphilis, which was a major concern of dermatologists back then. (There was some similarity: Syphilis is caused by a bacterium of the type known as spirochetes, the same group of corkscrewing creatures that includes *Borrelia burgdorferi*, the Lyme disease pathogen.) Afzelius didn't claim to know the cause of the woman's rash, but over the next dozen years he saw a similar pattern in five more patients. Other physicians in Europe also began noticing such annular rashes, each resembling a target with a tiny red dot as the bull's-eye. In some cases, the rash was associated with the bite of an unidentified arthropod (an insect, a spider, a tick?), and often it came with more serious symptoms. Sven Hellerstrom, another Swedish dermatologist, reported in 1930 seeing a man with the distinctive red rash, plus meningitis. As years passed, Hellerstrom found that annular rashes, resulting from tick bites and sometimes involving meningitis, were far from rare in the Stockholm area.

Almost two decades after his first report, Dr. Hellerstrom crossed the Atlantic to attend a medical conference in Cincinnati, where he described his continuing work. The cause of the rash-and-meningitis syndrome, he postulated, was a spirochete. Because the conference was sponsored by the Southern Medical Association, a printed version of Hellerstrom's 1949 talk appeared in the *Southern Journal of Medicine*, an otherwise unlikely outlet for a Swedish clinician. These were not high-profile publications, neither the papers of Afzelius nor of Hellerstrom nor the others, and of course there was no Internet, no Google, no PubMed, nor any other such means by which to summon obscure citations at the touch of a few keys. But a good memory, broad education, and luck could serve the same purpose.

And eventually did. Twenty more years passed before Rudolph J. Scrimenti, still another dermatologist, practicing in Milwaukee, had reason to recall Hellerstrom's paper, which he had read as a medical student. Scrimenti, in 1970, became the first physician to report a case of *erythema migrans* in America. His patient, a fellow physician, had been bitten by a tick while grouse hunting in central Wisconsin, and the rash grew outward from the site of the bite, eventually encircling much of his chest, right armpit, and back. Scrimenti treated the symptoms with penicillin. In his brief published report, he echoed Hellerstrom's guess that it might have been caused by a spirochete, but Scrimenti hadn't been able to find one.

This is all part of the medical groundwork that was available—though not conspicuously available—when doctors at the Yale School of Medicine heard about the cluster of juvenile arthritis cases in Lyme, Connecticut. One of those doctors was Allen C. Steere, a first-year fellow in the rheumatology division. Rheumatology is the science of joint disorders such as rheumatoid arthritis, which is an autoimmune condition, not an infectious disease. Juvenile rheumatoid arthritis, Steere recognized, should not be occurring in any such cluster. It didn't pass from one patient to another. It didn't infect people through their drinking water. It didn't fly on the wind like Q fever . . . did it?

Steere and his colleagues followed out the cases brought to their attention, did some further epidemiological legwork, found many more cases in roughly the same area, and began calling the syndrome "Lyme arthritis." Steere's group also took note of the associated symptom among a sizable fraction of the patients: a circular red rash. Other medical practitioners, in Connecticut and nearby parts of New York, also saw cases of this peculiar skin inflammation and began wondering. Was it caused by an insect bite? Was it the same condition, *erythema migrans,* that had been described in the literature from Europe? About that point, in the summer of 1976, a field biologist named Joe Dowhan, working in a forested area some miles east of Lyme, pulled a tick off his leg and dropped it into a jar. Dowhan had noticed the bite because, unlike most other tick attachments he'd experienced in his career, it registered as a small, painful nip. Three days later, he developed a rash. As the red circle grew, he remembered having seen an article about Allen Steere's work. So he called, got an appointment, sat through an exam, and then handed Steere the tick.

Dowhan's specimen was identified as *Ixodes scapularis*, commonly known as the deer tick, an arthropod widely distributed throughout the eastern and midwestern United States. This became an important but ambiguous clue in the Lyme disease story, leading both toward insight and into confusion. The insight came first. Fieldwork along the lower Connecticut River revealed that *Ixodes scapularis* ticks were far more numerous in small woodlands and brush on the east bank of the river—the bank on which sat the village of Lyme—than on the west bank. That finding, combined with the fact that human cases also were far more common on the east bank, pointed further suspicion at the "deer tick" as a vector of what even Steere and his rheumatologist colleagues, having dropped the term "Lyme arthritis," were now calling "Lyme disease."

The confusion grew more slowly. If the "deer tick" carried the pathogen (whatever it was) and infected people like Joe Dowhan by biting them, then the abundance of human cases must reflect the abundance of ticks, and the abundance of ticks must reflect the abundance of deer in those suburban woodlands of coastal Connecticut. Yes?

No. This was an ecological system with the intricacy of chess, not a board game with the clarity of checkers, and its cause-and-effect relations weren't nearly so simple. The "deer tick," as later research has shown, lives a complicated life.

49

Willy Burgdorfer meanwhile made his crucial discovery of the pathogen itself, giving a name and a biological identity to the agent responsible for the mysterious clusters of cases.

Burgdorfer was a Swiss-born and Swiss-trained microbiologist with a shovel-wide jaw, a cagey smile, a great domed head like Niels Bohr, and a deep interest in medical entomology. He did his doctorate on a tick-borne spirochete, *Borrelia duttonii*, which in Africa causes an illness called relapsing fever. By the time he finished that project, Burgdorfer had dissected thousands of ticks to scrutinize their innards. He had also invented a quick, practical technique for determining whether a given tick carries spirochetes: snip off a leg and look through a microscope at the body juice (hemolymph) that dribbles out. Emigrating to the United States, in 1952 he joined the Rocky Mountain Laboratory, in Hamilton, Montana, the same facility where Herald Cox and Gordon Davis had done their work on Q fever. In fact, Davis became his early sponsor there, and for a couple years Burgdorfer continued to work on *Borrelia* spirochetes (and the variants of relapsing fever they cause in America) among captive tick colonies that Davis had established. Some laboratory scientists work with fruit flies, some with carefully inbred mice; Davis and Burgdorfer nurtured teeming cagefuls of ticks.

Then the winds changed: A high administrator told young Willy Burgdorfer that relapsing fever was "a disease of the past," no longer justifying government-supported research, and advised him to pick a different specialty. By his own later account, Burgdorfer followed that advice only partially. He managed to stay at the Rocky Mountain Laboratory (which remained a leading research institution, despite its remote location), doing his primary work on plague, Rocky Mountain spotted fever, and other infamous diseases while pursuing his special interest in tick-borne spirochetes as "a moonlighting job." When Gordon Davis retired, Burgdorfer inherited the elder man's lab technician and his captive colonies of ticks. All of this qualified him well for the role he would eventually play with Lyme disease.

Almost three decades later, near the end of his own career, Burgdorfer's lifelong interest became urgently relevant. By the late 1970s, Allen Steere and others had begun to suspect that what they had first called "Lyme arthritis" was actually a tick-borne infectious disease, which had affected 512 patients, mostly along the northeastern seacoast and in Wisconsin. Hundreds more cases would soon be reported by the CDC. Around the same time, a family practitioner on Shelter Island, New York, just across the Long Island Sound from Lyme, was treating patients with similar histories—unusual feverish ailments that seemed to have been transmitted by ticks. Other tick-borne diseases also occurred on Shelter Island, a small but insalubrious place, so Lyme disease there was just one hypothesis among several. Then a batch of ticks, collected from low vegetation on Shelter Island, was sent out to Burgdorfer's lab in Montana, where he dissected their gut cavities and found more than 60 percent of them harboring some sort of spirochete. "No longer did we hear, 'get out of the spirochete business,'" Burgdorfer recalled later. Spirochetology was back in fashion. These ticks were alive with tiny corkscrewing forms.

When Burgdorfer and his colleagues allowed infected ticks to feed on white laboratory rabbits, the rabbits developed circular skin rashes that grew outward like ripples from the bites, replicating the telltale annular pattern seen so often in human cases. Burgdorfer's group also cultured the spirochete from ticks and then tested it against antibodies in blood sera from Lyme patients. The positive results in those tests, plus the rabbit reactions, constituted evidence that they had found the agent of Lyme disease. This was how Burgdorfer earned his place in what he later jovially called the "lymelight." When other researchers wrote up a formal identification of the spirochete, shortly afterward, they named it *Borrelia burgdorferi* in his honor. The only hitch in this tale of elegant lab science is that the identity of the ticks was still a matter of confusion.

50

It was confused in two ways, one of which is more interesting for our purposes than the other. The uninteresting confusion involved the scientific name. Was it *Ixodes scapularis* carrying the Lyme spirochete in those coastal New England habitats, or did the creature belong to a similar but undescribed species, which should be given its own scientific identity? For a while the Lyme-bearing tick became known as *Ixodes dammini*, until further taxonomic scrutiny invalidated that distinction, in 1993, and restored it to *Ixodes scapularis*. This back-and-forthing was merely a matter of taxonomic practice, reflecting the chronic tension between splitters (who like to delineate many species and subspecies) and lumpers (who prefer fewer). The splitters won a temporary victory; the lumpers prevailed.

A second sort of confusion, more consequential, derived from uncertainty over the tick's less formal label. As *Ixodes scapularis*, it had been familiarly known as the blacklegged tick. When it was mistakenly split off into a new species, it received also a new common (but not *very* common) name, "Dammin's northeastern deer ixodid." That clumsy phrase was later shortened to "deer tick." Name-giving influences perception, of course, and "deer tick" reinforced a misunderstanding about the little beast in question: that this blood-sucking, disease-transmitting arthropod is somehow uniquely associated with deer. Wrong.

Calling it the "deer tick" led to a mistake of circularity. If white-tailed deer are the host animals from which "deer ticks" draw their crucial sustenance, and "deer ticks" are the vectors that transmit Lyme disease to humans, it would seem to follow logically that high deer populations must contribute to high levels of human infection. It *does* follow logically—but erroneously. The syllogism would be sound, except that its first premise is oversimplified and misleading. "Deer ticks" of the species *Ixodes scapularis* do not draw their crucial sustenance from deer.

An ecologist named Richard S. Ostfeld has done much to

untangle this confusion. Ostfeld made a two-decade investigation of one ecosystem, in suburban New York, within which *Borrelia burgdorferi* lives. He also reviewed the research done elsewhere and the conclusions that had been (sometimes erroneously) drawn. White-tailed deer, he found, are a misleading distraction. Ostfeld's book on the subject, *Lyme Disease: The Ecology of a Complex System*, appeared in 2011. "The notion that Lyme disease risk is closely tied to the abundance of deer arose from field studies that began shortly after the discoveries of the bacterial agent of Lyme disease and the involvement of ticks as vectors of these bacteria," he wrote. Those studies were thorough and energetic, he noted, but perhaps driven too much by desire for a simple answer from which public health actions could be taken. Their context was "the hunt for the culprits—the *critical species*." One journal article had called white-tailed deer "the definitive host" of the tick. According to another study, the deer was "the one indispensable piece" of the Lyme disease puzzle in North America. An overview account, otherwise excellent and written by a doctor with an acute sense of the medical issues, had pounced on the same conclusion as a way of explaining why Lyme seemed to be a newly emergent disease: "If the Lyme spirochete had been around for so long, why did it begin to surface as a recognized medical entity only in the past few decades? This question can be answered in one word—deer." They all agreed: deer deer deer. The one-word answer seemed to point toward a pragmatic solution to the problem of Lyme disease: Reduce the number of infected ticks by reducing the number of white-tailed deer.

And so that was tried. In one early effort, on a small island off Cape Cod, state wildlife biologists shot 70 percent of the deer; then researchers assessed the effect on tick populations by counting tiny, immature ticks on one kind of mouse. Result: The abundance of ticks on the mice was at least as high as before deer eradication. In years since, heavy deer-hunting has been encouraged in some areas of Maine, Massachusetts, Connecticut, and New Jersey for the sake of drawing down deer populations, while researchers again monitored the effects, if any, on populations of ticks. The town of Dover, Massachusetts, for instance, recently announced its first deer hunt on open town land, reflecting recommendations from the local board of health and the Lyme Disease Committee. Nineteen deer (sixteen does and three bucks) were killed, after which a Dover newspaper explained confidently: "The higher the number of deer in an area, the higher the chances are of spreading Lyme disease to humans."

Well, actually, no. That simple formula is as false as the notion that swamp vapors bring malaria.

The premise behind such civic efforts is that the landscapes in question contain "too many" deer and that their overabundance accounts for the emergence of Lyme disease since 1975. And it's true enough that there are *lots* of deer out there. Populations in the northeastern United States have rebounded robustly (because of forest regrowth, absence of big predators, lessened hunting by meat-hungry humans, and other factors) since the hard times of the eighteenth and nineteenth centuries. There might be more deer in Connecticut today than at the time of the Pequot War in 1637. But that abundance of whitetails, as Ostfeld's work showed, is probably irrelevant to the chances you'll catch Lyme disease during a stroll in, say, Cockaponset State Forest. Why?

"Any infectious disease is inherently an ecological system," Ostfeld wrote. And ecology is complicated.

51

Rick Ostfeld, seated in his office at the Cary Institute of Ecosystem Studies, in Millbrook, New York, his walls and door decorated with tick humor, told me that he's a "heretic" on the subject of deer and Lyme disease. But he's a heretic with data, not one who listens to private voices of revelation.

Ostfeld is a fit, cheerful, fiftyish man with short brown hair and ovoid glasses. His primary research interest is small mammals. He studies the ways they interact, the factors affecting their distribution and abundance, the effects of their presence or absence, the things they carry. Since the early 1990s, he and his group at Cary have live-trapped tens of thousands of small mammals in the forest patches of Millbrook and neighboring areas—mainly mice, chipmunks, squirrels, and shrews, but also creatures as large as possums, skunks, and raccoons. Initially his research had nothing to do with Lyme; he was tracing population cycles of a native rodent, the white-footed mouse. Many kinds of small mammal tend to show such population cycles, passing from relative scarcity one year to abundance the next, even greater abundance the year after, and then crashing back to scarcity, as though governed by some mysterious rhythm. Many mammal ecologists have studied such cycles, trying to determine their causes. What drives the boom and the bust?

Ostfeld was more curious about the consequences. When animal A becomes inordinately plentiful, how might that affect the populations of animals B, C, and D? Specifically, he wondered whether high population levels of white-footed mice might control outbreaks of a certain pestiferous moth by eating up most of the caterpillars. As he trapped his animals, examined them, and marked them with ear tags before release back into the understory, he noticed that their ears were covered with tiny dark bodies, as small as the dots of a colon: baby ticks. The mice were infested. They were supplying blood meals to the immature stages of *Ixodes scapularis*, known to Ostfeld as the blacklegged (not "deer") tick. "Thus began my interest in Lyme disease ecology," he wrote in the preface of his book.

Over those twenty years, mammal by mammal, tick by tick, Ostfeld and his team collected an enormous body of information, and the work continues. They use Sherman live traps (from the H. B. Sherman company, of Tallahassee, a venerable supplier) baited with oats and set out on the forest floor. They release most of the captured animals alive, after a brief examination to check body condition and remove ticks. Small mammal biologists like him, for whom trap-and-release protocols are the daily routine of data gathering, tend to become highly adept—gentle but efficient—at handling live rodents. Ostfeld's group has found that, in about one minute of close scrutiny, they can detect 90 percent of the ticks on a mouse. (They measured their own field-exam thoroughness by taking some mice into captivity after the one-minute check-over, holding them captive, and waiting for all ticks to fall off into a pan of water beneath the cage. Then they sorted the ticks from the mouse shit and other detritus—"a messy and challenging task," Ostfeld testified—and counted this fuller total for comparison with what had been seen in the field.) For chipmunks, the method of quick visual inspection worked almost as well. On other small mammals, including squirrels and shrews, the tick burdens were higher and harder to count, but Ostfeld's group could still make well-informed estimates.

Larval ticks are minuscule and even a tiny masked shrew, weighing only five grams (about the same as two dimes), carried on average fifty-five ticks, the researchers found. That's a mighty burden of infestation for such a small, delicate creature. The short-tailed shrew, a larger animal, averaged sixty-three ticks per animal. Given Ostfeld's estimate (also derived from trapping data) of about ten short-tailed shrews resident in an acre of woodland around Millbrook, it begins to add up to quite a few ticks, whole forests a-crawl with sanguineous dots, a disquieting prospect, even if the blacklegged tick never fed on anything but the blood of shrews.

But it does. Its life cycle is complex. Like an insect, the blacklegged tick undergoes metamorphosis, passing through two immature stages (larva and nymph) on the way to adulthood. At each of those stages, it needs a single blood meal from a vertebrate host to nourish its transmogrification; an adult tick needs another blood meal to supply energy and protein for reproduction. In most cases the vertebrate host is a mammal, though it might also be a lizard, or a ground-nesting bird such as the veery, exposing itself to larval ticks on the forest floor. The blacklegged tick is such a generalist, in fact, that its menu of known hosts includes more than a hundred North American vertebrates, ranging from robins to cows, from squirrels to dogs, from skinks to skunks, from possums to people. "These ticks are unbelievably catholic in their tastes," Ostfeld told me.

An adult female tick spends her winter with a bellyful of blood and then in spring lays her eggs, which hatch into larvae by midsummer. Whether as immatures or as adults, ticks can't travel very quickly or very far. They don't fly. They're not so acrobatic as fleas or springtails. They lumber around like tiny tortoises. But they seem to be "exquisitely sensitive" to chemical and physical signals, according to Ostfeld, and thereby "able to orient toward safe locations for overwintering and toward hosts emitting carbon dioxide and infrared radiation." They smell out their food. They may not be agile, but they're opportunistic, alert, and ready.

The complete life cycle takes two years and entails three distinct episodes of parasitic drinking, each of which can involve a different kind of vertebrate host. Acarologists (tick biologists) have a wonderfully high-flown term for the behavior by which a tick seeks its next attachment, climbing to the top of a grass stem or out to the edge of a leaf, front legs extended, sniffing the signals, positioned to grab a new host; the word is "questing." The smaller the life stage, the more likely that questing occurs very low to the ground. One consequence of this, reflected in the data of Ostfeld and his colleagues, is that those two kinds of shrew supply about 30 percent of all the blood meals taken by larval ticks in the study area. White-footed mice are second in importance as blood hosts for the larval stage.

White-tailed deer seem to play a much different role. They are important mainly to adult ticks—not just for their blood, but also for providing a venue where male blacklegged ticks can meet females. A whitetail in the woods of Connecticut, during November, is like a teeming singles' bar in lower Manhattan on Friday night, crowded with lubricious seekers. One poor doe might be carrying a thousand mature blacklegged ticks. Mating occurs, somewhat gracelessly, when a male tick, prowling across the skin of the deer, encounters a preoccupied female—she is tapped in, drinking, immobile. Don't look for romance in arachnoid sex. Once the female has had her drink, and the male has had his congress, they drop off the deer, making way for others. Given such turnover, during a four-week season of tick procreation, a single whitetail can supply blood for the production of 2 million fertilized tick eggs. If half of those hatch, it's a million larvae from one deer.

Such data and calculations helped make Rick Ostfeld a heretic on the significance of deer in the Lyme disease system. The prevailing assumption was that more deer yield more ticks and therefore more risk of disease. "But it looks like all you need is a *few* deer to support a very abundant tick population," he told me. The more important risk factors, in an area like coastal Connecticut, might be local abundance of white-footed mice and shrews. Who knew?

But hold on. We're dealing with ecology, therefore complexity, and two additional factors must be considered. One is an unchanging fact and one is a variable. The unchanging fact is that *Borrelia burgdorferi* infection doesn't pass vertically between blacklegged ticks. In plainer language: It is not inherited. Of those million baby ticks, all derived from the female ticks that fed on a single deer, none will be carrying *B. burgdorferi* when they hatch—not even if every mother tick was infected and the deer was too. The youngsters will come into the world clean and healthy. Each generation of ticks must be infected anew. Generally what seems to happen is that a larval tick acquires the spirochete by taking its blood meal from an infected host—a mouse, a shrew, a whatever. It molts to become a nymph and then, if it gets its next meal from an uninfected host, the nymph passes the infection to that animal, by drooling spirochetes into the wound along with its anticoagulant saliva. "If mammals didn't make ticks sick," Ostfeld said, "ticks wouldn't make mammals sick later on." Such reciprocal infectivity helps keep the prevalence of *B. burgdorferi* high in both tick populations and hosts.

Related to the unchanging fact of noninheritability is a variable that Ostfeld and others call "reservoir competence." This is the measure of likelihood that a given host animal, if it's already infected, will transmit the infection to a feeding tick. Reservoir competence varies from species to species, most likely depending on differences in the strength of immune response against the pathogen. If the immune response is weak and the blood teems with spirochetes, that species will serve as a highly "competent" reservoir of *B. burgdorferi*, transmitting infection to most ticks that bite it. If the immune response is strong and effective, damping down the level of blood-borne spirochetes, that species will be a relatively less competent reservoir. Studies by Ostfeld's group, involving captive animals and the ticks feeding on them, showed white-footed mice to be the most competent of reservoirs for the Lyme disease spirochete. Chipmunks were a distant second in reservoir competence, with shrews close behind them.

Further complication: Besides being very competent as reservoirs, white-footed mice are also inefficient groomers, poor at clearing off the ticks, which target especially their faces and ears, so that a high percentage of their ticks survive into later stages. Shrews are also inefficient self-groomers, unfortunately for them, and therefore mice and shrews contribute disproportionally to the feeding, infecting, survival, and successful metamorphosis of larval ticks. By this standard, chipmunks were third in overall importance.

What matters perhaps less than their relative rankings is the more general point that these four little mammals together weigh so heavily in the system. Summary statistics compiled by Ostfeld and his gang indicate that up to 90 percent of the infected nymphal ticks "questing" for their next hosts, in a typical forest patch near Millbrook, New York, had taken their larval blood meal from (and therefore been infected by) either a white-footed mouse, a chipmunk, a short-tailed shrew, or a masked shrew. Those four hadn't fed 90 percent of all blacklegged nymphs but, because of the differences in reservoir competence and grooming efficiency, they had fed 90 percent of those that became infected and dangerous to people. Should I repeat that? Four kinds of small mammal fueled nine-tenths of the disease-bearing ticks.

So forget about deer abundance. White-tailed deer are involved in the Lyme disease system, yes, but involved like a trace element, a catalyst. Their presence is important but their numerousness is not. The littler mammals are far more critical in determining the scale of disease risk to people. Adventitious years of big acorn crops, yielding population explosions of mice and chipmunks, are more likely to influence the number of Lyme disease cases among Connecticut children than anything that deer hunters may do. Beyond helping the blacklegged tick (infected or uninfected) to survive, white-tailed deer are almost irrelevant to Lyme disease epidemiology. They don't magnify the prevalence of infection in the forest. They don't pass the spirochete to humans or to newly hatched ticks. They're dead-end hosts, Ostfeld told me.

Then again, he said, "We also happen to be dead-end hosts, in that, once we're infected, the infection goes nowhere. It stays in our body. It doesn't go back into ticks. So we're an incompetent reservoir." Mice and shrews make the ticks sick; the ticks make us sick; and we don't make anybody sick. The *Borrelia burgdorferi* spirochete, if a person catches it, stops there. It doesn't travel on a sneeze or a handshake. It doesn't move downwind. It's not an STD. This is interesting ecologically but probably cold consolation to anyone suffering from Lyme disease.

52

Ostfeld is sensitive to the human toll, not just to the wondrous intricacy of *Borrelia burgdorferi* dynamics in American forests. He showed me some figures from Dutchess County, New York, which includes Millbrook and the Cary Institute, between 1986 and 2005. The twenty-year trend of human infections was steeply upward, with especially high peaks in 1996 and 2002. People were suffering. In 1996 there were 1,838 reported cases of Lyme disease. After that came a sizable decline until, in 2002, again almost two thousand new cases were reported.

Still, it's best understood as an ecological phenomenon, not just a medical problem. "Lyme disease in humans exists because we are sort of unwitting victims of a wildlife-tick interaction," Ostfeld said. "We're interlopers into this system where ticks and these hosts—the reservoir hosts—pass bacterial infections back and forth." One way of construing those peaks in 1996 and 2002, he explained, is that they reflect autumns of bounteous production in the local forests. White-footed mice love acorns and, because the mice reproduce quickly and mature quickly, responding to food abundance with bursts of heightened fecundity, big masting events are often followed (after a two-year lag) by big increases in the mouse population. One pair of mice, given circumstances of plentiful food, could produce a net gain of fifty to seventy-five mice within a year. More acorns, more mice, more infected ticks, more Lyme.

Dutchess County is a halcyon Yankee getaway just east of the Catskills and only two hours from Manhattan via the Taconic State Parkway. It's a landscape of rolling hills, stone fences, small towns, old roadhouses, little gullies and streams carrying rain to the Hudson River, golf courses, and suburban neighborhoods, including some graceful homes with sizable yards shaded by hardwood trees and bordered by hedges or feral brush. The residential areas, even the commercial districts and malls, are well garnished with greenery. Scattered between and around the zones of concentrated human presence are parks, woodlots, and forest patches, dominated not by people but by oak and maple. The understory of those patches is rife with mosses, leaf litter, barberry, chickweed, acorn scraps, poison ivy, wild mushrooms, rotting logs, soggy swales, and the newts, frogs, salamanders, crickets, pill bugs, earthworms, spiders, and garter snakes that thrive in such places. Ticks too, of course—manymanymany ticks. During the year before my visit, Dutchess County health authorities had recorded another 1,244 cases of Lyme disease within a resident population of less than three hundred thousand people. It was enough to make you think twice about a stroll through the woods.

Ostfeld and his team can't afford to be squeamish, though, because those forest patches are where they gather their data. I had tagged along earlier that day, walking trap lines with him and some of his young colleagues. One of them was Jesse Brunner, a postdoc from Helena, Montana, bearded and balding, who was engaged in a multiyear study exploring the correlation, if any, between Lyme prevalence and species diversity on forest patches of various sizes. Another teammate was Shannon Duerr, a tech assistant employed in Ostfeld's lab, presently suffering a case of Lyme disease herself and under treatment with amoxicillin. Ostfeld, I noticed, wore his jeans tucked into his socks as we moved through the forest, and he worked in latex gloves while handling a captured animal. Jesse Brunner showed me his own technique with a white-footed mouse, and then handed the creature to me.

I held the mouse, as instructed, with a gentle pinch of the skin over its shoulders. Its eyes were dark and huge, protrusive with fear, gleaming like steel BBs. Its ears were large and velvety. Its fur was a soft brownish gray. Attached to one ear I could see several dark dots, each no bigger than a period. Those were larval ticks, Brunner explained; they had recently come aboard and scarcely begun to drink. In the other ear was a larger black lump, big as a pinhead. That larva had been attached longer and was now engorged with blood. At this time of the season, Brunner told me, the mouse was probably already infected with *B. burgdorferi* from the bite of a nymphal tick. The engorged larva had probably just become infected, in turn, from the mouse. So I was holding, most likely, two infected carriers. As I listened raptly to Brunner, the mouse sensed my lapse of attention, sprang free of my grip, hit the ground running, and disappeared in the undergrowth. And so the cycle continued.

That afternoon, during our chat in his office, I asked Ostfeld a practical question: Say you're a parent with young children, living here in your Millbrook dream house on three acres of beautiful lawn and shrubbery—what do you want for protection against Lyme disease? There might be a whole range of desperate options. Pesticide spraying by the county? Deer eradication by the state? Thousands of mousetraps (not Shermans but the lethal kind), deployed in the forest and baited with cheese, snapping away like brushfire? Do you pave your yard and ring it with an oil-filled moat? Do you put flea-and-tick collars on your kids' ankles before they go out to play?

No, none of those. "I would feel a lot more comfortable," Ostfeld answered, "if I knew that the landscape would support healthy populations of owls, foxes, hawks, weasels, squirrels of various kinds—the components of the community that could regulate mouse populations." In other words, biological diversity.

This was his offhand way of expressing the most notable conclusion that has emerged from twenty years of research: Risk of Lyme disease seems to go up as the roster of native animals, in a given area, goes down. Why? Probably because of the differences in reservoir competence between mice and shrews (both with high competence) and almost all other vertebrate hosts (low competence) that may share habitat with them. The effect of the most competent reservoirs is diluted by the presence, when there is such presence, of less-competent alternatives. In forest patches containing a full cast of ecological players—medium-sized predators such as hawks, owls, foxes, weasels, and possums, as well as smallish competitors such as squirrels and chipmunks—the populations of white-footed mice and shrews are relatively small, held in check by predation and competition. The average reservoir competence is therefore low. In forest patches with little diversity, on the other hand, white-footed mice and shrews are almost certainly there, flourishing inordinately. And where they flourish, transmitting infection efficiently to the ticks that bite them, *Borrelia burgdorferi* flourishes also.

This insight had led Ostfeld to another interesting question, one with direct implications for public health. Which forest patches contain less species diversity than others? In practical terms: Which woodlots and green zones and parks harbor the greatest risk of exposure to Lyme disease?

Bear in mind that any patch of forest, surrounded by pavement and buildings and other forms of human impact, is to some degree an ecological island. Its community of land animals is insularized because, when individuals try to leave or to enter, they get squashed. (Birds are a special case, though they tend to conform to the same pattern.) Be aware too that big islands generally support more diversity than small islands do. Madagascar is more richly diverse than Fiji, which in turn is more richly diverse than Pohnpei. Why? The simple answer is that greater land area and greater habitat diversity allow the survival of more kinds of creatures. (The complicated details behind that simple answer are addressed by a field of science called island biogeography, familiar to Rick Ostfeld because it so heavily influenced ecological thinking during the 1970s and 1980s, and familiar to me because I wrote a book about it in the 1990s.) Apply that principle to Dutchess County, New York, and it yields a prediction that small forest patches, postage-stamp woodlands, contain fewer kinds of animal than larger forest tracts. That's what Rick Ostfeld did—applied the prediction of area-related diversity as a rough hypothesis and then studied real sites to test it. By the time of my visit to Millbrook, he could say that the pattern did seem to hold true, while Jesse Brunner's postdoctoral work probed further into the same topic.

Then time passed. Five years after I spoke with him, Rick Ostfeld could state the matter more confidently based on two decades' worth of continuous investigation. It became an important theme in his *Lyme Disease* book. With his increasing confidence in the general principles has come increasing appreciation for the various ways those principles play out in differing circumstances. All his conclusions are now carefully modified with conditionals. But the basic findings are clear.

A tiny patch of woodland in a place such as Dutchess County is likely to harbor only a few kinds of mammal, one of which is the white-footed mouse. The mouse is a good colonizer, a good survivor, a fecund breeder, an opportunist; it is there to stay. Restrained by few predators and few competitors, its population fluctuates around a relatively high average level and, in summers following a big acorn crop, goes much higher still. A plague of mice will infest the little woodland, like rats on the road out of Hamelin. There will also be plenty of ticks. The ticks drink heartily of mouse blood and have a high rate of survival, because white-footed mice (unlike possums, catbirds, or even chipmunks) are not very good at grooming themselves clear of larvae. And because the mouse is such a competent reservoir of *Borrelia burgdorferi*—so efficient at harboring and transmitting it—most ticks carry the infection.

In a larger area of forest, with a more richly diverse community of animals and plants, the dynamics are different. Facing a dozen or more kinds of predators and competitors, the white-footed mouse is less numerous; the other mammals are less competent as hosts for the spirochete and less tolerant of thirsty tick larvae; the net effect is fewer infected ticks.

Although it's an intricate system, as Ostfeld warned in his title, certain points about Lyme disease emerge plainly. "We know that walking into a small woodlot," he wrote, "is riskier than walking into a nearby large, extensive forest. We know that hiking in the oak woods two summers after a big acorn year is much riskier than hiking in those same woods after an acorn failure. We know that forests that house many kinds of mammals and birds are safer than those that support fewer kinds. We know that the more opossums and squirrels there are in the woods, the lower the risk of Lyme disease, and we suspect that the same is true of owls, hawks, and weasels." As for white-tailed deer: They're involved, yes, but far from paramount, so don't believe everything you've heard.

Some people take "All life is connected" to be the central truth of ecology, Ostfeld added. It's not. It's just a vague truism. The real point of the science is understanding which creatures are more intimately connected than others, and how, and to what result when change or disturbance occurs.

53

One of the signal lessons of Lyme disease, as Rick Ostfeld and his colleagues have shown, is that a zoonosis may spill over more readily within a disrupted, fragmented ecosystem than within an intact, diverse ecosystem. Another lesson, though, has little to do with Ostfeld's work and can't be addressed at the scale of Sherman traps baited with oats. This one derives from a more basic fact—the fact that *Borrelia burgdorferi* is a bacterium.

It's a bacterium, admittedly, with some peculiar traits. When assaulted with antibiotics, for instance, *B. burgdorferi* seems to retreat into a defensive, impervious form, a sort of cystlike stage known as a "round body." Round bodies are resistant to destruction and very difficult to detect. A patient who seems cured of Lyme disease by the standard two-to-four-week course of amoxicillin or doxycycline might still be harboring round bodies and therefore subject to relapse. Round bodies might even explain the "chronic Lyme disease" syndrome so hotly contested by suffering patients, maverick physicians, and the IDSA. Or not.

Don't confuse the round bodies of *Borrelia burgdorferi* with the small form of *Coxiella burnetii*, the agent of Q fever, also cystlike but found adrift on the Dutch breezes, carrying infection downwind from a birthing goat. Nobody is claiming, not so far, anyway, that Lyme disease can likewise travel on the wind. Both the round bodies of *B. burgdorferi* and the small form of *C. burnetii* merely illustrate that, even in the age of antibiotics, bacteria can be sneaky and tough. These microbes remind us that you don't have to be a virus to cause severe, intractable, mystifying outbreaks of zoonotic disease in the twenty-first century. Although it helps.

VI

GOING VIRAL

54

Viruses were an invisible mystery, like dark matter and Planet X, until well into the twentieth century. They were momentously consequential but undetectable, like the neutron. Anton van Leeuwenhoek's microbial discoveries hadn't encompassed them, nor had the bacteriological breakthroughs of Pasteur and Koch, two hundred years later. Pasteur worked on rabies as a disease, true, and even developed a vaccine, but he never laid eyes on the rabies virus itself nor quite understood what it was. Likewise, in 1902, William C. Gorgas eliminated yellow fever from Cuba, by a program of mosquito eradication, without ever knowing just what infectious agent those mosquitoes carried. It was like a blindfolded hunter shooting ducks by the sound of their quacking. Even the influenza virus of 1918–1919, having killed up to 50 million people around the world, remained a ghostly cipher, unseen and unidentified at the time. Viruses couldn't be viewed with an optical microscope; they couldn't be grown in a culture of chemical nutrients; they couldn't be captured, as bacteria could, with a porcelain filter. They could only be inferred.

Why so elusive? Because viruses are vanishingly minuscule, simple but ingenious, anomalous, economical, and in some cases fiendishly subtle. Expert opinion even divides on the conundrum of whether viruses are alive. If they aren't, then at the very least they're mechanistic shortcuts on the principle of life itself. They parasitize. They compete. They attack, they evade. They struggle. They obey the same basic imperatives as all living creatures—to survive, to multiply, to perpetuate their lineage—and they do it using intricate strategies shaped by Darwinian natural selection. They evolve. The viruses on Earth today are well fit for what they do because only the fittest have survived.

The word "virus" has a much longer history than the study of what we now call by that name. It comes directly from the Latin *virus*, a term meaning "poison, sap of plants, slimy liquid." You can even find the Latin word rendered as "poisonous slime." Its earliest known use in English to denote a disease-causing agent was in 1728, though for the rest of the eighteenth century, throughout the nineteenth, and for several decades beyond, there was no clear distinction between "virus" as a vague term, applicable to any infectious microbe, and the very particular group of entities we know as viral today. As late as 1940, even Macfarlane Burnet sometimes called the Q fever microbe a "virus" in casual usage, though by then he knew perfectly well it was a bacterium.

The effects of viruses were detected long before viruses themselves. Smallpox and rabies and measles were excruciatingly familiar at the clinical level for centuries, millennia, although their causal agents weren't. Acute disease and epidemic outbreaks were understood in a variety of inventive ways—as caused by miasmal vapors and "effluvia," by decaying matter and filth, by poverty, by the whim of God, by bad magic, by cold air or wet feet—but the recognition of infectious microbes came slowly. Around 1840, a German anatomist named Jakob Henle began to suspect the existence of noxious particles—creatures or things—that were too small to be seen with a light microscope and yet able to transmit specific diseases. Henle had no evidence, and the idea didn't immediately take hold. In 1846, a Danish physician named Peter Panum witnessed a measles epidemic on the Faroe Islands, a remote archipelago north of Scotland, and drew some keen inferences about how the ailment seemed to pass from person to person, with a delay of about two weeks (what we'd now call an incubation period) between exposure and symptoms. Robert Koch, who had been a student of Jakob Henle's at Göttingen, advanced beyond observation and supposition with his experimental work of the 1870s and 1880s, identifying the microbial causes of anthrax, tuberculosis, and cholera. Koch's discoveries, along with those of Pasteur and Joseph Lister and William Roberts and John Burdon Sanderson and others, provided the empirical bases for a swirl of late-nineteenth-century ideas that commonly get lumped as "the germ theory" of disease, which marked a movement away from older notions of malign vapors, transmissible poisons, imbalanced humors, contagious putrefaction, and magic. But the germs with which Koch, Pasteur, and Lister mainly concerned themselves (apart from Pasteur's brilliant guesswork on rabies) were bacteria.

And bacteria weren't quite so ineffable. They could be seen with a normal microscope. They could be cultured in a Petri dish (the invention of Julius Petri, Koch's assistant) containing a nutrient-rich medium of agar. They were bigger and easier to grasp than viruses.

The next crucial insight came from agronomy, not medicine. During the early 1890s, a Russian scientist named Dmitri Ivanofsky, in St. Petersburg, studied tobacco mosaic disease, a problem on plantations within the empire. The "mosaic" spots on the leaves led eventually to stunting and shriveling, which lowered productivity and cost growers money. Earlier work had shown that this disease was infectious—it could be transferred experimentally from one plant to another by applying sap drawn from infected leaves. Ivanofsky repeated the transmission experiment, with one added step. He put the juice through a Chamberland filter, a device made from unglazed porcelain, with tiny pores, for purifying water by screening out bacteria. Ivanofsky's report, that "the sap of leaves infected with tobacco mosaic disease retains its infectious properties even after filtration," constituted the first operational definition of viruses: infectious but "filterable," meaning so small they would pass through where bacteria wouldn't. Soon afterward, a Dutch researcher named Martinus Beijerinck arrived independently at the same result and then pushed one step farther. By diluting the filtered sap from an infected plant and using that tincture to infect another plant, Beijerinck found that the infectious stuff, whatever it was, regained its full strength even after dilution. That meant it was reproducing itself in the second plant's living tissues, which meant in turn that it wasn't a toxin, a poisonous excretion, of the sort that some bacteria produce. A toxin, diluted in volume, is reduced in effect—and it doesn't spontaneously recover its strength. This stuff did. But in a container of filtered sap alone, it wouldn't grow. It needed something else. It needed the plant.

So the cumulative work of Martinus Beijerinck, Dmitri Ivanofsky, and a few colleagues showed that tobacco mosaic disease is caused by an entity smaller than a bacterium, invisible by microscope, and capable of multiplication within—only within—living cells. That was the basic profile of a virus, though still nobody had seen one. Beijerinck guessed that the tobacco-mosaic agent was liquid and labeled it *contagium vivum fluidum*, a contagious living fluid. Later work, including the invention of the electron microscope in the 1930s, proved him wrong on that point. A virus is not liquid but solid: minute particles.

This was all about plants. The first animal virus discovered was the one causing foot-and-mouth disease, another sore problem to agriculture. Cattle and swine passed it to one another, like a sneeze on the breeze, and died from it or else had to be culled. Friedrich Loeffler and Paul Froesch, at a university in northern Germany, using the same techniques of filtering and dilution as Beijerinck, proved in 1898 that the foot-and-mouth agent is also a filter-passing entity capable of replication only in living cells. Loeffler and Froesch even noted that it might be just one of a whole class of disease agents, so far undiscovered, possibly including some that infected people, causing phenomena such as smallpox. But the first viral infection recognized in humans wasn't smallpox; it was yellow fever, in 1901. Around the time William Gorgas was solving the practical problem of yellow fever in Cuba, by killing off all those mosquitoes, Walter Reed and his little team of microbiologists showed that the causative agent was indeed mosquito-transmitted. Still, they couldn't see it.

Scientists then began using the label "filterable virus," which was a clumsy but more precise application of the old poisonous-slime word. Hans Zinsser, for example, in his 1934 book *Rats, Lice and History*, a classic chronicle of medical groping and discovery, declared himself "encouraged by the study of the so-called 'filterable virus' agents." Many epidemic diseases, Zinsser wrote, "are caused by these mysterious 'somethings'—for example, smallpox, chicken pox, measles, mumps, infantile paralysis, encephalitis, yellow fever, dengue fever, rabies, and influenza, to say nothing of a large number of the most important afflictions of the animal kingdom." Zinsser realized, too, that some of those animal afflictions might overlap with the first category, human epidemics. He added a crucial point: "Here, as in bacterial disease, there is a lively interchange of parasites between man and the animal world." Zinsser was a panoramic thinker as well as an acutely trained microbiologist. Eight decades ago he sensed that viruses, only lately discovered, might be among the most nefarious of zoonoses.

55

The difficulty of cultivating viruses *in vitro* made them obscure to early researchers, elusive in the laboratory, but it was also a clue to their essence. A virus won't grow in a medium of chemical nutrients because it can only replicate inside a living cell. In the technical parlance, it's an "obligate intracellular parasite." Its size is small and so is its genome, simplified down to the bare necessities for an opportunistic, dependent existence. It doesn't contain its own reproductive machinery. It mooches. It steals.

How small is small? The average virus is about one-tenth the size of the average bacterium. In metric terms, which are how science measures them, roundish viruses range from around fifteen nanometers (that's fifteen *billionths* of a meter) in diameter to around three hundred nanometers. But viruses aren't all roundish. Some are cylindrical, some are stringy, some look like bad futuristic buildings or lunar landing modules. Whatever the shape, the interior volume is minuscule. The genomes packed within such small containers are correspondingly limited, ranging from 2,000 nucleotides up to about 1.2 million. The genome of a mouse, by contrast, is about 3 *billion* nucleotides. It takes three nucleotide bases to specify an amino acid and on average about 250 amino acids to make a protein (though some proteins are much larger). Making proteins is what genes do; everything else in a cell or a virus results from secondary reactions. So a genome of just two thousand code letters, or even thirteen thousand (as for the influenzas) or thirty thousand (the SARS virus), is a very sketchy set of engineering specs. Even with such a small genome, though, coding for just eight or ten proteins, a virus can be wily and effective.

Viruses face four basic challenges: how to get from one host to another, how to penetrate a cell within that host, how to commandeer the cell's equipment and resources for producing multiple copies of itself, and how to get back out—out of the cell, out of the host, on to the next. A virus's structure and genetic capabilities are shaped parsimoniously to those tasks.

Sir Peter Medawar, an eminent British biologist who received a Nobel Prize the same year as Macfarlane Burnet, defined a virus as "a piece of bad news wrapped up in a protein." The "bad news" Medawar had in mind is the genetic material, which so often (but not always) inflicts damage on the host creature while exploiting its cells for refuge and reproduction. The protein wrap is known as a capsid. The capsid serves two purposes: It protects the viral innards when they need protection and it helps the virus lever its way into cells. The individual viral unit, one particle, standing intact outside a cell, is called a virion. The capsid also defines the exterior shape of a virus. Virions of Ebola and Marburg, for instance, are long filaments, which is why they've been placed in a group known as filoviruses. Other viruses have particles that are spherical, or ovoid, or helical, or icosahedral (twenty-sided, like a soccer ball designed by Buckminster Fuller). HIV-1 particles are globular. Rabies virions are shaped like bullets. A plate of Ebola virions mixed with Hendra virions would resemble capellini in a light sauce of capers.

Many viruses are wrapped with an additional layer, known as an envelope, comprising not only protein but also lipid molecules drawn from the host cell—in some cases, pulled from the wall of the cell when the virion made its exit. Across the outer surface of the envelope, the virion may be festooned with a large number of spiky molecular protuberances, like the detonator stubs on an old-fashioned naval mine. Those spikes serve a crucial function. They're specific to each kind of virus, with a keylike structure that fits molecular locks on the outer surface of a target cell; they allow the virion to attach itself, docking like one spaceship to another, and they open the way in. The specificity of the spikes not only constrains which kinds of host a given virus can infect but also which sorts of cell—nerve cells, stomach cells, cells of the respiratory lining—the virus can most effectively penetrate, and therefore what sort of disease it may cause. Useful as they are to a virus, though, the spikes also represent points of vulnerability. They are the primary targets of immune response by an infected host. Antibodies, produced by white blood cells, are molecules that glom onto the spikes and prevent a virion from grabbing a cell.

The capsid shouldn't be mistaken for a cell wall or a cell membrane. It's merely analogous. Viruses, from the beginning of virology, have been defined in the negative (*not* captured by a filter, *not* cultivable in chemical nutrients, *not* quite alive), and the most fundamental negative axiom is that a virion is not a cell. It doesn't function the way a cell functions; it doesn't share the same capacities or frailties. That's reflected in the fact that viruses are impervious to antibiotics—chemicals valued for their ability to kill bacteria (which are cells) or at least impede their growth. Penicillin works by preventing bacteria from building their cell walls. So do its synthetic alternatives, such as amoxicillin. Tetracycline works by interfering with the internal metabolic processes by which bacteria manufacture new proteins for cell growth and replication. Viruses, lacking cell walls, lacking internal metabolic processes, are oblivious to the effects of such killer drugs.

Inside the viral capsid is usually nothing but genetic material, the set of instructions for creating new virions on the same pattern. Those instructions can only be implemented when they're inserted into the works of a living cell. The material itself may be either DNA or RNA, depending on the family of virus. Both types of molecule are capable of recording and expressing information, though each has its advantages and its drawbacks. Herpesviruses, poxviruses, and papillomaviruses contain DNA; so do half a dozen viral families you've never heard of, such as the iridoviruses, the baculoviruses, and the hepadnaviruses (one of which causes hepatitis B). Others, including filoviruses, retroviruses (most notoriously, HIV-1), coronaviruses (SARS-CoV), and the families encompassing measles, mumps, Hendra, Nipah, yellow fever, dengue, West Nile, rabies, Machupo, Junin, Lassa, chikungunya, all the hantaviruses, all the influenzas, and the common cold viruses, store their genetic information in the form of RNA.

The different attributes of DNA and RNA account for one of the most crucial differences among viruses: rate of mutation. DNA is a double-stranded molecule, the famed double helix, and because its two strands fit together by way of those very specific relationships between pairs of nucleotide bases (adenine linking only with thymine, cytosine only with guanine), it generally repairs mistakes in the placement of bases as it replicates itself. This repair work is performed by DNA polymerase, the enzyme that helps catalyze construction of new DNA from single strands. If an adenine is mistakenly set in place to become linked with a guanine (not its correct partner), the polymerase recognizes that mistake, backtracks by one pair, fixes the mismatch, and then moves on. So the rate of mutation in most DNA viruses is relatively low. RNA viruses, coded by a single-strand molecule with no such corrective arrangement, no such buddy-buddy system, no such proofreading polymerase, sustain rates of mutation that may be thousands of times higher. (For the record, there's also a smaller group of DNA viruses that code their genetics on single strands of DNA and suffer high mutation rates, as in RNA. And there's a little group of double-stranded RNA viruses. To every rule, an exception. But we're going to ignore those minor anomalies because this stuff is already complicated enough.) The basic point is so important I'll repeat it: RNA viruses mutate profligately.

Mutation supplies new genetic variation. Variation is the raw material upon which natural selection operates. Most mutations are harmful, causing crucial dysfunctions and bringing the mutant forms to an evolutionary dead end. But occasionally a mutation happens to be useful and adaptive. And the more mutations occurring, the greater chance that good ones will turn up. (More mutations also mean more chance of harmful ones, lethal to the virus; this puts a cap on the maximum sustainable mutation rate.) RNA viruses therefore evolve quicker than perhaps any other class of organism on Earth. It's what makes them so volatile, unpredictable, and pesky.

Notwithstanding the quip by Peter Medawar, not every virus is "a piece of bad news wrapped up in a protein"—or at least, it's not bad news for every host infected. Sometimes the news is merely neutral. Sometimes it's even good; certain viruses perform salubrious services for their hosts. "Infection" need not always entail any significant damage; the word merely means an established presence of some microbe. A virus doesn't necessarily achieve anything by making its host sick. Its self-interest requires just replication and transmission. The virus enters cells, yes, and subverts their physiological machinery to make copies of itself, yes, and often destroys those cells as it exits, yes; but maybe not so many cells as to cause real harm. It may inhabit a host rather quietly, benignly, replicating at modest levels and getting transmitted from one individual to another without producing any symptoms. The relationship between a virus and its reservoir host, for instance, tends to involve such a truce, sometimes reached after long association and many generations of mutual evolutionary adjustment, the virus becoming less virulent, the host becoming more tolerant. That's in part what defines a reservoir: no symptoms. Not every virus-host relationship evolves toward such amicable relations. It's a special form of ecological equilibrium.

And like all forms of ecological equilibrium, it's temporary, provisional, contingent. When spillover occurs, sending a virus into a new kind of host, the truce is canceled. The tolerance is nontransferable. The equilibrium is ruptured. An entirely new relationship occurs. Freshly established in an unfamiliar host, the virus may prove to be an innocuous passenger, or a moderate nuisance, or a scourge. It all depends.

56

The virus known informally as herpes B (and more precisely now as Macacine herpesvirus 1, referring to its natural reservoirs, macaques) sprang from obscurity to medical attention in 1932, after a laboratory mishap at New York University. A young scientist named William Brebner was doing research toward a polio vaccine. Monkeys were important for such work, and the animal of choice was the rhesus macaque (*Macaca mulatta*), which belongs to the cercopithecine family. Because poliovirus hadn't yet been cultured in glass (that would eventually be possible, but only when living cells could be maintained in the medium as viral hosts), rhesus macaques typically served both as incubators of the virus and as test subjects. Poliomyelitis is not a zoonosis; it doesn't naturally affect any animals other than humans; but with the help of a hypodermic needle, it could be made to grow in monkeys. An experimenter would take the poliovirus from one animal, which had been artificially infected, and inject that into the brain or the spinal cord of another, keeping the chain of infection continuous and observing effects on the monkeys along the way. One day, handling a monkey, William Brebner got bitten.

It wasn't a bad bite, just a nip across the ring finger and the pinkie of his left hand. Brebner dosed the wounds with iodine, then with alcohol, and kept working. The monkey seemed normal and healthy, though understandably cantankerous, and if it was already carrying polio, that doesn't seem to have concerned Brebner. Soon afterward the monkey died (under ether, during another experimental procedure), and it wasn't necropsied.

Three days later, Brebner noticed "pain, redness, and slight swelling" around the bite. Another three days passed and he was admitted to Bellevue Hospital. His symptoms developed slowly—tender lymph nodes, abdominal cramps, paralysis of his legs, inability to urinate, tingling numbness in his arms, and then a high fever and hiccupping—until, after two weeks, he was very sick indeed. His breathing became labored and he turned blue. Put into a respirator, he convulsed and lost consciousness. Frothy liquid came wheezing out of his mouth and nostrils. Five hours later, William Brebner was dead at the age of twenty-nine.

What killed him? Was it polio? Was it rabies? A fellow researcher in the same NYU lab, just out of medical school but bright and ambitious, assisted at Brebner's autopsy and then made a further investigation, using bits of Brebner's brain, spinal cord, lymph nodes, and spleen. This man was Albert B. Sabin, decades before his fame as creator of an oral polio vaccine. Sabin and a colleague injected an emulsion from Brebner's brain back into monkeys; they also injected some mice, guinea pigs, and dogs. None of those animals showed signs of what Brebner had suffered. But rabbits, likewise injected, did. Their legs went limp, they died of respiratory failure, their spleens and livers were damaged. From the rabbits, Sabin and his partner extracted a filtered essence capable of causing the same course of infection again. They called it simply "the B virus," after Brebner. Other work showed that it was a herpesvirus.

Herpes B is a very rare infection in humans but a nasty one, with a case fatality rate of almost 70 percent among those few dozen people infected during the twentieth century (before recent breakthroughs in antiviral pharmaceutics) and almost 50 percent even since then. When it doesn't kill, it often leaves survivors with neurological damage. It's an occupational hazard of scientists and technicians who work with laboratory macaques. Among the macaques themselves it's common, but merely an annoyance. It abides within nerve ganglia and emerges intermittently to cause mild lesions, usually in or around the monkey's mouth, like cold sores or canker sores from herpes simplex in humans. The monkey sores come and go. Not so with herpes B in people. In the decades since Brebner's death, forty-two other human cases have been diagnosed, all involving scientists or laboratory technicians or other animal-handlers who had contact with macaques in captivity.

The number of human cases rose quickly during the era of fervid research toward a polio vaccine, in the 1950s, probably because those efforts entailed such a sharp increase in the use of rhesus macaques. Conditions of caging and handling were primitive, compared with standards for medical research on primates today. Between 1949 and 1951, a single project within the overall effort financed by the National Foundation for Infantile Paralysis (aka the March of Dimes) consumed seventeen thousand monkeys. The foundation maintained a sort of clearinghouse for imported monkeys in South Carolina, from which one leading researcher had a standing order of fifty macaques per month, at $26 apiece, delivered. Nobody knows exactly how many macaques were "sacrificed" in the labs of Albert Sabin and Jonas Salk, let alone other researchers, but the incidence of herpes B infections peaked in 1957–1958, just as the polio vaccine quest came to its crescendo. Most of those cases occurred in the United States, the rest in Canada and Britain, places where rhesus macaques were thousands of miles removed from their natural habitat but medical research was intensive.

From that 1950s peak, the rate of accidental infections declined, possibly because lab techs began taking better precautions, such as wearing gloves and masks, and tranquilizing monkeys before handling them. In the 1980s came a small second uptick in herpes B incidents, correlated with another increase in the use of macaques, this time for research on AIDS.

The most recent case occurred at the Yerkes National Primate Research Center, in Atlanta, in late 1997. On October 29, a young woman working among the captive monkeys was splashed in the eye with some sort of bodily gook from a rhesus macaque. It may have been urine, or feces, or spit; nobody seems to know. She wiped her eye with a paper towel, soldiering on through her chores, and almost an hour later found time to rinse the eye briefly with water. That wasn't enough. She filed no incident report, but ten days later the eye was red and swollen. She went to an ER, where the physician on duty prescribed antibiotic eyedrops. Thanks a lot. When the eye inflammation worsened, she saw an ophthalmologist. More days passed, and another ophthalmologist examined her, before she was hospitalized for suspected herpes B. Now they put her on strong antiviral drugs. Meanwhile, cultures taken from swabbing her eyes were quietly retrieved from the commercial laboratories to which they had been sent for analysis—um, never mind, we'll just take those back. Her cultures had belatedly been deemed too dangerous for ordinary lab workers to handle.

The young woman seemed to improve slightly and left the hospital. But she woke the next morning with worsening symptoms—abdominal pain, inability to urinate, weakness in her right foot—and went back. At the end of the month, she began having seizures. Then came pneumonia. She died of respiratory failure on December 10, 1997. Despite the fact that her own father was an infectious-disease doctor, her mother was a nurse, and Yerkes was full of people who knew about herpes B, modern medicine hadn't been able to save her.

This pathetic mishap put some people on edge. The probability of cross-species transmission might be low—very low, under normal circumstances—but the consequences were high. Several years later, when eleven rhesus macaques at a "safari park" in England tested positive for herpes B antibodies, management decided to exterminate the entire colony. This decision was driven by the fact that Britain's Advisory Committee on Dangerous Pathogens had lately reclassified herpes B into biohazard level 4, placing it in the elite company of Ebola, Marburg, and the virus that causes Crimean-Congo hemorrhagic fever. National regulations specified that any animals infected with a level-4 agent had to be either handled under BSL-4 containment (meaning space suits, triple gloves, airlock doors, and all the rest, not quite practicable at a tourist attraction for viewing wildlife) or destroyed. Of course, positive results on antibody tests meant only that those eleven monkeys had been exposed to the virus, not that they were presently infected, let alone shedding herpes B. But that scientific distinction didn't stop the cull. Hired shooters killed all 215 animals at the safari park, using silenced .22 rifles, in a single day. Two weeks later, another animal park in the English countryside followed suit, killing their hundred macaques after some tested positive for herpes B antibodies. The law was the law, and macaques (infected or not) were probably now bad for business. A more sensitive question, raised by primatologists who considered such cullings grotesque and unnecessary, was whether herpes B does or doesn't belong in level 4. Some arguments suggest that it doesn't.

The rhesus macaque isn't the only monkey that carries herpes B. The same virus has been found in other Asian monkeys, including the long-tailed macaque (*Macaca fascicularis*) within its native range in Indonesia. In the wild, though, neither rhesus macaques nor the others have passed any known herpes B infections to humans, not even in situations where the monkeys come into close contact with people. For this there's no easy explanation, because the opportunities do seem to exist. Both rhesus macaques and long-tailed macaques are opportunistic creatures, largely unafraid of humans or human environments. As the chainsaws and machetes of humanity's advance guard have driven them out of their native forest habitats—in India, Southeast Asia, Indonesia, and the Philippines—they have been only more willing to take their chances scavenging, stealing, and panhandling at the edges of civilization. They live anywhere they can find food and a modicum of tolerance. You can see rhesus macaques lurking along the parapets of government buildings in Delhi. You can glimpse long-tailed macaques scrounging garbage from the corridors of dormitories at a university not far from Kuala Lumpur. And because both the Hindu and Buddhist religions embrace gentle attitudes toward animals in general, toward nonhuman primates in particular, macaques have become abundantly, boldly present at many temples around their native regions, especially where any such temple stands near or within a remnant of forest.

At Hindu sites, they have the advantage of their resemblance to the monkey god Hanuman. Buddhism, at least as practiced in Japan, China, and India, also carries ancient threads of monkey veneration. You can see it in iconic art and sculpture, such as the famed three-monkeys carving (see no evil, hear no evil, speak no evil) on the Toshogu Shrine, north of Tokyo. Over generations, over centuries, macaques within these landscapes have come in from the wild and habituated themselves to human proximity. Now they're mascot monkeys at many temples and shrines, indulged like acolytes of Hanuman or the Shinto deity Sanno, living largely on handouts from pilgrims and tourists.

One such place is the Sangeh Monkey Forest in central Bali, amid the green volcanic slopes and the shapely rice paddies of the world's most decorous island. There at Sangeh, two hundred long-tailed macaques wait to cadge handouts from the thousands of visitors who traipse through the temple and its little woodland every month. That's why an anthropologist named Lisa Jones-Engel, of the University of Washington, and her husband, Gregory Engel, a physician, chose Sangeh as a place to study human exposure to monkey-borne herpes B. They knew that the circumstances would be very different from those in a laboratory.

Bali, with a population of almost 4 million in an area barely larger than Delaware, is one of the more crowded human habitats on Earth—but gracefully crowded, ingeniously built up and terraced and irrigated and partitioned, not so squashed and squalid as other densely populous tropical states. Bali is home to most of the Hindus of Indonesia, otherwise a predominantly Muslim country. The little forest at Sangeh amounts to about fifteen acres of hardwoods, providing shade and cover for the macaques but not much natural food. They live instead on peanuts, bananas, cold rice, flower petals, and other treats and offerings, all supplied by temple workers, tourists, and Hindu worshippers. The lane leading into the forest is lined with shops selling souvenirs, clothes, and monkey food. The monkeys aren't shy about accepting, even demanding, those handouts. They have lost their wild instincts about personal space. Enterprising local photographers run a brisk trade in photos of tourists posed with macaques. *And here's me in Bali, with a monkey on my head. Cute little guy, just wanted that Snickers bar.* But the cute little guys sometimes bite and scratch.

Engel, Jones-Engel, and their colleagues gathered two interesting sets of data from this place. They surveyed the monkey population, by way of blood samples; and they surveyed the human workforce at Sangeh, by way of interviews and also blood samples. What they found says a lot about the scope of opportunity for virus spillover between Asian monkeys and people.

The team drew blood from thirty-eight macaques, of which twenty-eight were adults, the rest youngsters. They screened the blood serum for evidence of antibodies to herpes B, the same virus that killed William Brebner and most of the other people ever infected with it. The results of the lab work were chilling: Among adult long-tailed macaques at Sangeh, the prevalence of herpes B antibodies was 100 percent. Every mature animal had been infected. Every mature animal had either once carried the virus or (more likely, because it's a herpesvirus, capable of long-term latency) still did. Among juveniles the rate was lower, presumably because they are born free of the virus and acquire it, as they get older, by social interaction with adults.

Matched against that was the human survey, measuring opportunities for the virus to cross between species. The team found that almost a third of the shopkeepers, photographers, and other local people they interviewed had been bitten at least once by a macaque. Almost 40 percent had been scratched. Some people had been bitten or scratched more than once.

This study, focused on workers, didn't even attempt to count bites and scratches among the tourists who come and go. The researchers merely estimated that there must be thousands of monkey-bitten tourists walking away from Sangeh each year—and Sangeh is just one such Balinese monkey temple among a handful. The odds of a human contracting herpes B under these circumstances seem vast.

But it hasn't happened, so far as anyone knows. Engel, Jones-Engel, and their coauthors wrote that "no case" of human infection with the virus has been reported in Bali, "either in association with monkey forests or in any other nonlaboratory context." Thousands of bites, thousands of scratches, thousands of opportunities, and zero cases (anyway, zero reported cases) of humans sickened by herpes B. If that sounds like good news, rather than a spooky enigma, you're more of an optimist than I am. When I finished reading their paper, still puzzled, I wanted to hear more in person.

57

Before I knew it, I was helping Lisa Jones-Engel and Gregory Engel trap monkeys at a shrine in northeastern Bangladesh.

We had come to a city called Sylhet, along the banks of the Surma River, an area where the Bangladesh lowlands begin to wrinkle up into hills. The hills rise northward into mountains, beyond which lie Assam, Bhutan, and Tibet. Sylhet is a district capital, home to a half million people and an indeterminate number of other primates. Its streets are flooded with traffic that somehow manages to move despite a near-total absence of stoplights. Hundreds of green motorbike taxis, powered by natural gas, and thousands of brightly decorated bicycle rickshaws, powered by longsuffering men with skinny brown legs, jockey for position alongside the bashed-up busses and creeping cars. In early morning, two-wheeled pushcarts also roll through the streets, moving vegetables to market. At the bigger intersections loom shopping complexes and upscale hotels behind gleaming glass. It's a thriving city, one of the richest in this poor country, thanks much to investment and spending by emigrant families, with roots here, who have thrived in Great Britain. They often return home, or at least send money back. Many of the curry shops in London, a man told me, are run by expat Bangladeshis from Sylhet.

Religious tourism also helps fuel the local economy. There are quite a few shrines. And those shrines, besides bringing pilgrims from all over Bangladesh, are what had brought us.

On our first afternoon in Sylhet, we scouted a holy place known as Chashnipeer Majar. It's a small domed structure atop a hillock that looms above a crowded neighborhood, surrounded below by concrete walls, small shops, blank-faced houses fronting the street, and sinuous alleys. A long staircase led us to the shrine, which was overarched by five or six scraggly trees, one with dead limbs where monkeys perched, shaking the branches like mad sailors in a ship's rigging. The hillsides around the shrine were covered with ragged brush, trash, and the graves of Sylhetian ancestors. It wasn't a verdant spot, this little island of sacred ground at the heart of an urban neighborhood, but the resident wildlife didn't seem to mind. There were macaques on the shrine roof, macaques in the trees, macaques on rooftops of the houses below, macaques climbing drainpipes, macaques crossing power lines, macaques loitering on the staircase and walking its railings, macaques scampering among the graves. Having scouted the place on that first afternoon, we came back two days later, in early morning, to disturb the peace.

Our monkey trap was assembled and ready. It was a frame cube of aluminum tubing and nylon mesh, big as a closet, custom built for this purpose, with a falling door controlled by a remote tripwire. You sat at a distance, you watched, you saw monkeys enter, you pulled the line—and the door came down. But don't pull too soon. Don't settle for the first animal that ventures inside. Part of optimal technique for trapping macaques, I'd been told, was to catch as many as possible on the first go, because these critters are smart and they learn quickly. They become trap-shy after seeing the trick worked on their comrades. So whoever holds the tripwire must be patient, waiting for just the right moment, when as many animals as possible are inside the trap.

My assignment was minor: When the door fell, I should get there immediately and lock it down with my foot, so the captured macaques couldn't widget their way out. Gregory Engel would then do the hard part, tranquilizing them one by one with a hypodermic full of Telazol, a fast-acting veterinary anesthetic. How do you inject a hysterical monkey? In this case, by jabbing into its thigh through the mesh of the trap. Professor Mohammed Mustafa Feeroz, Engel and Jones-Engel's principal Bangladeshi collaborator, would stand as defense. Four of Feeroz's students would help. Defense was important because the uncaptured monkeys might charge, frantic to free their comrades. They could be a formidable platoon. Lisa Jones-Engel, chief genius of the whole project but prohibited from entering this shrine because of her gender, would be waiting in a courtyard nearby, along with several female assistants, to begin drawing blood. One, two, three: trap, tranquilize, draw. What could be simpler?

Lots of things, let me tell you, could be simpler.

The trap was baited with puffed rice and bananas. Within moments of seeing the bait placed, a few monkeys came to inspect. They climbed all over the trap, inside and out. Most of the others held back. Word seemed to pass among them, excitement rose, more animals arrived across the rooftops; there must have been a hundred, all nervously curious about our presence and tantalized by the bait. We loitered discreetly, on the steps, on the slope, looking casual and averting our eyes. Feeroz held the trip line. He had the patience of a fisherman watching a bobber jiggle. He waited, he waited, as several of the biggest macaques entered the trap to investigate. One of them, a great male with a Schwarzenegger physique and very long canines, may have been the alpha of the troop. He was bold. Greedy for his share. A few more animals entered behind him. Feeroz pulled.

The door fell, trapping Schwarzenegger plus six others, and all hell broke loose.

58

Maybe it has occurred to you to wonder: sacred monkeys in an Islamic country? Bangladesh's population is 90 percent Muslim, mostly composed of traditional Sunnis. Doesn't Islam forbid graven images and totemism? Aren't those monkey temples supposed to be Hindu or Buddhist?

Right enough, but with an exception: the Sufi shrines of northeastern Bangladesh, including Sylhet. Chashnipeer Majar is a Sufi site.

Sufism in the region traces back seven hundred years, to a pious invader named Hazrat Shah Jalal. It may be practiced by either Shiites or Sunnis, but it's a more mystical, esoteric brand of Islam than mainstream Shia or Sunni observance. As the story goes, Shah Jalal came out of the west, from Mecca by way of Delhi, with his army of 360 disciples. Sylhet was a Brahmin kingdom in those years, but a kingdom of faded strength, ruled by a tribal chieftain. Shah Jalal either conquered the chieftain or (depending on which version you hear) scared him into retreat. One among Shah Jalal's entourage was a man named Chashnipeer, a sort of wizard geologist, charged with finding just the right place for a new kingdom of Sufi believers, where the soil would match Mecca's sacred soil. Sylhet was it. Shah Jalal and his followers settled in the region and converted much of the populace to Sufism. After a long rule, Shah Jalal died and was buried there. His mausoleum, now encompassed within a large mosque complex in a north neighborhood of the city, still attracts pilgrims from all over Bangladesh. I don't believe it welcomes monkeys.

But other sites of worship were also established, taking their names from the lesser founding heroes. These were different from normal Islamic mosques; they were *majars*, shrines, implying veneration of a holy personage, whose body might be entombed (like Shah Jalal's) on the spot. Because this recognition of saintliness can be construed as idolatry—implicitly comparing a mortal individual to God—such Sufi majars may offend against the letter of Islam as understood by Sunni or Shia. They are heterodox. You won't find them down south in the capital, Dhaka.

Then too, in more recent times, some of the Sylhet majars underwent another stage of transformation. With macaque habitat shrinking as the landscape became farmed and urbanized, monkeys found refuge at the shrines. At first they may have stolen food or picked garbage. Gradually they became half-tame. They learned how to beg food and were accommodated, tolerated, eventually indulged, by the men who looked after those sites. Several majars, including Chashnipeer, became monkey shrines.

People arrived to worship, enjoyed seeing the macaques, gave alms, and came back again, occasionally in great number and from long distances for festivals that involved feasting and prayer. The macaques were novel. They were popular. It was a good business model, pardon my secular soul, for a religious establishment. Some pilgrims believed that if a monkey took food from your hand, your prayers would be answered. The whole arrangement might seem sacrilegious in most parts of the Islamic world, but in Sylhet it became holy tradition.

59

Mustafa Feeroz is professor of zoology at Jahangirnagar University in Savar, just north of Dhaka. He's a sweet-spirited fellow, a careful scientist, and an observant Muslim, though not a Sufi. He and Dr. Jones-Engel had of course sought permission to trap monkeys at Chashnipeer Majar, explaining their scientific purposes and their concern that no animals be harmed. That satisfied the committee in charge but not the macaques themselves, who went ballistic when they saw that we had trapped one of their honcho males and a half dozen others, including a female with an infant.

Inside the trap, the captives panicked, bouncing and scrambling across the mesh walls and ceiling. Outside the trap, about eighty other macaques came down from their tree limbs and wires and rooftops, screaming and chattering, surging around us, making moves to attack in support of the hostages. Feeroz and the students had prepared for this moment by picking up large sticks. Now they brandished those weapons, swinging, threatening, smacking the ground, shouting to drive the macaques back. I pinned the door with my foot, as instructed, so that nimble monkey fingers couldn't unlatch it. The loose animals weren't easily cowed. They dodged the sticks, backed off, jumped around, screeched all the more, and came forward again, like those infernal winged monkeys in *The Wizard of Oz*. Gregory Engel meanwhile moved to the trap with his syringe and, through the mesh, managed to jab the Schwarzenegger macaque in its thigh; in the same motion, he rammed down the plunger. It was a nifty move, somewhat outside the usual duties of a family-practice physician from Seattle.

Within a few seconds, Schwarzenegger's ferocity started to wilt. The animal went clumsy, then limp. Lights out, for at least half an hour.

Working quickly, Engel tried to get each of the others. But it was difficult with six monkeys still ricocheting around the cage and others at his back. He poked a couple more and then reloaded his syringes with Telazol. Nobody wanted to get clawed or bitten. Grab a tail if you can! he hollered to me. Pin one against the mesh! Yeah, right. I made a lame tail-grab attempt, but I was the amateur here, and I found little zeal for exposing my hands to the flying claws and teeth of animals well known for carrying herpes B.

Somehow, within a few minutes, Engel injected all five adults in the trap. When we opened the door, one juvenile and the infant skittered away, but the others were down like drunks.

We loaded them into a duffel bag. Go, go fast, said Engel, and two students carried the bag down the staircase and then hoisted it gingerly over a wall, below which Jones-Engel stood ready to help catch the bundle of doped monkeys. She was dressed in traditional Bangladeshi attire—a camise and salwar pants plus a veil over her shoulders, which was her usual field garb, worn in deference to local sensibilities—but now she also wore exam gloves and a surgical mask. She guided the monkey-bearers down an alley to the private courtyard, where women were welcome, where tables had been prepared, where swabs and vials and clipboards and more syringes had been laid out in readiness. The gathering of data began.

Lisa Jones-Engel is a forceful, direct person with years of experience among Asia's nonhuman primates. She loves her subject animals but doesn't romanticize them. As she and her assistants started drawing blood and taking oral swabs, her husband and Feeroz, followed by the male students and me, headed back to the shrine for another round of trapping. Now that we had shown our methods, and our devious intentions, it was dicey to say how the troop might behave. "If the monkeys in the last half hour have figured out their plan of attack," Lisa commanded us, "you just retreat."

60

"Herpes B scares the shit out of people," she told me a few days later. We had returned to Dhaka, and after another long day she and Gregory and I were sharing wee drams of Balvenie in my hotel room. Lisa was adamant. "Herpes B gets populations of monkeys shot in the head and . . ."—she had in mind the safari park culling as well as other such events—"just eradicated. Herpes B is like Ebola that way." It's not only frightful and potent, she meant, but profoundly misunderstood.

Herpes B and Ebola, of course, are very different sorts of bug. But she was right; there are similarities worth noting. In both cases, the virus is often lethal to humans but not nearly so consequential as it would be if not constrained by the limits of its transmissibility. It has no preternatural powers. It finds humans a dead-end host. People are ignorant about its actual properties and inclined to imagine an unreal breadth of risk. Among differences between the two, there's this: Ebola is infamous and herpes B is largely unknown. It's unknown, that is, unless you work in a monkey lab or run a safari park.

Killing off captive macaques is uncalled for, Lisa insisted, even in populations that might carry the virus, so long as their likelihood of passing it to a human is extremely low. And a positive test for antibodies doesn't even prove that the virus is still present.

She mentioned a recent case, just three months earlier, in which a research colony of macaques at a university in France was condemned to extermination. Some of those individual animals were known to and studied by attentive ethologists for twenty-five years. The colony was notable for expressing some fascinating behavioral patterns. A thousand primatologists, from the International Primatological Society and other scientific groups, signed petitions challenging the logic of wholesale condemnation. "Look, don't do this," they argued. "You don't really understand what these results mean." The university council made its decision anyway and, on a Sunday in August, before the scientists and the keepers could protest further, the macaques were all killed.

However dangerous herpes B might be when infecting a person, the chances of monkey-human transmission seem to be extremely small. That's what those research results from the Sangeh Monkey Forest in Bali suggest. Lisa and Gregory found a high prevalence of the virus among the macaques there, and a high incidence of macaque bites and scratches among the people, but no evidence of herpes B transfer. If cases do sometimes occur in Bali, they must escape medical notice, or else get taken for some other dreadful disease, such as polio, or rabies, which is a serious problem in Bali because of its prevalence among the island's dogs. Nobody knows whether any undetected herpes B infections have come out of Sangeh. Possibly, none have.

Other data, published almost a decade earlier by a different team, support the impression that herpes B doesn't leap readily to humans. This study looked at blood samples from 321 laboratory workers—scientists and technicians who handled live primates or else primate cells in culture. Most of those people worked with macaques. Many of them had been bitten, scratched, or splashed. Yet none of the 321 workers tested positive for exposure to herpes B. Evidently the virus is not easily transmitted, and evidently it's not causing subtle, asymptomatic infections among people in close contact with monkeys.

The medical record notes just forty-three cases, beginning with William Brebner, in which contact between a macaque and a person led to infection. True, those forty-three infections often brought dire results. But over the same period of time, during untold thousands or millions of other such contacts—in laboratories, in the wild, from monkey temples to Petri dishes, via scratching or biting or drool or needlestick accident or splashed urine—herpes B didn't make the monkey-human leap. Why not? Apparently this virus isn't ready.

Another way of saying that: Ecology has provided opportunities, but evolution hasn't yet seized them. Maybe it never will.

61

The blood drawn from the macaques we trapped at Chashnipeer Majar would be screened for evidence of another virus too. Lisa Jones-Engel and her team had lately shifted their attention to this one. It's a favorite of mine because of its lurid name: simian foamy virus. No, infected hosts don't foam at the mouth. The "foamy" part derives from its tendency to cause cells in a host to fuse with one another, forming gigantic, nonfunctional megacells that, under a microscope, resemble bubbles of foam.

There's actually a whole gaggle of foamy viruses, all lodged within the genus *Spumavirus*. Some of them infect cows, cats, and horses. They have also been found among gorillas, chimpanzees, orangutans, baboons, macaques, and other primates, in each of which they seem to be ancient infections, having coevolved with their hosts for as long as 30 million years, one species of simian foamy virus (SFV) per species of simian. Maybe that's why, nowadays, they seem so benign. One team working in Central Africa reported evidence of SFV passing from primates that are hunted for bushmeat (mandrills, gorillas, and guenons) into people who hunt those animals. Whether SFV makes the hunters sick is another question, not addressed by that study. If it does, the effects must be slow and subtle. Then again, the HIVs are slow and subtle. And SFV, like the HIVs, is a retrovirus. Jones-Engel isn't the only researcher who feels that simian foamy virus bears watching.

Thirty years ago, scientists believed that we humans have our own foamy virus, our own endemic version, distinct from the zoonotic foamies we may acquire while feeding rice to a sacred monkey or cutting open a gorilla with our machete. Destructive in cell cultures but apparently harmless in a living person, human foamy virus was called "a virus in search of a disease." Later research with advanced molecular methods—most notably, genetic sequencing—showed that it was probably just a variant of the foamy virus endemic to chimpanzees. Anyway, that one isn't what interests Lisa Jones-Engel and her husband. They're more concerned with the versions that dwell in Asian macaques.

Like the African SFVs, those Asian viruses seem to be innocuous when they get into human hosts. During our talk in Dhaka, Lisa stated the point a little more guardedly. "There's no known disease in nonhuman primates infected with simian foamy virus. Now when the virus jumps the species barrier to humans . . . "—when that happens, well, it's hard to say what may occur, because of limited data. "The number of people that we've had to look at so far is so small that we really can't speak yet to whether it does cause disease in humans." The cases observed have been too few, and the time of observation has been too short. As retroviruses, the SFVs might conceivably have a long, sneaky period of latency and slow replication within the body, before emerging from their secret lairs to wreak havoc.

For Engel and Jones-Engel, this particular line of investigation had its origin at the Sangeh temple, in Bali, where they screened for simian foamy virus as well as for herpes B. And like herpes B, simian foamy seemed to be widespread throughout the population; they found antibodies against it in most macaques tested. A common infection, then, probably passed from monkey to monkey by social contact, again like herpes B. But how often does it spill into humans?

Besides trapping and sampling monkeys, the researchers drew blood from more than eighty people and screened those samples by the same method used for the monkeys. All the humans tested negative except one, a forty-seven-year-old Balinese farmer. This man lived near Sangeh, visited the temple often, and had been bitten once and scratched several times. No, he told them, he had never eaten a monkey. No, he did not keep a pet monkey. If the virus was in him, it must have come from those aggressive animals at the temple. In retrospect, the most notable aspect of what Jones-Engel and Engel found among their eighty-some test subjects in Bali was that *only* the farmer had been infected. Since then, further sampling in other Asian countries (Thailand, Nepal, and Bangladesh) has shown that simian foamy gets into humans more readily than the early results suggested.

But if it causes no known disease, so what?

Beyond the obvious point that it might cause an *unknown* disease, Engel and Jones-Engel have another reason for studying this virus. "It's a marker," Gregory told me. "We caught a marker for transmission," Lisa echoed. What they meant is that the presence of SFV within a human population marks opportunities having occurred for cross-species infection of all kinds. If simian foamy has made the leap from a half-tame macaque to a person—to several people, maybe to thousands of people passing through sites such as Sangeh—then so could other viruses, their presence still undetected, their effects still unknown.

"And why is that important?" I asked.

"Because we're looking for the Next Big One," she said.

62

The Next Big One, as I mentioned at the start of this book, is a subject that disease scientists around the world often address. They think about it, they talk about it, and they're quite accustomed to being asked about it. As they do their work or discuss pandemics of the past, the Next Big One (NBO) is at the back of their minds.

The most recent big one is AIDS, of which the eventual total bigness (the scope of its harm, the breadth of its reach) cannot even be predicted. About 30 million deaths, 34 million living people now infected, with no end in sight. Polio was a big one, at least in America, where it achieved special notoriety by crippling a man who would become president despite it. Polio also, during its worst years, struck hundreds of thousands of children and paralyzed or killed many, captured public attention like headlights freezing a deer, and brought drastic changes to the way large-scale medical research is financed and conducted. The biggest of the big ones during the twentieth century was the 1918–1919 influenza. Before that, on the North American continent, the big one for native peoples was smallpox, arriving from Spain about 1520 with the expedition that helped Cortez conquer Mexico. Back in Europe, two centuries earlier, it was the Black Death, probably attributable to bubonic plague. Whether the plague bacterium or another, more mysterious pathogen caused the Black Death (as several historians have recently argued), there's no question of its bigness. Between the years 1347 and 1352, this epidemic seems to have killed at least 30 percent of the people in Europe.

Moral: If you're a thriving population, living at high density but exposed to new bugs, it's just a matter of time until the NBO arrives.

Note that most of these big ones but not all of them (plague the exception) were viral. Now that modern antibiotics are widely available, vastly reducing the lethal menace of bacteria, we can guess confidently that the Next Big One will be a virus too.

To understand why some outbreaks of viral disease go big, others go *really* big, and still others sputter intermittently or pass away without causing devastation, consider two aspects of a virus in action: transmissibility and virulence. These are crucial parameters, defining and fateful, like speed and mass. Along with a few other factors, they largely determine the gross impact of any outbreak. Neither of the two is an absolute constant; they vary, they're relative. They reflect the connectedness of a virus to its host and its wider world. They measure situations, not just microbes. Transmissibility and virulence: the yin and yang of viral ecology.

You've heard a bit already about transmissibility, including the simple statement that viral survival demands replication and transmission. Replication can occur only within cells of a host, for the reasons I've mentioned. Transmission is travel from one host to another, and transmissibility is the packet of attributes for achieving it. Can the virions concentrate themselves in a host's throat or nasal passages, cause irritation there, and come blasting out on the force of a cough or a sneeze? Once launched into the environment, can they resist desiccation and ultraviolet light for at least a few minutes? Can they invade a new individual by settling onto other mucous membranes—in the nostrils, in the throat, in the eyes—and gaining attachment, cell entry, another round of replication? If so, that virus is highly transmissible. It goes airborne from one host to another.

Fortunately, not every virus can do that. If HIV-1 could, you and I might already be dead. If the rabies virus could, it would be the most horrific pathogen on thc planct. Thc influenzas are well adapted for airborne transmission, which is why a new strain can circle the world within days. The SARS virus travels this route too, or anyway by the respiratory droplets of sneezes and coughs—hanging in the air of a hotel corridor, moving through the cabin of an airplane—and that capacity, combined with its case fatality rate of almost 10 percent, is what made it so scary in 2003 to the people who understood it best. But other viruses employ other means of transmission, each with its own advantages and limitations.

The oral-fecal route sounds disgusting but is really quite common. It works well for some viruses because host creatures (including humans) are often forced, especially when living at high densities, to consume food or water contaminated by excrement from other members of their population. This is one of the reasons why children die of dehydration in rainy refugee camps. The virus goes in the mouth, replicates in the belly or the intestines, causes gastrointestinal distress, may or may not spread to other parts of the body, and comes gushing out the anus. Diarrhea, for such a virus, is part of an effective strategy for dispersal. Viruses transmitted this way tend to be fairly hardy in the environment, because they may need to linger in that polluted sump for a day or two before some desperate person comes to drink from it. There's an entire group of such viruses, known as the enteroviruses, including polio and about seventy others, that attack us in the gut. Most of those enteroviruses are uniquely human infections, not zoonoses. Evidently they don't *need* other animal hosts for maintaining themselves in a crowded human world.

For blood-borne viruses, transmission is more complicated. Generally it depends on a third party, a vector. The virus must replicate abundantly in the blood of the host to produce severe viremia (that is, a flood of virions). The vector (a blood-sucking insect or some other arthropod) must arrive for a meal, bite the host, slurp up virions along with the blood, and carry them away. The vector itself must be a hospitable host, so that the virus replicates further within it, producing many more virions that make their way back to the mouth area and stand ready for release. Then the vector must drool virions (as anticoagulant saliva) into the next host it bites. The yellow fever virus, West Nile, and dengue transmit this way. It has an upside and a down.

The downside is that vector transmission requires adaptations for two very different sorts of environment: the bloodstream of a vertebrate and the belly of an arthropod. What works well in one may not work at all in the other, so the virus must carry genetic preparedness for both. The upside is that a vector-borne virus has a vehicle that can carry it some distance, searching thirstily for new hosts. A sneeze travels downwind, more or less at random, but a mosquito can fly upwind toward a victim. That's what makes vectors such effective modes of transmission.

Blood-borne viruses can also spread to new hosts by way of hypodermic needles and transfusions. But those opportunities are adventitious addenda, recent and accidental, patched onto ancient viral strategies shaped by evolution. Ebola and HIV-1, two viruses of very different character, very different adaptive strategies, both happen to move well via needles. So does hepatitis C virus.

In the case of Ebola, transmission from human to human occurs also by blood-to-blood contact in intimate situations, as when one person takes care of another. For a nursing sister in a Congolese clinic with small cracks on her chapped hands, a few minutes spent wiping bloody diarrhea off the floor could be exposure enough. This is extraordinary transmission, so far as the virus is concerned. Ordinary transmission is however Ebola gets from one individual to another within whatever animal host—identity still unknown—serves as its reservoir. Ordinary transmission allows the virus to perpetuate itself. Extraordinary transmission gives it a burst of high replication, high notoriety, but soon brings it to a dead end. Passing between people via bloody rags or reused needles, in this or that African clinic, is not a strategy that serves Ebola for long-term survival. It's just an occasional anomaly that has little or no significance (so far, anyway) within Ebola's broader evolutionary history. Of course, that could change.

Ordinary transmission, for Ebola, need not be blood-borne. If the virus resides in fruit bats of the Central African forest, as suspected but not yet proven, then it might pass from bat to bat during sex, or suckling of infants, or mutual grooming among adults, or breathing on one another, or biting and scratching, or any other form of close contact. At this point in Ebola research, we can only guess. Drops of urine, falling from one bat into the eyes of another? Saliva on shared fruit? Blood-sucking bat bugs? Saliva on fruit would explain how Ebola gets into chimpanzees and gorillas. Bat bugs (yes, there are such things, related to bedbugs) would allow us to imagine a specialist parasite I'll call *Cimex ebolaensis*. It's all speculation. We might even come to learn that Ebola is a natural infection of African ticks, who carry it among fruit bats, gorillas, and chimps. Merely a thought. But please remember that I've just invented tick-borne Ebola from zero evidence.

Sexual transmission is a good scheme for viruses with low hardiness in the external environment. It's a mode of passage that doesn't require them to go outside. They're virtually never exposed to daylight or dry air. The virions pass from one body to another by way of direct, intimate contact between host cells lining delicate genital and mucosal surfaces. Rubbing (not just pressing) those surfaces together probably helps. Transmission during coitus is a conservative strategy, reducing risk to such viruses, sparing the need for hardening against desiccation or sunlight. But it has a downside too—notably, that opportunities for transmission are fewer. Even the most lubricious humans don't have sex as often as they exhale. So the sexually transmitted viruses tend toward patience. They cause persistent infections and endure long periods of latency, punctuated by recurrent outbreaks (like herpesviruses); or else they replicate slowly (like HIV-1 and hepatitis B) up to a critical point at which things get bad. Such patience within a host gives the virus more time and therefore more sexual encounters by which to get itself passed along.

Vertical transmission, meaning mother-to-offspring, is another slow, cautious mode. It can occur during pregnancy, during birth, or (in the case of mammals) by way of milk while an infant nurses. HIV-1, for instance, can be transmitted from mother to fetus across the placenta, or to a newborn in the birth canal, or through breastfeeding; but each of those outcomes is far from inevitable, and the likelihood of their occurrence can be lowered with medical precautions. Rubella (loosely known as German measles) is caused by a virus capable of vertical as well as airborne transmission, and it can kill a fetus or inflict severe damage, including heart disorders, blindness, and deafness. That's why young girls were counseled, in the era before rubella vaccine, to get themselves infected with the virus—suffer a mild bout and be done with it, permanently immune—before they reached childbearing age. From a strictly evolutionary perspective, though, vertical transmission is not a strategy upon which rubella virus could depend for long-term success. A miscarried fetus or a blinded baby with heart troubles will most likely be a dead-end host, just as terminal for the virus as a Congolese nun with Ebola.

Whatever mode of transmission a virus favors—airborne, oral-fecal, blood-borne, sexual, vertical, or just getting itself passed along in the saliva of a biting mammal, like rabies—the common truth is that this factor doesn't exist independently. It functions as half of that ecological yin-yang.

63

And the other half, virulence, is more complicated. In fact, virulence is such an iridescent, relativistic concept that some experts refuse to use the word. They prefer "pathogenicity," which is nearly a synonym but not quite. Pathogenicity is the capacity of a microbe to cause disease. Virulence is the measurable degree of such disease, especially as gauged against other strains of similar pathogen. To say that a virus is virulent almost sounds tautological—the noun and the adjective come from a single Latin root, after all. But if "virus" hearkens back to "poisonous slime," the point of virulence is to ask, *How* poisonous? The virulence of a given virus within a given host tells you something about the evolutionary history between the two.

Just what does it tell you? That's the tricky part. Most of us have heard an old chestnut on the subject of virulence: The first rule of a successful parasite is Don't kill your host. One medical historian has traced this idea back to Louis Pasteur, noting that the most "efficient" parasite, in Pasteur's view, was one that "lives in harmony with its host," and therefore latent infections should be considered "the ideal form of parasitism." Hans Zinsser voiced the same notion in *Rats, Lice and History*, observing that a long period of association between one species of parasite and one species of host tends to lead, by evolutionary adaptation, to "a more perfect mutual tolerance between invader and invaded." Macfarlane Burnet agreed:

> In general terms, where two organisms have developed a host-parasite relationship, the survival of the parasite species is best served, not by destruction of the host, but by the development of a balanced condition in which sufficient of the substance of the host is consumed to allow the parasite's growth and multiplication, but not sufficient to kill the host.

It does seem logical, at first consideration, and it's still often taken as dogma—at least by people who don't happen to study the evolution of parasites. But even Zinsser and Burnet, to their credit, hedged their endorsements of this idea. They must have recognized that the "rule" was just a generalization with important, revealing exceptions. Some very successful viruses do kill their hosts. Lethalities of 99 percent, and persisting at that level over time, aren't unknown. Case in point: rabies virus. Case in point: HIV-1. What matters more than *whether* a virus kills its host is *when*.

"A disease organism that kills its host quickly creates a crisis for itself," wrote the historian William H. McNeill, in his landmark 1976 book *Plagues and Peoples*, "since a new host must somehow be found often enough, and soon enough, to keep its own chain of generations going." McNeill was right, and the key word in that statement is "quickly." Timing is all. A disease organism that kills its host slowly but inexorably faces no such crisis.

Where's the balance point in that dynamic interplay between transmission and virulence? It differs from case to case. A virus can succeed nicely in the long term, despite killing every individual infected, if it manages to get itself passed onward to new individuals before the death of the old. Rabies does that by traveling to the brain of an infected animal—commonly a dog, a fox, a skunk, or some other mammalian carnivore, with flesh-biting habits and sharp teeth—and triggering aggressive changes of behavior. Those changes induce the mad animal to go on a biting spree. In the meantime, the virus has traveled to the salivary glands as well as the brain, and therefore achieves transmission into the bitten victims, even though the original host eventually dies or is killed with an old rifle by Atticus Finch.

Rabies also occurs sometimes in cattle and horses, but you seldom hear about that, probably because herbivores are less likely to pass the infection along with a furious bite. A poor rabid cow may let out a piteous bellow and bump into a wall, but it can't easily skulk down a village lane, snarling and nipping at bystanders. Reports occasionally filter out of eastern Africa about rabies outbreaks in camels, which are especially worrisome to pastoralists who tend them because of the dromedary's notorious tendency to bite. One recent dispatch from the northeastern Uganda borderlands told of a rabies-infected camel that ran mad and "started jumping up and down, biting other animals, before it died." Another, from Sudan, mentioned that rabid camels get excitable, sometimes attacking inanimate objects or biting their own legs—which can't do the camels much harm, not at that stage, but does reflect the strategy of the virus. Even a human in the last throes of rabies infection could potentially transmit the virus with a bite. No such case has ever been confirmed, according to WHO, but precautions are sometimes taken. There was a farmer in Cambodia, several years ago, who broke with the disease after being bitten by a rabid canine. In his late stages, the man hallucinated, he convulsed, and worse. "He barked like a dog," his wife recalled later. "We put a chain on him and locked him up."

HIV-1, like rabies, seems almost invariably to kill its host. It did, anyway, during the gruesome decades before combined antiretroviral therapy became available, and possibly does (time will tell) even now. Death rates have slowed among some categories of HIV-positive people (mainly those with access to the expensive drug cocktails), though this doesn't mean that the virus itself has mellowed. The HIVs by their nature are very slow-acting creatures, which is why they are lumped within the genus *Lentivirus* (from the Latin *lentus*, meaning "slow") along with such other laggardly agents as visna virus, feline immunodeficiency virus, and equine infectious anemia virus. HIV-1 may circulate within a person's bloodstream for ten years or more, replicating gradually, evading the body's defenses, fluctuating in abundance, doing its damage bit by bit to the cells that mediate immune functions, before full-blown AIDS arrives with its fatal results. During that period, the virus has ample time and opportunity to get from one person to another; in the early stage of infection (when viremia goes high, before falling back down), its chances of onward transmission are especially good. More on this below, when we come to the subject of how the HIVs originally spilled over. The point here is that evolution may coax the human immunodeficiency viruses toward various changes, various adaptations, various new proclivities, but a reduction in lethality will not necessarily be one of them.

The most famous instance of a virus becoming less virulent is the case of myxoma virus among Australian rabbits. This one is literally a textbook example. Myxomatosis isn't a zoonotic disease but it played a small, important role in helping scientists understand how virulence can be adjusted by evolution.

64

The story began in the mid-nineteenth century, when a misguided white landholder named Thomas Austin had the bright idea of introducing wild European rabbits to the Australian landscape. Austin was an "ardent acclimatizer," meaning a willful introducer of nonnative animals and plants, who had also given Australia the gift of sparrows. In 1859, a shipment of twenty-four rabbits from England reached him by boat. He wasn't the first to bring rabbits to Australia, but he was the first to seek out wild rabbits, in preference to docile, hutch-bred representatives of the species (*Oryctolagus cuniculus*), which had long been domesticated. He released them on his property in Victoria, the southernmost state of Australia's mainland. Liberated from the problems of home, still capable of life in the wild, and having a naturally high reproductive rate (they were *rabbits*, after all), Austin's imports and their offspring multiplied crazily. If he had brought them over for the joy of shooting them, or hunting them with dogs, he got more than his wish. Within just six years, twenty thousand rabbits had been killed on his estate, and others had gone hopping away in all directions.

By 1880, they had crossed the Murray River into New South Wales and were still headed north and west, the rabbitty front advancing at about seventy miles per year, a formidable pace, considering that it included occasional pauses to drop and rear offspring. Decades passed, with the situation only getting worse. By

1950 there were about 600 million rabbits in Australia, competing with native wildlife and livestock for food and water, and Australians were desperate for action.

That year, the government approved introduction of a poxvirus from Brazil, myxoma, which was known to infect but not greatly harm Brazilian rabbits. There, in its native land, in its accustomed host, it caused small sores on the skin, which remained small or gradually healed. But the Brazilian rabbit was an animal of the Americas that belonged to a genus (*Sylvilagus*) of the Americas, and experimental work suggested that European rabbits might be affected more drastically by this American bug.

In the European rabbits of Australia, sure enough, myxoma turned out to be pestilential, killing about 99.6 percent of the individuals it infected, at least during the first outbreak. In them too it caused sores—not just small ones but big ulcerous lesions, and not just on the skin but also on organs throughout the body, severe enough to kill an animal within less than two weeks. It was carried from rabbit to rabbit mainly by mosquitoes, of which Australia had a more than adequate supply, thirsty for blood and quite willing to drink from a new kind of mammal. The transfer of virus seems to have been mechanical, not biological—meaning that virions traveled as a smear on mosquito mouthparts, not as replicating contaminants within a mosquito's gastric and salivary organs. It's a clumsier mode of vector transmission, such mechanical transfer, but it's simple and in some cases effective.

After a few experimental releases, myxoma caught hold in the Murray River valley, causing what was called a "spectacular epizootic," which for its speed and its scale "must be almost without parallel in the history of infections." Thanks to mosquitoes and the breezes they rode, the virus spread quickly. Dead rabbits began piling up by the thousands in Victoria, New South Wales, and Queensland. Everybody was happy except bunny sympathizers and people who made their livings from cheap fur. Within a decade, though, two things happened: The virus became inherently less virulent and the surviving rabbits became more resistant to it. Mortality fell and the rabbit population began climbing back. This is the short, simple version of the story, with its facile lesson: Evolution lowers virulence, tending toward that "more perfect mutual tolerance" between pathogen and host.

Well, not quite. The real story, teased out through careful experimental work by an Australian microbiologist named Frank Fenner and his colleagues, is that virulence declined quickly from its original extreme, north of 99 percent, and then stabilized at a lower level that was still pretty damn high. Would you consider a kill rate of "just" 90 percent to be mutual tolerance? Me neither. That's as lethal as Ebola virus, at its most extreme, in a Congolese village. But it's what Fenner found. He and his co-workers studied the changes in virulence by collecting samples of virus from the wild and testing those samples against naïve, healthy rabbits in captivity, comparing one sample against others. They detected a wide diversity of strains and, for purposes of analysis, they grouped those strains into five distinct grades of Australian myxoma, on a descending scale of lethality. Grade I was the original strain, with its case fatality rate of nearly 100 percent; grade II killed upward of 95 percent; grade III, the intermediate among all five, still killed between 70 and 95 percent of infected rabbits. Grade IV was milder, and grade V milder still (though far from harmless), killing less than 50 percent of the rabbits it infected.

What was the relative mix of these five grades among infected rabbits? By sampling from the wild, measuring the presence of each grade, and tracing changes in their proportional dominance over time, Fenner and his co-workers hoped to answer some basic questions, chief of which was: Did the virus trend steadily toward becoming innocuous? Did the evolutionary interaction between rabbit and microbe progress toward Zinsser's "more perfect mutual tolerance," as represented by grade V, the mildest grade? Did myxoma learn not to kill its host?

The answer was no. After a decade, Fenner and his partners discovered, grade III myxoma had come to predominate. It was still causing upward of 70 percent mortality among the rabbits, and it constituted more than half of all the samples collected. The most lethal strain (grade I) had nearly disappeared, and the most benign strain (grade V) was still rare. The situation seemed to have stabilized.

But had it? A ten-year span is an eyeblink in the timescale of evolution, even for creatures that reproduce as quickly as viruses and rabbits. So Frank Fenner kept watching.

After another twenty years, he saw a significant change. By 1980, grade III myxoma accounted for two-thirds, not just half, of all collected samples. Highly lethal but not *always* lethal, it was thriving in the wild, an evolutionary success. And the mild strain, grade V, had vanished. It wasn't competitive. For one reason or another, it seemed to have flunked the Darwinian test; the unfit don't survive.

What explains this unexpected result? Frank Fenner guessed astutely that it was the dynamic between virulence and transmission. His tests of one grade versus another, using captive rabbits and captive mosquitoes, revealed that the efficiency of transmission correlated with the amount of virus available on a rabbit's skin. More lesions, or lesions that lasted longer, meant more available virus. More virus smeared on mosquito mouthparts, more chance of transmission to the next rabbit. But "available virus" assumed that the rabbit was still alive, still pumping warm blood, and therefore still of interest to the vector. Dead, cold rabbits don't attract mosquitoes. Between the two extreme outcomes of infection—healed rabbits and dead rabbits—Fenner found a point of balance.

"Laboratory experiments showed that all field strains produced lesions that provided sufficient virus for transmission to occur," he wrote. But the strains of very high virulence (grades I and II) killed rabbits "so quickly that infectious lesions were only available for a few days." The milder strains (grades IV and V) produced lesions that tended to heal quickly, he added—and then the payoff, "whereas strains of grade III virulence were highly infectious for the lifetime of the rabbits that died and for a much longer period in those that survived." Grade III, at that point, was still killing around 67 percent of the rabbits it touched. Myxoma virus, thirty years after its introduction, had found this level of virulence—being pretty damn lethal—to maximize its transmission. It was still capable of killing most of the rabbits it infected, but capable also of assuring its own survival with a continuous chain of infections.

The first rule of a successful parasite? Myxoma's success in Australia suggests something different from that nugget of conventional wisdom I mentioned above. It's not Don't kill your host. It's Don't burn your bridges until after you've crossed them.

65

Who makes these rules? Unless you're a creationist, you'll likely recognize that the answer is nobody. Where do they come from? Evolution. They are life-history strategies, carved by evolutionary chisels from a broader universe of possibilities. They persist because they work. You can find it in Darwin: descent with modification, natural selection, adaptation. The only surprise, if it is a surprise, is that viruses evolve just as surely as creatures that are unambiguously alive.

Around the time that Frank Fenner published his thirty-year retrospective on myxoma, two other scientists started developing a theoretical model of parasite-host interactions. They meant to codify not just the first rule but various others. They proposed to do it with mathematics. Their names were Anderson and May.

Roy M. Anderson is a parasitologist and ecologist of mathematical bent, employed in those days at Imperial College, in London. He did his dissertation on the flatworms that infect bream. Robert M. May is an Australian, like Frank Fenner, like Macfarlane Burnet—but then again, very different. He took a doctorate in theoretical physics, migrated to Harvard to teach applied mathematics, and somewhere along the way became interested in animal population dynamics. He came under the influence of a brilliant ecologist named Robert MacArthur, then at Princeton, who had applied new levels of mathematical abstraction and manipulation to ecological thinking. MacArthur died young in 1972. May moved to Princeton as his handpicked successor, became a professor of zoology there, and continued the project of applying mathematics to theoretical ecology. His first published paper on parasites was titled "Togetherness among Schistosomes," describing transmission dynamics in another form of flatworm.

Brought together by their common interests (ecology, math, flatworms) and their complementing strengths, Robert May and Roy Anderson teamed up, like Watson and Crick, like Martin and Lewis, and presented the earliest form of their disease model in 1978. Over the following dozen years, they elaborated on that and related subjects in a series of papers that were verbally lucid, bestrewn with math, and widely noticed by other scientists. Then in 1991 they put it all, and more, into a thick tome titled *Infectious Diseases of Humans*. They had built their work on the same sort of conceptual schema used by disease theorists for sixty years, the *SIR* model, representing a flow of individuals, during the course of an outbreak, through those three classes I mentioned earlier: from susceptible (*S*) to infected (*I*) to recovered (*R*). Anderson and May improved the *SIR* model in several ways, making it more complex and more realistic. Their most telling improvement involved a fundamental parameter: population size of the hosts.

Almost all earlier disease theorists, such as Ronald Ross in 1916, Kermack and McKendrick in 1927, and George MacDonald in 1956, had treated population size as a constant. The math was simpler that way, and it seemed a practical shortcut for dealing with real situations. For instance: If the population of a city is two hundred thousand and measles strikes, then as the outbreak progresses the sum of those people still susceptible, plus those now infected, plus those recovered, will always equal two hundred thousand. This assumes that the population is inherently stable, with births balanced by deaths, and that its inherent stability continues despite the epidemic. Epidemiologists and other medical people, even the mathematically adept ones, had generally taken such an approach.

But that was too simple, too static, for Anderson and May. They came from the realm of ecology, where population sizes are always changing in complex, consequential ways. Let's treat population size as a dynamic variable, they proposed. Let's get beyond assuming any artificial, inherent stability and recognize that a disease outbreak itself may affect population size—by killing a large fraction of the populace, say, or by lowering the birth rate, or by increasing societal stresses (such as overcrowding in hospitals) that might raise the rate of death from other causes. Maybe all three of those factors together, plus others. Their aim, wrote Anderson and May, was to "weave together" the two approaches, the medical and the ecological, into a single, savvy method for understanding (and predicting) the course of infectious diseases through populations.

"That got a whole bunch of ecologists interested in the phenomenon," one senior member of the guild told me. This was Les Real, of Emory University, whose work on Ebola among gorillas I mentioned earlier. "Ecologists who were looking for what to do in population ecology suddenly got interested in infectious diseases," he said. As an afterthought, Les qualified his statement: Of course, May and Anderson hadn't *invented* the ecological approach to diseases. That had been around for a long time, at least since Macfarlane Burnet. They had done something else. "Bob and Roy mathematized it. And they mathematized it in an interesting way."

Math can be correct but boring. It can be elaborate, impeccable, and sophisticated yet at the same time stupid and useless. Anderson and May's math wasn't useless. It was nifty and provocative. Don't take my word for it, but you can trust Les Real on this point. Or consult *Science Citation Index*, the authoritative scoreboard of scientific influence, and see how frequently the papers of Anderson and May (or May and Anderson, as they occasionally signed) have been cited by other scientists down through the years.

Some of those papers appeared in august journals such as *Nature, Science*, and *Philosophical Transactions of the Royal Society of London*. My own favorite saw print in a more specialized organ called *Parasitology*. This one, titled "Coevolution of Hosts and Parasites," appeared in 1982. It began by dismissing those "unsupported statements" in medical and ecological textbooks "to the effect that 'successful' parasite species evolve to be harmless to their hosts." Bosh and nonsense, said Anderson and May. In reality the virulence of a parasite "is usually coupled with the transmission rate and with the time taken to recover by those hosts for whom the infection is not lethal." Transmission rate and recovery rate were two variables that Anderson and May included in their model. They noted three others: virulence (defined as deaths caused by the infectious agent), deaths from all other causes, and the ever-changing population size of the host. The best measure of evolutionary success, they figured, was the basic reproductive rate of the infection—that cardinal parameter, R_0.

So they had five crucial variables and they wanted to understand the net effect. They wanted to trace the dynamics. This led them to a simple equation. There will be no math questions in the quiz at the end of this book, but I thought you might like to cast your eyes upon it. Ready? Don't flinch, don't worry, don't blink:

$$R_0 = \beta N/(\alpha + b + v)$$

In English: The evolutionary success of a bug is directly related to its rate of transmission through the host population and inversely but intricately related to its lethality, the rate of recovery from it, and the normal death rate from all other causes. (The clunky imprecision of that sentence is why ecologists prefer math.) So the first rule of a successful parasite is slightly more complicated than Don't kill your host. It's more complicated even than Don't burn your bridges until after you've crossed them. The first rule of a successful parasite is $\beta N/(\alpha + b + v)$.

The other thing that makes Anderson and May's 1982 paper vivid is its discussion of myxoma in Australian rabbits. That brought their modeling to an empirical case and allowed them to test theory against fact. They described Frank Fenner's five grades of virulence. They saluted his methodical combination of field sampling and lab experiments. They mentioned the mosquitoes and the open sores. Then, using Fenner's data and their own equation, they plotted a relationship between virulence and success. Their result was a model-generated prediction: Given *this* rate of transmission, given *that* rate of recovery, given *those* unrelated mortalities, then . . . an *intermediate* grade of virulence should come to predominate.

Son of a gun, it matched what had happened.

The match showed that their model, though still crude and approximate, might help predict and explain the course of other disease outbreaks. "Our major conclusion," wrote Anderson and May, "is that a 'well-balanced' host-parasite association is *not necessarily* one in which the parasite does little harm to its host." Their italics: *not necessarily*. On the contrary, *it depends*. It depends on the specifics of the linkage between transmission and virulence, they explained. It depends on ecology and evolution.

66

Anderson and May were theoreticians who worked much with other people's data. So is Edward C. Holmes. Unlike them, he's a specialist in viral evolution, one of the world's leading experts. He sits in a bare office at the Center for Infectious Disease Dynamics, which is part of Pennsylvania State University, in a town called State College, amid the rolling hills and hardwoods of central Pennsylvania, and discerns patterns of viral change by scrutinizing sequences of genetic code. That is, he looks at long runs of those five letters, A, C, T, G, and U, strung out in unpronounceable streaks as though typed by a manic chimpanzee. Holmes's office is tidy and comfortable, furnished sparsely with a desk, a table, and several chairs. There are few bookshelves, few books, few files or papers. A thinker's room. On the desk is a computer with a large monitor. That's how it all looked when I visited, anyway.

Above the computer hung a poster celebrating "the Virosphere," meaning the unplumbable totality of viral diversity on Earth. Beside that, another poster showed Homer Simpson as a character in Edward Hopper's famous painting "Nighthawks." I'm not sure what that one was celebrating, unless perhaps donuts.

Edward C. Holmes is an Englishman, transplanted to central Pennsylvania from London and Cambridge. His eyes bug out slightly when he discusses a crucial fact or an edgy idea, because good facts and ideas impassion him. His head is round and, where not already bald, shaved austerely. He wears wiry glasses with a thick metal brow, as in old pictures of Yuri Andropov. Though shaved, though brilliant, though Andropovian at first glance, Edward C. Holmes isn't austere. He's lively and humorous, a generous soul who loves conversation about what matters: viruses. Everyone calls him Eddie.

"Most emerging pathogens are RNA viruses," he told me, as we sat beneath the two posters. RNA as opposed to DNA viruses, he meant, or to bacteria, or to any other type of parasite. He didn't need to cite the particulars about RNA viruses because I already had that list in my mind: Hendra and Nipah, Ebola and Marburg, West Nile, Machupo, Junin, the influenzas, the hantas, dengue and yellow fever, rabies and its cousins, chikungunya, SARS-CoV, and Lassa, not to mention HIV-1 and HIV-2. All of them carry their genomes as RNA. The category does seem to encompass much more than its share of dastardly zoonoses, including most of the newest and the worst. Some scientists have begun asking why. To say Eddie Holmes wrote the book on this subject wouldn't be metaphorical. It's titled *The Evolution and Emergence of RNA Viruses*, published by Oxford in 2009, and that's what had brought me to his door. Now he was summarizing some of the highlights.

Granted, Eddie said, there are an *awful* lot of RNA viruses generally, which might seem to raise the odds that many would come after humans. RNA viruses in the oceans, in the soil, in the forests, and in the cities; RNA viruses infecting bacteria, fungi, plants, and animals. It's possible that every cellular form of life on the planet supports at least one RNA virus, he had said in the book, though we don't know for sure because we've just begun looking. A glance at his virosphere poster, which portrayed the universe of known viruses as a brightly colored pizza, was enough to support that point. It showed RNA viruses accounting for at least half the slices. But they're not merely common, Eddie said. They're also highly evolvable. They're protean. They adapt quickly.

Two reasons for that, he explained. It's not just the high mutation rates but also the fact that their population sizes are huge. "Those two things put together mean you'll produce more adaptive change."

RNA viruses replicate speedily, generating their big populations (high titers) of virions within each host. Stated another way, they often produce acute infections, severe for a short time and then gone. Either they soon disappear or they kill you. Eddie called it "this kind of boom-bust thing." Acute infection also means lots of viral shedding—by way of sneezing or coughing or vomiting or bleeding or diarrhea—which facilitates transmission to other victims. Such viruses try to outrace the immune system of each host, taking what they need and moving onward before a body's defenses can defeat them. (Lentiviruses, including the HIVs, are exceptional here, following a different strategy.) Their fast replication and high rates of mutation supply them abundantly with genetic variation. Once an RNA virus lands in another host—maybe even another *species* of host—that abundant variation serves the virus well, giving it many chances to adapt to the new circumstances, whatever those circumstances might be. In some cases it fails to adapt; in some it succeeds well.

Most DNA viruses embody the opposite extremes. Their mutation rates are low and their population sizes can be relatively small. Their strategies of self-perpetuation "tend to go for this persistence route," Eddie said. Persistence and stealth. They lurk, they wait. They hide from the immune system rather than trying to outrun it. They go dormant and linger within certain cells, replicating little or not at all, sometimes for many years. I knew he was talking about things like varicella zoster virus, a classic DNA virus that begins its infection of humans as chickenpox and can recrudesce, decades later, as shingles. The downside for DNA viruses, Eddie said, is that they can't adapt so readily to a new species of host. They're just too stable. Hidebound. Faithful to what has worked in the past.

The stability of DNA viruses derives from the structure of the genetic molecule and how it replicates, using DNA polymerase to assemble and proofread each new strand. The enzyme employed by RNA viruses, on the other hand, is "error prone," according to Eddie. "It's just a really crappy polymerase," which doesn't proofread, doesn't backtrack, doesn't correct erroneous placement of those nucleotide bases, A, C, G, and U. Why not? Because the genomes of RNA viruses are tiny, ranging from about two thousand nucleotides to about thirty thousand, which is much less than what most DNA viruses carry. "It takes more nucleotides," Eddie said—a larger genome, more information—"to make a new enzyme that works." One that works as neatly as DNA polymerase does, he meant.

And why are RNA genomes so small? Because their self-replication is so fraught with inaccuracies that, given more information to replicate, they would accumulate more errors and cease to function at all. It's sort of a chicken-and-egg problem, he said. RNA viruses are limited to small genomes because their mutation rates are so high, and their mutation rates are so high because they're limited to small genomes. In fact, there's a fancy name for that bind: Eigen's paradox. Manfred Eigen is a German chemist, a Nobel winner, who has studied the chemical reactions that yield self-organization of longer molecules, a process that might lead to life. His paradox describes a size limit for such self-replicating molecules, beyond which their mutation rate gives them too many errors and they cease to replicate. They die out. RNA viruses, thus constrained, compensate for their error-prone replication by producing huge populations and achieving transmission early and often. They can't break through Eigen's paradox, it seems, but they can scoot around it, making a virtue of their instability. Their copying errors deliver beaucoup variation, and beaucoup variation allows them to evolve fast.

"DNA viruses can make much bigger genomes," Eddie said. Unlike the RNAs, they're not limited by Eigen's paradox. They can even capture and incorporate genes from the host, which helps them to confuse a host's immune response. They can reside in a body for longer stretches of time, content to get themselves passed along by slower modes of transmission, such as sexual and vertical. Most crucially, they can repair copying errors as they replicate, thus lowering their mutation rates. "RNA viruses can't do that." They face a different set of limits and options. Their mutation rates can't be lowered. Their genomes can't be enlarged. "They're kind of stuck."

What do you do if you're a virus that's stuck, with no long-term security, no time to waste, nothing to lose, and a high capacity for adapting to new circumstances? By now we had worked our way around to the point that interested me most. "They jump species a lot," Eddie said.

VII

CELESTIAL HOSTS

67

From where do these viruses jump? They jump from animals in which they have long abided, found safety, and occasionally gotten stuck. They jump, that is, from their reservoir hosts.

And which animals are those? Some kinds are more deeply implicated than others as reservoirs of the zoonotic viruses that jump into humans. Hantaviruses jump from rodents. Lassa too jumps from rodents. Yellow fever virus jumps from monkeys. Monkeypox, despite its name, seems to jump mainly from squirrels. Herpes B jumps from macaques. The influenzas jump from wild birds into domestic poultry and then into people, sometimes after a transformative stopover in pigs. Measles may originally have jumped into us from domesticated sheep and goats. HIV-1 has jumped our way from chimpanzees. So there's a certain diversity of origins. But a large fraction of all the scary new viruses I've mentioned so far, as well as others I haven't mentioned, come jumping at us from bats.

Hendra: from bats. Marburg: from bats. SARS-CoV: from bats. Rabies, when it jumps into people, comes usually from domestic dogs—because mad dogs get more opportunities than mad wildlife to sink their teeth into humans—but bats are among its chief reservoirs. Duvenhage, a rabies cousin, jumps to humans from bats. Kyasanur Forest virus is vectored by ticks, which carry it to people from several kinds of wildlife, including bats. Ebola, very possibly:

from bats. Menangle: from bats. Tioman: from bats. Melaka: from bats. Australian bat lyssavirus, it may not surprise you to learn, has its reservoir in Australian bats. And though the list already is long, a little bit menacing, and in need of calm explanation, it wouldn't be complete without adding Nipah, one of the more dramatic RNA viruses to emerge within recent decades, which leaps into pigs and via them into humans: from bats.

68

The debut appearance of a new zoonotic disease is often confusing as well as alarming, and Nipah was no exception. In September 1998, people began getting sick in a northern district of peninsular Malaysia, near the city of Ipoh. Their symptoms included fever, headache, drowsiness, and convulsions. The victims were pig farmers or somehow associated with pig processing. One was a pork seller, who died of a brain inflammation. In December, after the northern outbreak seemed to be tapering off, a new cluster of cases appeared southwest of the capital, Kuala Lumpur, in a pig-farming area of the state of Negri Sembilan. By the end of the year, ten workers had fallen ill, gone comatose, and died. The government reacted quickly but with imperfect comprehension. At first it was all about mosquitoes and pigs.

Mosquitoes were implicated as the presumed vectors; pigs, as the presumed reservoir hosts. But vectors and reservoirs of what? Japanese encephalitis virus was the presumed cause.

Japanese encephalitis (JE) is an endemic disease in Malaysia and much of southeastern Asia, tallying upward of thirty thousand human cases (mostly nonfatal) throughout the region each year. The JE virus belongs to the same family as West Nile, dengue, and yellow fever virus. It's vector-borne, traveling by mosquito from its reservoirs in domestic pigs and wild birds. Antibodies found in some of the sickened Malaysian pig workers seemed to confirm its responsibility for the 1998 outbreak, and so Japanese encephalitis became the object of rising public concern and government action. Health officials started pondering how many people—or how many pigs—they should vaccinate against it.

In early January, a story ran in the *New Straits Times*, Malaysia's leading English-language newspaper, under the headline: GIRL IS FOURTH PERSON IN NEGRI TO DIE OF ENCEPHALITIS. The girl in question, thirteen years old and unnamed in the article, had been helping her family with their pig business. Below the piece about her was another, a small item, reporting that Malaysia's Health Minister had ordered a campaign of fogging to kill mosquitoes. Kill mosquitoes, eliminate the vector, stop JE transmission, yes? Yes but no. One day later, in the same newspaper: GIRL DIES OF SUSPECTED JE IN IPOH. That brought the death count, between Negri Sembilan in the south and Ipoh in the north, to thirteen. This child was only a toddler. She expired at her family home, a half mile from the nearest pig farm. "Pigs are a common host for the virus," the story added—meaning the JE virus, of course. Was there any other?

Maybe. While the news media flamed over Japanese encephalitis, and the government took steps to control it, scientists in the Department of Medical Microbiology at the University of Malaya (not "Malaysia," because it has preserved its historical name), in Kuala Lumpur, grew increasingly skeptical. They knew JE about as well as anyone, and some aspects of what was happening now just didn't seem to fit the pattern. Apart from the two young girls so conspicuously mourned in the newspapers, almost all other recent victims had been adult males, men with hands-on involvement in the farming, transport, or butchery of pigs. In fact, most of them were not only male and adult but ethnic Chinese, a group that dominated the Malaysian pig industry. Japanese encephalitis as previously known, on the other hand, was notorious for affecting mainly children. Professor Sai Kit Lam ("Ken" Lam, to his Anglophone friends), then head of Medical Microbiology at the university, stated publicly that this outbreak was killing too many adults to fit the normal profile of JE. The case fatality rate of the current outbreak, too, seemed weirdly high. It was running at more than 54 percent. Maybe this was a new strain of the JE virus, more virulent than usual, more aggressive against adults, less widely spread to the general populace by its insect vector.

Or a different virus altogether, with a different mode of transmission. Mosquito vectoring didn't seem to fit. What sort of mosquito chooses to bite only adult male Chinese pig farmers?

Meanwhile the pigs of Malaysia were sick too, suffering their own epizootic outbreak of something or other. Again, the familiar form of Japanese encephalitis didn't explain it, since pigs usually tolerate that infection without showing clinical signs like this. They can be amplifier hosts as well as reservoirs of JE, in that their prevalence of infection may help increase the prevalence of the virus in mosquitoes, which then may bite humans. Pregnant sows infected with JE may also abort or deliver stillborn young; but it doesn't cause conditions such as were now being seen in Malaysia. And there were other problems with the JE hypothesis. Whereas the new human disease among pig-industry workers was neurological, causing encephalitis and other problems of the nervous system, the pig ailment was both neurological and respiratory. It seemed very contagious from pig to pig, evidently moving by airborne transmission. One after another, first in the big sties of the Ipoh region and then down into Negri Sembilan, animals started coughing, shuddering, barking, wheezing piteously, collapsing off their feet, and in some cases dying.

The lethality among pigs, though, was much lower than the rate among human cases. Their symptoms at first suggested something called classical swine fever, a viral infection also known as hog cholera. But that guess was soon dismissed. Hog cholera, which isn't zoonotic, couldn't account for the human illnesses. Then maybe Japanese encephalitis of a nasty new sort? The outbreak spread from one pig farm to another in almost a rolling chorus of porcine hacking—people could hear it coming and wait with dread. "It became known as a one-mile barking cough," according to a visiting expert from Australia, "because you could hear it a mile away. People would know that the disease had arrived in their area." It traveled on the sneeze of a pig. It traveled too by truck, when animals were moved from one farm to another. And it traveled across borders, as in early 1999, when Malaysian pigs were exported to Singapore and the disease struck abattoir workers there. Eleven Singaporeans got sick. In the excellent medical facilities of the city-state, only one died.

Still no one knew what this thing was. Most of the early laboratory diagnostics, in Malaysia, had been done either by the Ministry of Health or, for the pig samples, by the national veterinary research institute up in Ipoh. Scientists at the University of Malaya, especially in Ken Lam's Department of Medical Microbiology, followed the crisis closely but quietly. Paul Chua was the department's chief clinical virologist. His work involved wet-lab methods, such as viral culturing and microscopy. Sazaly AbuBakar was the molecular virologist, meaning that he looked at viral genomes as Eddie Holmes does: in blurps of dry code, ACCAAACAAGGG, letter by letter. For a while, neither Chua nor AbuBakar could do much more than read the newspaper accounts, talk with colleagues, and speculate, because they didn't have samples of blood, tissue, or cerebrospinal fluid, the raw evidence for lab diagnostics.

And then suddenly they did. As the outbreak continued in Negri Sembilan, not far from the capital, patients began arriving at the University of Malaya Medical Center. These patients were treated, some died, and Paul Chua received samples taken from three of the bodies. One of those victims had been a fifty-one-year-old pig farmer from a village called Sungai Nipah. This man had come to the hospital feverish, confused, with a twitchy left arm. Six days later he was dead.

Chua and his trusted lab technician isolated virus from the Sungai Nipah sample, growing it in a line of tame laboratory cells derived originally from the kidney of an African monkey. Immediately the virus in culture started causing damage. The damage didn't look like JE. Individual cells were enlarged, merging into big membranous bubbles peppered with multiple nuclei. Chua called in his colleague AbuBakar to look.

"Really unusual," AbuBakar said, recalling the sight of those cells, when I stopped by his office in Kuala Lumpur. I had met him at a Nipah conference and he'd welcomed further chat. Paul Chua by then had left for a job in the Ministry of Health, but AbuBakar (his young students called him Prof. Sazaly) was now chair of Medical Microbiology himself. "We all concluded it is something unusual that we see in the cell culture."

The logical next step, Prof. Sazaly told me, was to get a look at this virus under a good electron microscope. Although cell cultures reveal the collective action of the virus, visible to the naked eye as reflected in ravaged cells, it takes electron microscopy to show individual virions. "But unfortunately, at that time, we don't have good electron microscopes anywhere in the country." The one at the university was old and bleary. Malaysia is an Asian tiger, with many keen and well-educated scientists, but still short on certain technological resources.

So the department head, Ken Lam, called on old contacts in the United States, making arrangements for Paul Chua to visit. Chua tucked some frozen samples into a bag and got on a plane for America. Many hours later, he was in Fort Collins, Colorado. At the CDC's satellite center there, which houses its Division of Vector-Borne Diseases, he and CDC scientists examined the Sungai Nipah samples under a topnotch electron microscope. What they saw wasn't Japanese encephalitis virus. It looked more like a scrum of paramyxovirus, containing long filaments with a sort of herringbone structure. Malaysian measles? Murderous porcine mumps? Based on that tentative identification, Chua was redirected to CDC headquarters in Atlanta, where his new contacts were paramyxovirus researchers. They doused his samples with various assays, testing for antibody reaction, and got a provisional positive from the assay for Hendra antibodies. Sequencing part of the viral genome, though, they found that this was an entirely new bug: not Hendra, something similar but distinct. Paul Chua and his colleagues named it Nipah virus, after that little village of the fifty-one-year-old farmer. The disease eventually became known as Nipah virus encephalitis.

69

There's a convergence of stories here. Once the Malaysian microbiologists knew that their outbreak was caused by a virus closely resembling Hendra, Ken Lam phoned another colleague, this time in Australia. "Look, we've got something," he said. That was an understatement. The worrisome part was that he didn't know where this "something" had come from or where it might go. He wanted expert help. No one was an expert on Nipah virus, not yet, but an expert on Hendra might be the next best thing. Through an intermediary, Lam's request reached Hume Field, the lanky former veterinarian who had discovered Hendra in fruit bats. Field saddled up quickly. He got the call on a Thursday, to the best of his recollection, and by Monday he was on a plane to Kuala Lumpur.

Field joined an international team, led by a senior man from the CDC, which had convened from Atlanta and elsewhere to help the Malaysian professionals deal with the crisis. Their first task was to halt the immediate risk to people. "At that time, the human case rate was escalating," Field told me later, during one of our talks in Brisbane. "Something like fifty new cases a week. So there was huge pressure—social, political—to stop the source of infection." To do that, he added, the team had to understand the virus and learn how it behaved in pigs.

They began at what he called "hot farms," where the infection was still burning its way through resident pigs. You could tell a hot farm by ear; it was Field whom I quoted above, describing the "one-mile barking cough." He and the rest of the team wanted sick pigs from which to collect samples, hoping those samples might yield a virus matching the one Paul Chua had isolated from his pig farmer. "And that's what happened," Field said. They dispatched samples to the Australian Animal Health Laboratory, in Geelong, where colleagues isolated a virus that matched Paul Chua's. Final proof of that match came from AbuBakar's team in Kuala Lumpur. All this confirmed pigs as an amplifier host of the same Nipah virus that was killing humans. But it said nothing about where Nipah might ultimately reside.

The Malaysian government in the meantime had ordered a mass culling—that is, the extermination of every pig, infected or uninfected, on every farm that the outbreak had touched. Some of those piggeries had been abandoned by their operators, panicky and bewildered, even before the discovery of the new virus. People in certain areas even fled their homes; Sungai Nipah became a ghost town. By the end of the outbreak, at least 283 humans had been infected and 109 had died, for a case fatality rate of almost 40 percent. Nobody wanted to eat pork, or to handle it, or to buy it. Pigs were left starving in their pens. Some broke out to roam the roadways like feral dogs, foraging for food. Malaysia at that time contained 2.35 million pigs, half of them from Nipah-affected farms, so this could have become an almost medieval problem, like a scene from the Black Death: herds of infected pigs stampeding ravenously through empty villages. A phalanx of cullers, including soldiers from the army as well as police and veterinary officers, moved into the countryside wearing protective suits, gloves, masks, and goggles. Their assigned task was to shoot, bury, or otherwise dispose of more than a million animals, and to do it quickly, without splashing virus everywhere. Despite all precautions, at least half a dozen soldiers did get infected. Hume Field noted: "There's no easy way to kill a million pigs."

Later in conversation he corrected himself: It was in fact 1.1 million pigs. The difference might seem like just a rounding error, he told me, but if you ever had to kill an "extra" hundred thousand pigs and dispose of their bodies in bulldozed pits, you'd remember the difference as significant.

Field and the international team, racing ahead of the cullers, also visited farms that had been, but no longer were, hot—where the infection had come and gone. What they found at those sites, by drawing blood from surviving pigs and testing for antibodies, was that the virus seemed to be extraordinarily contagious, at least among swine, even when it wasn't extraordinarily virulent. The prevalence of antibodies in the animals on recovered farms was

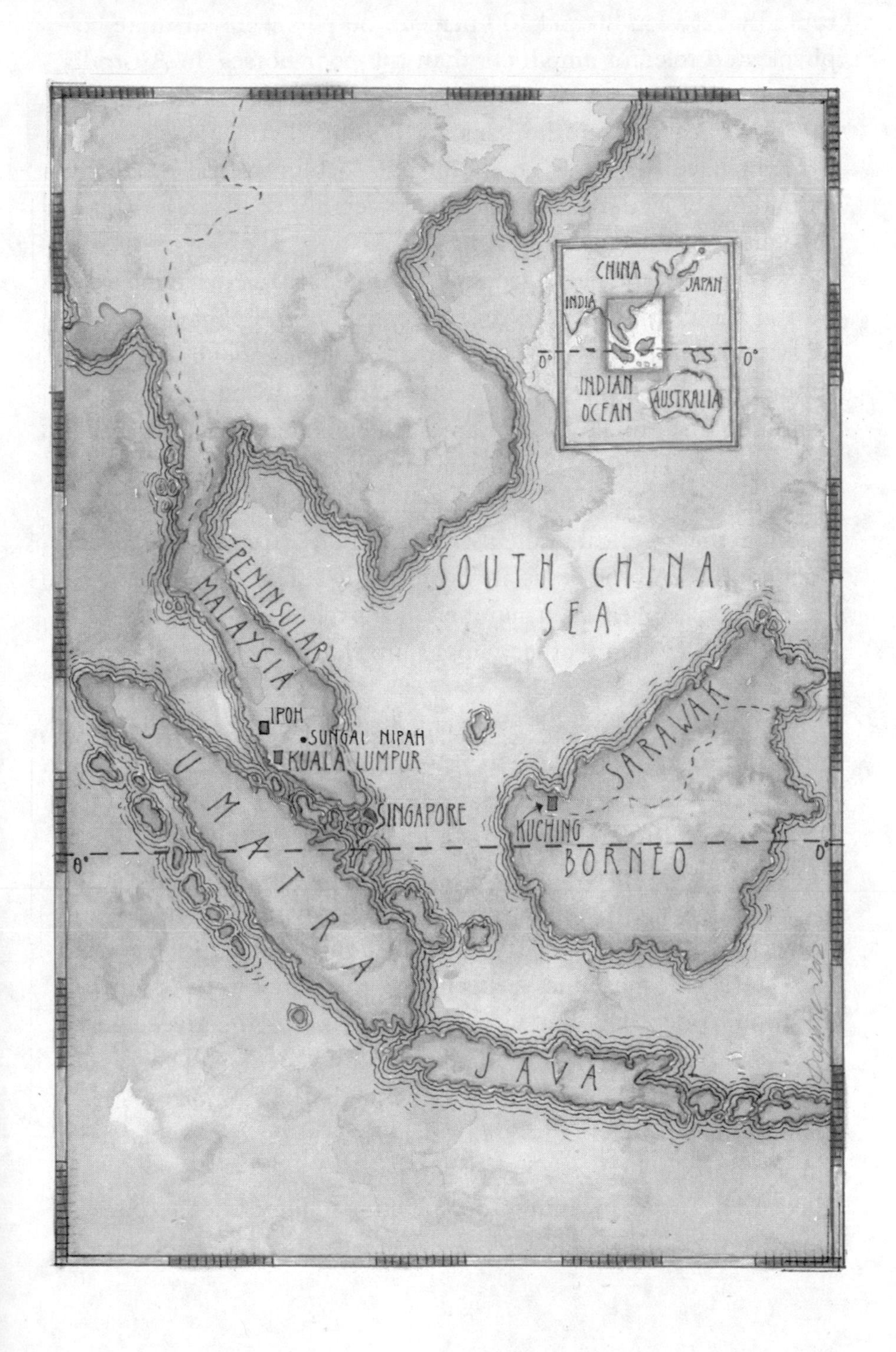
CHINA
JAPAN
INDIA
0°
0°
INDIAN
OCEAN
AUSTRALIA
SOUTH CHINA
SEA
PENINSULAR
MALAYSIA
IPOH
SUNGAI NIPAH
KUALA LUMPUR
SINGAPORE
SUMATRA
SARAWAK
KUCHING
BORNEO
0°
0°
JAVA

typically between 80 and 100 percent. So pigs were far more hospitable and tolerant amplifiers than the poor horses, in Australia, who came down with Hendra. If Nipah virus hadn't been zoonotic, capable of leaping into humans and killing them, Field told me, it might have passed as no more than "a blip on the productivity output" of Malaysian pig farming overall. "That's an intriguing thought," he added.

I wasn't sure, and neglected at that moment to ask, what intrigued him so about this alternate-universe version of Nipah. One possibility is that Field had in mind other potential zoonoses that are simmering, unrecognized, presently harmless to humans, among domesticated animals. How many such bugs may be working their way through large-scale livestock operations around the globe? How many RNA viruses may be achieving high rates of evolution (because they replicate quickly, they mutate often, their populations are big, and the herds are big too) in our factory farms? What are the odds, given such numbers, of a mutation that facilitates spillover? How many other Nipahs are slouching toward Bethlehem to be born?

Maybe the Next Big One will emerge from a Malaysian piggery, travel to Singapore in exported sows, and then from Singapore to the world (riding airplanes, as SARS did) in the lungs of a tourist or a flight attendant who ate a lunch of mu shu pork at one of those trendy, overpriced cafes along the waterfront near the Raffles Hotel. Forget about palm civets, for a moment, and consider mass-production animal husbandry. It's almost impossible to screen your pigs, cows, chickens, ducks, sheep, and goats for a virus of any sort until you've identified that virus (or at least a close relative), and we have only begun trying. The larger meaning of Nipah, in accord with Hume Field's "intriguing thought," is that tomorrow's pandemic zoonosis may be no more than "a blip on the productivity output" of some livestock industry today.

Nipah has other meanings too, not quite so large but also intriguing. One takes us back to the subject of bats.

70

After three weeks in Malaysia, Hume Field split away from the pig investigation and, along with a Malaysian veterinarian named Mohd Yob Johara and a few other colleagues, began searching for the origin of the virus. That was why he had been asked to join the international response team, after all—because of his experience in tracing a closely related virus, Hendra, to its reservoirs.

Drawing on the Hendra parallel, Field's little group now focused mainly on bats, of which Malaysia contains a high diversity, including thirteen species of fruit-eating bats and about sixty species of small insectivorous bats. Two of the native fruit bats are flying foxes, big animals with broad wingspans, belonging to the same genus, *Pteropus*, as the Hendra reservoirs in Australia. The small bats were caught using mist nets erected near their feeding and roosting sites. For flying foxes, the team used a more opportunistic method. Bat hunting is legal in most parts of Malaysia, so Field and Johara accompanied sport hunters into the woods and, with the hunters' indulgence, took samples from bagged animals. Some hunters were shooting wild boar, so the researchers snipped bits from boar carcasses also, to test whether the virus had gotten from domestic pigs into wild ones. Another group from the international team, around the same time, sampled domestic dogs, rats, house shrews, chickens, ducks, and pigeons. Both groups wanted answers to the same urgent question: Where was this virus lurking in the bigger world beyond the piggeries?

The wild boar, the rats, the shrews, and the birds all tested negative—no signs of Nipah nor of antibodies against it. Some of the dogs tested positive for antibodies, probably because they had been living closely with sick pigs or eating dead ones. The dogs didn't seem to be spreading the virus much, neither from one canine to another nor to humans (though some evidence suggests that dog-to-human transmission *did* happen occasionally). Most of the bats tested negative, except for a few species, two of which stood apart from all others, showing significant prevalence of Nipah antibodies within their populations. Those two were the variable flying fox (*Pteropus hypomelanus*) and the large flying fox (*Pteropus vampyrus*). This wasn't surprising, given the other similarities between Nipah and Hendra. But it didn't constitute final proof of the bats as reservoirs. Antibodies merely suggested exposure, which could mean one thing or another, and the samples taken by Field and Johara didn't yield any live virus.

That task remained for Paul Chua, back in Malaysia following his mission to Fort Collins and Atlanta. Later in 1999, after the furor, after the 1.1 million pigs had been killed and the outbreak among humans stifled, Chua and his own team visited one of the flying fox colonies and tried a new technique. Instead of shooting bats and dissecting out tissues, they spread big plastic sheets beneath the roosting sites and collected a few precious drops of bat urine. Beneath the feeding sites, too, they collected samples—in the form of masticated fruit. Some of the fruit was mango; there was also a local delectable known as *jambu air* (in English, water apple). The water apple is an unprepossessing little thing, bell-shaped, usually pinkish or red, sweet and succulent enough to quench the thirst of children. Culturing those samples sedulously, Chua's group grew three isolates of Nipah virus, two from urine and one from a gobbet of water apple. The virus closely matched strains found in Nipah-sickened humans. This proved that flying foxes are reservoirs of Nipah virus, capable of spilling it into pigs that spill it into people.

But more. Chua's work established a plausible scenario for spillover. How did the virus go from bats to pigs? All it required was a mango or water apple tree, laden with ripe fruit, overhanging a pigsty. An infected bat feeds on a water apple, discarding the pulp (as bats do), which is besmeared with virus; the pulp drops down among the pigs; one pig snarfs it up and gets a good dose of virus; the virus replicates in that pig and passes to others; soon the whole herd is infected and human handlers begin to fall sick. It wasn't a far-fetched scenario. Amid the diversified agriculture of Malaysia at the time, wherein marketable fruit could supplement revenue from livestock, there were more than a few pigsties with mango, water apple, and other fruit trees growing nearby. Nipah virus may have been falling in sweet little packets. What pig could resist?

71

Malaysia acted firmly, tightening its agricultural regulations, closing some farms, getting the pigsties out from under the fruit trees, and producing a blitz of cautionary public education. Watch out for Nipah! Watch out for asthmatic swine! Still, eliminating all threat of this virus wasn't so simple. Two years later, it re-emerged in Bangladesh, Malaysia's regional neighbor, a Muslim country containing very few pigs.

Bangladesh is at special risk from infectious disease outbreaks for several reasons, most obvious of which is the density of its population. Within its fifty-seven thousand square miles of territory it contains almost 150 million people, making it the most densely populated country in the world (apart from tiny city-states such as Singapore and Malta). Its generally low elevation (barely thirty feet above sea level in most areas) and its regular cycles of flooding (because of monsoonal rains and high rivers) exacerbate the problem of waterborne diseases such as cholera and bacterial diarrhea, which kill tens of thousands of Bangladeshis (especially children) each year. Although the numbers for Nipah are much smaller and the mechanism very different, the emergence of this virus in Bangladesh and the fact that (as you'll see) it can sometimes be transmitted human-to-human have caused researchers and health officials to take the situation very seriously. Any infectious disease that achieves highly efficient airborne transmission might rampage through greater Dhaka (with its 17 million people), the other major cities, and the continuous crowded sprawl of villages to devastating result. And such a vast epidemic in Bangladesh, besides killing Bangladeshis, would also give the virus in question abundant opportunity to adapt still better to human hosts.

The first Nipah outbreak in Bangladesh, during April and May 2001, occurred in a place called Chandpur, a village of six hundred souls in the southern lowlands. Thirteen people got sick, nine of them died, blood samples confirmed the presence of Nipah, and then the problem seemed to go away. People die all too frequently in Bangladesh, from one cause or another, and this cluster didn't provoke any panic or rigorous investigation. From where had the virus come? Unknown. If bats again were the reservoir, what had caused the spillover? Unknown. Was there an amplifier host? Unknown. Pigs, anyway, weren't implicated.

When considered in retrospect by a team of epidemiologists, several years later, the Chandpur cases seemed to share only two risk factors worth mentioning. Some of the victims had lived with or cared for other victims, suggesting the possibility of person-to-person transmission, which was new. And more than a few of them had had contact with a sick cow. A cow? The epidemiologists' published report, conscientious, exact, groping for leads, mentioned that animal several times. If the virus thrives in Malaysian pigs, couldn't it flourish in a Bangladeshi cow? Maybe. The cow's role remains undetermined.

In January 2003 another outbreak began, up in Naogaon District, about a hundred miles north of Chandpur. Again febrile illnesses, befuddlement, encephalitis, hospitalizations, a high death rate; and no good explanation of how the virus arrived. One suggestive fact was that a herd of pigs had passed through the area, presumably attended by nomadic drovers, and some of the Nipah encephalitis patients had been exposed to them. Aha. Reports didn't suggest that the pigs were sneezing and wheezing and stumbling and dying, as in Malaysia, but they may have been infected and infectious nonetheless. Disease scientists in Bangladesh were still puzzling over outbreaks one and two when the third began, in January 2004. It struck a couple of villages within Rajbari District, just west of the Padma River (an outlet finger of the Ganges), across from Dhaka. Again the case numbers were small, only a dozen; but ten of the dozen died. One other pattern in the data seemed curious: Most of these victims were children—boys, below the age of fifteen.

Another squad of epidemiologists arrived, including an American named Joel M. Montgomery, on a postgraduate training fellowship with the CDC. They came with their clipboards and questionnaires and phlebotomy tools, as epidemiologists do, hoping to make sense of what had happened. They did a case-control study, meaning that they tried to identify the source of the outbreak, and its spread, by identifying behavioral differences between those who had gotten sick and those who hadn't. What were the risky activities that made one a candidate for infection?

Of course, young boys in Bangladesh, like young boys anywhere, engage in lots of risky activities, many of which could result in cracked skulls, broken arms, drowning, snakebite, getting arrested, or being hit by a train. But which kinds of risky behavior could give you Nipah? Montgomery and his colleagues ticked through some possibilities: fishing, hunting, touching dead animals, playing cricket, playing soccer, playing hide-and-seek, picking fruit off the ground and eating it. Among that list, as data accumulated, "touching dead animals" looked like it might be important; several of the sickened children, a week earlier, had helped bury some dead chickens and ducks. Evidently the kids had been acting out funeral rites with deceased poultry. Then again, more than a few *un*infected village children had also touched the dead animals. The ducks and chickens turned out to be a false lead. See how tricky it is to do epidemiology in a Bangladesh village? None of those innocent childhood pastimes I've mentioned, from duck burial to cricket, was significantly more associated with the infected boys (whether recovered or dead) than with their healthy peers. But one was: climbing trees.

Climbing trees? That was puzzling. Although the Montgomery group documented a strong correlation, their results didn't explain *why* tree climbing might expose young Bangladeshis to Nipah infection. They could only make a calculated guess: It put the boys closer to bats.

Three months afterward, in April 2004, health officials in Bangladesh learned of still another outbreak. Faridpur District, just adjacent to Rajbari along the Padma River's right bank, was the latest site. Faridpur and Rajbari, reachable only by slow ferry, are where the urban clamor of greater Dhaka, groping upward in concrete and rebar, gives way to the silty, deltaic lowlands of southern Bangladesh. Rice paddies line the road. Palms and banana trees grow like weeds in a vacant lot. Among thirty-six patients in Faridpur, twenty-seven died. And the pattern of social connectedness among cases suggested another concern, which had also arisen regarding the Chandpur outbreak: that some people had caught the infection from other people. A team of investigators noted that such person-to-person transmission "increases the risk for wider spread of this highly lethal pathogen. In an impoverished, densely populated country such as Bangladesh, a lethal virus could rapidly spread before effective interventions are implemented." Judicious language, by which they meant: It could go like wildfire in dry grass.

Then came still another Bangladesh outbreak, the fifth within four years, this time in Tangail District, about sixty miles northwest of Dhaka. Twelve cases, eleven deaths, all during January 2005. Now it began to seem that Bangladesh was uniquely, persistently tormented by this killer disease, recurring in the early months of each year. Malaysia had had no further outbreaks. India, just north of the northwestern Bangladesh border, had had one. Elsewhere in the world, Nipah was unknown. Again a team went out from Dhaka and did a case-control study, looking for the cause of the spillover. The team leader was Stephen Luby, an American physician and epidemiologist from the CDC, seconded to Dhaka as a program director within the International Center for Diarrheal Disease Research, Bangladesh (fastidiously initialed as the ICDDR,B but commonly known as the Cholera Hospital), where he worked closely with his Bangladeshi counterpart from the Ministry of Health, Mahmudur Rahman.

Luby's group, like Montgomery's earlier, questioned people about potentially risky activities—things done by patients who sickened and died, or sickened and recovered, that might not have been done by neighbors who remained healthy. For the fatalities, they got their answers from surviving relatives or friends. Had the person climbed a tree? Some had, most hadn't, both among the patients and the healthy controls. Had the person touched a pig? No, nobody in Tangail was in the habit of touching pigs. Touched a fruit bat? No, nobody. Touched a duck? Yes, but so what, lots of people did that. Touched a sick chicken? Eaten a guava? Eaten a banana? Eaten an animal that was ill at the time of slaughter? Eaten a star fruit? Touched someone who was feverish, confused, and who later died?

The questions themselves are like pen strokes on a sketch of Bangladeshi village life. But none of those questions—not even the one about tree climbing, this time—yielded any statistically significant distinction between those who had gotten sick and those who hadn't. Only one question asked by the Luby team did: Have you drunk any raw date-palm sap recently?

Gulp, um, yeah. Date-palm sap is a seasonal delicacy in the villages of western Bangladesh. It flows in the veins of a certain palm tree, the sugar date palm (*Phoenix sylvestris*), and if the tree is tapped, sap will drain into a carefully placed clay pot. Like the sap of a maple tree, it's sugary—even more sugary than maple, evidently, because it needn't be rendered down with hours of cooking. Some people are ready to pay good takas, scarce cash, for date-palm sap offered fresh and raw. Tappers sell it door-to-door in the nearby villages, or else on the roadside, like a neighbor kid with a lemonade stand. Customers usually bring a glass or a jar of their own. They drink it down on the spot or carry it home to share with the family. The best quality sap is red, sweet, and clear. Natural fermentation sets in quickly, and the price plummets after 10 a.m., when the sap is no longer so fresh. Impurities also lower the value. Impurities, as you'll see, have another result too.

The investigation at Tangail found that single distinction between the sick and the well: Among those infected, most had drunk raw date-palm sap. Their healthy neighbors mostly hadn't. It suggested a more intricate story.

72

So I went to see Steve Luby, at the ICDDR,B. He's a tall, gaunt man with short brown hair and glasses, serious but not pompous, a former philosophy major who turned to medicine and epidemiology, and then chose to focus on infectious diseases in low-income countries. He has been in Bangladesh since 2004. He knows the place pretty well. He hears a steady tolling of preventable deaths and tries hard to prevent as many as possible. Much of his work involves familiar and mundane diseases, such as pneumonia, tuberculosis, and diarrhea, which cause far greater mortality than Nipah. Bacterial pneumonia, for instance, accounts for about ninety thousand deaths annually just among Bangladeshi children under age five. Bacterial diarrhea kills about twenty thousand newborn infants every year. Given those numbers, I asked Luby, why divert any attention at all to Nipah?

To be prudent, he said. Classic case of the devils you know versus the devil you don't know, none of which can you afford to ignore. Nipah is important because of what *might* happen and because we understand little about *how* it might happen. "This is a horrible pathogen," he said, reminding me that the lethality among Nipah cases in Bangladesh is more than 70 percent. "Of those who survive, a third of them have marked neurological deficits. This is a bad disease." And about half of all known cases in Bangladesh, he added, have acquired it by person-to-person transmission, a worrisome development that hadn't appeared during the Malaysian outbreak of Nipah.

Why has person-to-person spread been a major factor in some of the outbreaks but not others? How stable is the virus? What's the chance that it might evolve into a form that's even more readily transmissible? Bangladesh, as I've mentioned, is very densely populous, with about a thousand humans per square kilometer, and still increasing. That population, dispersed rather evenly across a crowded but rural landscape, with low levels of income and medical care, pressing relentlessly against the last remnants of native landscape and wildlife, puts the country at special risk of epidemics, whether from old mundane pathogens or strange new ones. So of course Nipah is an important part of our work, Luby said, even though the numbers (so far) are small.

And there's another reason, he added. No one in the world knows much about this virus. "If we do not study it in Bangladesh, it will not get studied." Malaysia has seen only one outbreak. India, one in 2001, and another recently. Bangladesh, he pointed out, citing the count as of 2009, has had eight outbreaks in eight years (and more since my conversation with him). Lab work can be done anywhere, but lab work won't solve the mysteries of how Nipah behaves in nature. "If we really want to understand how it moves from its wildlife reservoir into people, what happens in terms of human disease transmission, this is the place we're going to do it," he said.

To understand how it moves from its wildlife reservoir into people requires that one basic point of reference: the identity of the reservoir. Bats were logical suspects, of course—flying foxes in particular—based on what had been learned in Malaysia, and on the parallel findings for Hendra in Australia. The only flying fox native to Bangladesh is a big thing called the Indian flying fox (*Pteropus giganteus*). Luby and his team knew from earlier work that members of this species too had tested positive for Nipah antibodies. But how did the virus get from bats into people, if not by way of pigs? Well, it happens that Indian flying foxes enjoy date-palm sap. Tree owners complained of hearing bats in their palms at night. As the Luby team reported, after their work in Tangail: "Owners viewed the fruit bats as a nuisance because they frequently drink the palm sap directly from the tap or the clay pot. Bat excrement is commonly found on the outside of the clay pot or floating in the sap. Occasionally dead bats are found floating in the pots." But that's not enough to eliminate the demand for raw sap.

On a long list of possible risk factors that Luby's team took to Tangail, sap drinking was just another hypothesis, added to the interview protocols almost as a hunch. The first investigators on the scene were social anthropologists, Luby told me; they were very simpatico with the local people, very low-key, asking open-ended questions, not so formal and quantitative as epidemiologists. "And the anthropologists said, 'Everybody with a case drank date-palm sap.'" He meant everybody with a case of Nipah, not a case of bottled sap. The epidemiologists came next, confirming that hypothesis with hard data. "The Tangail outbreak was the epiphany moment for us," he said. The epiphany seems obvious in retrospect, as epiphanies often do: Yes, drinking raw date-palm sap is an *excellent* way to infect yourself with Nipah.

He explained the context. That western area of Bangladesh, in which most of the outbreaks occurred, could be considered the Nipah Belt. Possibly that's because it's the Date-Palm Belt. The bats range widely, but the west is where sugar date palms grow well and are much prized for their sap. The harvest begins in mid-December, with the first cold night of what passes for winter in Bangladesh. The tappers are known as *gachis*, tree people, from the Bangla word *gach*, meaning "tree." Other people own the palms, and the owners typically get a half share of the product. The gachis are poor, independent operators, generally agricultural laborers who do this as a seasonal sideline. To harvest sap, a gachi climbs a tree, shaves away a large patch of bark near the top to create a V-shaped bare patch (from which sap oozes out), places a hollow bamboo tap at the base of the V, and hangs his small clay pot beneath the tap. The sap flows overnight; the pot fills. Just before dawn, the gachi climbs up again and brings down a pot of fresh sap. Maybe he gets two liters per tree. Bounty! Those two liters are worth about twenty takas (US $0.24) if he can sell them before 10 a.m. He empties the clay pot into a larger aluminum vessel, mixing the sap and the bat feces (if any) and the bat urine (if any) and the virus (if any) from one tree with the sap (and its impurities) from others. Then off he goes to sell his product. Some gachis are complacent about the risk of adulteration. One told a colleague of Luby's: "I do not see any problem, if birds drink sap from my trees. Because birds drink a slight amount of sap. I would get God's grace by giving bats and other animals a chance to drink sap." He gets God's grace and the customer gets Nipah. Other gachis do care, because clear reddish sap brings a better price than foamy, gunky sap full of drowned bees, bird feathers, and bat shit.

The whole investigation, for Steve Luby, leads in two very different directions—one practical and immediate, the other farsighted and scientific. On the practical side, he and his people have been exploring low-cost methods for helping gachis keep bats away from their clay pots. A simple screen made of woven bamboo scraps, costing about ten cents, can be placed around a tapping wound and its clay pot, fencing the bats out. That's a simple fix, and probably more humane than passing a law against harvesting date-palm sap. On the scientific side, Luby told me, there are crucial unanswered questions about Nipah virus. How does it maintain itself in the bat population? Why does it spill over when it does? Is it readily capable of human-to-human transmission, or just under special circumstances? Has it emerged recently, a new pathogen, or is it something that's been killing Bangadeshis, unnoticed, for millennia?

Those questions lead to another. How have changes to Bangladesh's landscape, and the density of people upon it, affected the fruit bats, the virus they carry, and the likelihood of spillover? In other words: What's new in Nipah ecology? For more a more eloquent answer to that, Luby said, you could talk with Jon Epstein.

73

Eloquence is good but field time is better. I left Dhaka with Jon Epstein the next morning, headed west toward the river crossing that would take us into the southwestern Bangladesh lowlands.

Epstein is a veterinary disease ecologist, based in New York. He was employed at the time by an organization called Wildlife Trust, under its Consortium for Conservation Medicine (the same organization as Aleksei Chmura, and more recently rebranded as Eco-Health Alliance). In addition to his DVM, Epstein has a master's in public health and a lot of experience handling big Asian bats. He worked with Paul Chua in Malaysia, capturing flying foxes amid the coastal mangroves, sometimes while chest-deep in seawater. He led the team that found evidence of Nipah among flying foxes in India, after the first outbreak there, and was part of a multinational group that identified bats as the reservoir of the SARS virus in China. He's a large sturdy fellow with a crew cut and lozenge glasses, looking like a former high school quarterback grown fortyish and serious. He was in Bangladesh, not for the first time, to gather data toward understanding when, where, and how the Indian flying fox carries and sheds Nipah.

He brought along Jim Desmond, another American veterinarian, newly recruited to the organization, whom Epstein would train in the particular delicacies of searching for Nipah virus in bats as big as crows. The fourth member of our party was Arif Islam, also a veterinarian, one of very few in Bangladesh who works with wildlife and zoonotic diseases, and the only member of our group who spoke fluent Bangla. Arif was crucial because he could draw blood from a bat's brachial artery, negotiate with local officials, and order curried fish for us in a local restaurant.

It was almost 9 a.m. by the time we cleared the traffic of Dhaka, where the busses grind against one another like chummy elephants and the green motorbike taxis dodge through the gaps, seeming ever at peril of getting squashed. Finally the road opened. We rolled westward toward the river, relieved to be away. Behind us, the low sun shone feebly through the smog of the city, orange as a bloodied yolk.

We made the ferry crossing into Faridpur District—dry season, the Padma River was low—and proceeded on a two-lane between the rice paddies. We stopped in Faridpur city to pick up more personnel, a pair of field assistants named Pitu and Gofur, with special skills. Both were small men, as compact and agile as jockeys, expert climbers and bat catchers who had worked intermittently with Epstein for several years. Their bat-catching expertise came from an earlier career in poaching, but now they were on the side of the angels. With them aboard, we turned south, snacking on oranges and spicy cracker mix along the way. We eased through small towns clogged with rickshaws and busses and motorbikes; down here in the southwest, I noticed few private cars. One community seemed to specialize in the quarrying, bagging, and shipping of sand, a resource available in abundance. It was transplanting time for the new rice crop, and we could see men and women bent double, digging the dark green shoots from their thick nursery patches along the river bottoms, bundling them, moving them, replanting them carefully in flooded paddies. On drier ground grew small patches of other crops—corn, beans, grain—and the occasional cluster of banana trees or coconut palms. Drier ground, though, was becoming more scarce as we moved farther south. Straight ahead was the Sundarbans swamp, where the Ganges delta dissolves into mangrove islands and braiding channels and crocodiles and wet-footed tigers, but we weren't going that far. Already the land was so flat and low, the water table so high, that sumps of stagnant water surrounded every village and town we passed.

Along here we started to see more date palms, their smooth trunks scarred with barber-pole striations showing where gachis had tapped them in years past. It was mid-January now and the sap harvest was on, perfect timing in case we wanted to sample a glassful. We didn't. Bangladeshis call the stuff *kajul*, I learned from Arif. They believe that it's a salubrious beverage, killing parasites in the gut. But you've got to drink it fresh, Arif said. Boiling the sap ruins not just its taste but also the medicinal effect. He drank it himself as a boy, yeah, sure—but not anymore, no way, not since he's been working on Nipah.

In midevening we reached a city called Khulna, found rooms in a decent hotel, and the next day went out looking for bat roosts, of which Arif had prescouted several during an earlier trip. West of the city, the land seemed lower still, and water was plentiful—water in paddies, in sumps, in lagoons, in shrimp-raising ponds. Village people and their livestock lived on patches of dirt reached by footpath causeways, and the road itself ran along an embankment, material for which had presumably come from borrow pits that were now the funky greenish and brownish pools alongside. If you wanted high ground here, you had to build it. There were plenty of trees but nothing to call forest, just a scattering of coconut palms, bananas, papayas, tamarinds, a few hardwoods, and many more date palms, into one of which I saw a gachi climbing. He was barefoot, using his hands and feet plus a belt rope to ascend, like a Wichita lineman. He wore a *lungi* (a sarong, knotted at his waist), a turban, and over his shoulder a woven quiver, which held two long, curved knives. Nearby, a small boy on the roadside carried four red clay pots, empty and ready for placement to catch tonight's drippings.

The bats would be ready too. Meanwhile they were sleeping. Flying foxes, unlike insectivorous bats and some fruit bats, do not roost in caves, mines, or old buildings. They prefer trees, from the branches of which they dangle upside down, wrapped in their wings, like the weirdest of tropical fruit. We visited four or five sites. We gazed up into treetops at aggregations of sleeping bats, talked with locals, and inspected the lay of the land beneath each roost, none of which met Epstein's exacting standards. Either the bats were too few (a hundred here, a hundred there), the nearby trees or lack of them allowed no way to erect a net, or circumstances were wrong on the ground below. In one village, several hundred bats had established their roost in some legume trees, a tempting cluster, except that they dangled just above a big green puddle that seemed to serve as drain tank and garbage dump for the village. Lowering the net after captures would drop tangled bats into that water, Epstein foresaw, and oblige him to plunge in and untangle them before they drowned. Nope, he muttered. I'd rather take my chances with Nipah than with whatever's in that bilge.

So we returned to a site we had spotted along the road into Khulna: a derelict storage depot within a walled compound of several acres, government-owned and once used as a repository for road-building materials. From a grassy courtyard there, among the sheds and warehouses, towered a handful of great *karoi* trees in which dangled four or five thousand bats. It was an especially favored roost site, evidently, because the trees were so large, the walled compound protected them from village hubbub and boys with slingshots, and each evening around dusk they could drop

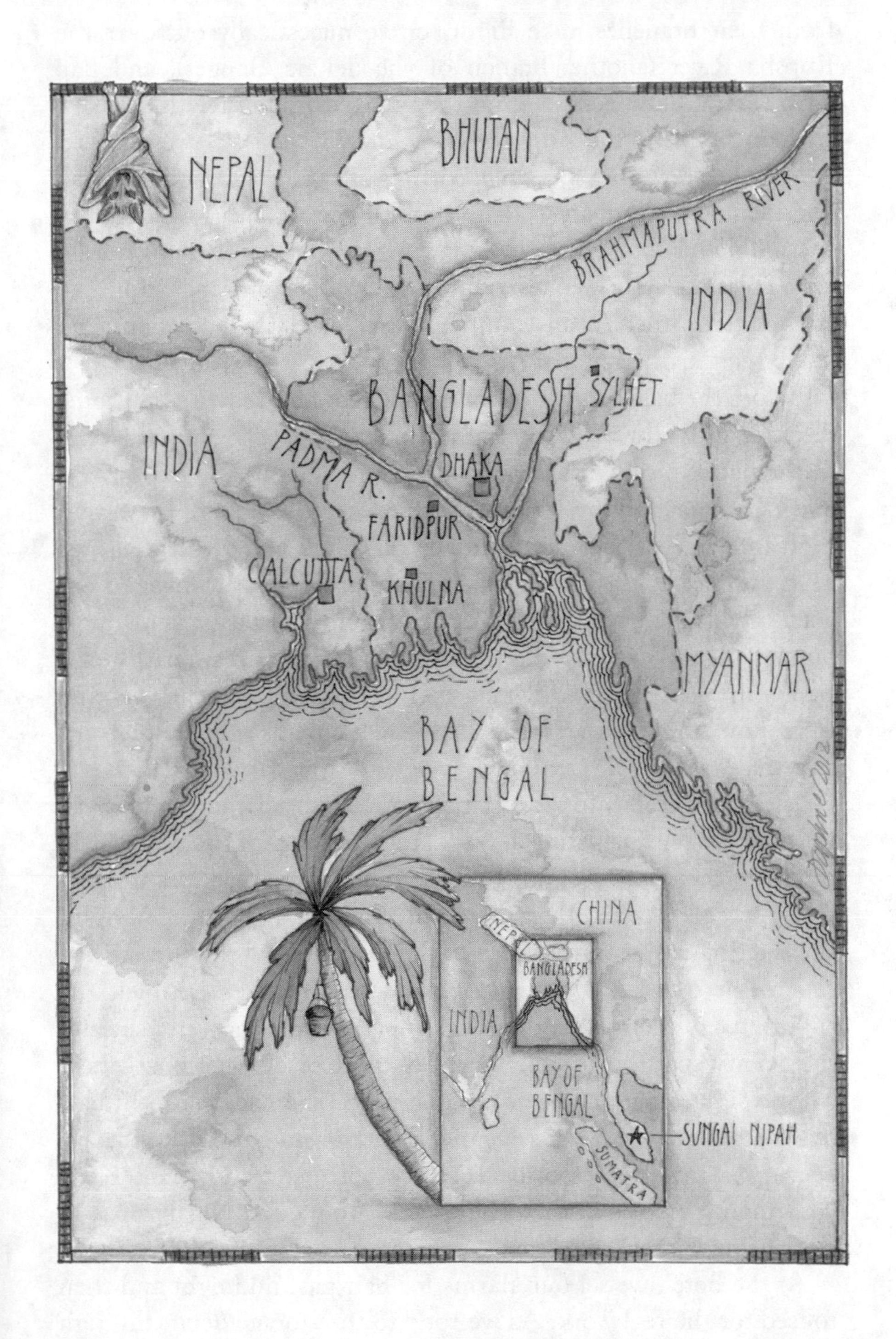

NEPAL
BHUTAN
BRAHMAPUTRA RIVER
INDIA
BANGLADESH
SYLHET
INDIA
PADMA R.
DHAKA
FARIDPUR
CALCUTTA
KHULNA
MYANMAR
BAY OF
BENGAL
CHINA
BANGLADESH
INDIA
BAY OF
BENGAL
SUNGAI NIPAH
SUMATRA

from their branches, take flight, circle majestically out over the Rupsha River (another branch of the deltaic Ganges), and flap away for a night's feeding amid the villages around Khulna. Okay, Epstein decided, this is it.

Within a day, after meetings with local officials, he and Arif had obtained permission for us to go spooking around this old depot in the middle of the night. *That's* why I like working in Bangladesh, Epstein said. Simple request, reasonable people, prompt action. Go into certain other Asian countries with similar expectations and you'll see the difference.

Before the bat catching could begin, though, we had to do some daytime groundwork. We climbed a long rickety bamboo ladder to the flat roof of a disused warehouse, just beside the karoi trees, and from that rooftop Gofur and Pitu kept climbing. They went high into one of the trees, nimble as sailors going to the crow's nest, and lashed a bamboo mast into place so that it towered out vertically above an uppermost limb. Atop that mast was a simple homemade pulley. They did the same in another tree, near the far side of the warehouse, and when their clambering and their rigging were done, they could raise and lower a huge mist net between the two trees.

Their intrusion into a roost tree, of course, disrupted the bats. Hundreds of animals stirred, woke, took flight, and circled out over the river, then back around, and then out again, like flotsam adrift on a great eddy of air. They looked big as geese against the daylight sky, soaring easily on thermals or flapping in slow rhythm. When they came over us, passing low, their features were visible—the auburn fur of their bodies, the big umber wings almost translucent, the pointy snouts. Although they didn't like being waked, there was no sign of panic. They were magnificent. I had seen fruit bats in Asia before, but never so many in motion so close. I must have been gawking like a fool because Epstein gently advised, "Keep your mouth closed when you look up." They shed Nipah virus in their urine, he reminded me.

At the hotel, we set our alarms for half past midnight and then roused for the real work. As we rode to the storage depot through slumbering Khulna, Epstein gave us what he called The Safety Briefing. Goggles and leather welder's gloves for the bat handlers, he said. Medical gloves underneath. Keep your hat on, keep your long sleeves down. When you take hold of such a large bat, you want to grasp it firmly around the back of its head, your fingers and thumb beneath its mandible so it can't bite you. Avoid being bitten. Avoid being scratched. If a bat hooks a claw into your arm, raise that hand high, over your head; the animal's instinct is to climb upward, and you don't want it to climb across your face. Pitu and Gofur will untangle captured bats from the net and then place them into your grasp. Take the head with one hand, get its limbs with the other, clamping each of its strong little ankles and wrists in the gaps between your fingers—one, two, three, four—and your thumb. Four pinch slots, just enough. Trust Pitu and Gofur, they'll help. That's how you control a flying fox so that nobody gets hurt. Drop each bat into its pillowcase—which Arif will be holding open—then knot the pillowcase, hang it from a limb, and come back for another bat. If you get scratched or bitten, we treat that as an exposure—possibly to Nipah, possibly also to rabies. We wash the wound for five minutes with soap and then douse it with benzalkonium chloride, a potent antiviral. Immediately after that, jab, you get a rabies booster. Are you vaccinated for rabies, David? (Yes.) When was your last booster, how are your titers? (Um, don't know.) As for Nipah exposure, never mind, because there's no vaccine, no treatment, no cure. (What a relief.) Have I said, Don't get bit? Our first principles are, one, safety for us; two, safety for the bats. Let's do take good care of the bats, Epstein said. (He's a veterinarian and a conservationist, before all.) Any questions?

Most of this, thank goodness, was for Jim Desmond's benefit, not mine. Arif and Pitu and Gofur were seasoned pros; they didn't need another briefing. Desmond was the real trainee, and I was along to watch. I didn't intend to let anyone hand me a Nipah-dripping bat if I could reasonably avoid it.

Just outside the compound wall, in another empty building, Epstein had established his field lab. There, in the early wee hours, he and his crew readied their equipment for later tasks: anaesthetizing captured bats, taking blood samples and urine swabs from each animal, centrifuging the blood tubes to allow aliquoting off the serum, and freezing all the samples in a liquid-nitrogen shipper tank. This room had a concrete floor, barred windows, a wooden table now covered with plastic sheeting, and a sterilizing footbath at the door, through which we would come and go in our rubber boots. Epstein issued respirator masks, safety goggles, and medical gloves (not latex, not rubber, but the latest material of choice: nitrile) to everyone, and we suited up. He and Desmond both donned old coveralls. Arif had a nice new Tyvek one-piece suit, like gleaming white footie pajamas. Get something else when you can, Epstein told him gently; these bats are visual, remember, not echolocators, and they can see you. Desmond tried on his respirator, and after a moment Epstein asked, "Can you breathe?"

"Yeah."

"That's good. You're not allowed to pass out. That's rule number five." I tried to remember the other four.

Just before pulling his own mask into place, Epstein noted cheerily: "With new and emerging viruses, it's all about prevention. Once you *have* the virus, there's not much you can do." He handed me a small packaged wipe, like the alcohol-laced face fresheners you get on an airplane, except instead of alcohol this thing contained benzalkonium chloride. Ooo, thanks. It was now 2:40 a.m., time to go to the roof.

"All right," he said. "Are we ready?"

74

There was no moon. We marched out through the darkness like Ghostbusters and took turns climbing the long bamboo ladder. The roof of the warehouse was a little spooky in itself, an expanse of tarpaper with a few patches and cracks, old and neglected, not guaranteed to support a person's weight. Wearing safety glasses that quickly became fogged with vapor leaking up from my respirator, I could scarcely see where I was walking. Worse still, I could scarcely see where the building ended and open space began. About all I could see was Arif, moving around in his Tyvek, pale and diaphanous as Casper the Friendly Ghost. Okay, him we won't bust. But don't get distracted, and watch where you step. Rule number six, I realized, is Don't fall off the roof.

The bats were all out for their nightly feeding. We would lurk here to catch them as they returned, sometime before daylight. Gofur and Pitu had already hoisted the net into place, an invisible wall of delicate mesh in the blackness somewhere above us, big as the screen for a drive-in movie. We hunkered down to wait. The night grew chilly—the first time in my limited Bangladesh experience I'd had occasion to get cold. I lay on my back upon the tarpaper, bundled as best I could be in a light jacket, and went to sleep. The first bat hit the net at 4:22 a.m.

Headlamps came alight, people jumped up, Gofur lowered the net on its pulleys while Epstein and Pitu converged on the animal and I staggered forward after them, safely blinded behind my safety glasses. Pitu untangled the bat and Epstein accepted it, using just the technique he had described: grabbing its head firmly, taking its legs and arms into his finger gaps—binga, binga, binga, binga—and then jouncing the bat into its bag. Close the bag's neck, tie firmly with a piece of twine. Captured bats, like captured snakes, evidently relax better if you confine them in soft cloth. Reraise the net and repeat. I was impressed by the proficiency of Epstein's team.

Between the first bat and daylight, before call to prayer even sounded from the local mosques, they bagged five more. Six bats for a night's work was below par for Epstein—he liked to average about ten—but it was a good start for a new location. Adjustments to the net placement, to the height of the masts, would improve the yield here in coming days. For now, enough. As dawn filtered in, we climbed down the ladder and repaired to the laboratory room. Here again, everybody had an assigned role. Mine was to stay the hell out of the way, and occasionally to assist with a swab.

Three hours later, blood samples drawn, swab samples taken, tubes in the freezer tank, it was time to release the bats. Each of them first received a drink of fruit juice to help restore bodily fluids lost in the blood draw. Then we all walked back to the grassy courtyard, beneath the karoi trees, where a small crowd of men, women, and children from the neighborhood had gathered. (The walls of the old depot compound were permeable to locals when something interesting was afoot.) Epstein, again now wearing welder's gloves, released the first five bats one by one from their bags, holding each animal high so it wouldn't crawl up his face, letting it free its legs and its wings, then relaxing his grip gently just as the wing beats began to find purchase on air, and watching—all of us watching—the animal catch itself short of the ground, rise slowly, circle languidly, and fly away. Eventually, after a circuit or two of the compound, a few minutes of befuddled relief, it would find its way back to the communal roost, sadder but wiser and no great harm done.

Before releasing the last bat, Epstein gave a brief talk to the assembled citizens, translated by Arif, congratulating them on their good fortune as a village at harboring so many wonderful bats, which are helpful to fruit trees and other plants, and assuring them that he and his colleagues had taken great care not to harm the animals while studying their health. Then he let the final bat drop. It climbed through the air, from knee level, and flew away.

Later he said to me: "Any one of those six bats could have been infected. That's what it looks like. They look totally healthy. There's no way to distinguish Nipah virus. That's why we take all these precautions." He dipped his boots again in the sterile footbath, as we left the lab, and washed up at the village pump. A little girl brought soap.

75

"The key is connectivity," Epstein told me, during a quiet chat the following afternoon. "The key is to understand how animals and people are interconnected." We were back at the hotel, showered and fed, after another full night of trapping, another fifteen bats sampled and released. You can't look at a new bug or a reservoir host, he said, as though they exist in a vacuum. It's a matter of contact with humans, interaction, opportunity. "Therein lies the risk of spillover."

Repeatedly over the next half hour he returned to the word "opportunity." It kept knocking. "A lot of these viruses, a lot of these pathogens that come out of wildlife into domestic animals or people, have existed in wild animals for a very long time," he said. They don't necessarily cause any disease. They have coevolved with their natural hosts over millions of years. They have reached some sort of accommodation, replicating slowly but steadily, passing unobtrusively through the host population, enjoying long-term security—and eschewing short-term success in the form of maximal replication within each host individual. It's a strategy that works. But when we humans disturb the accommodation—when we encroach upon the host populations, hunting them for meat, dragging or pushing them out of their ecosystems, disrupting or destroying those ecosystems—our action increases the level of risk. "It increases the opportunity for these pathogens to jump from their natural host into a new host," he said. The new host might be any animal (the horse in Australia, the palm civet in China) but often it's humans, because we are present so intrusively and abundantly. We offer a wealth of opportunity.

"Sometimes nothing happens," Epstein said. A leap is made but the microbe remains benign in its new host, as it was in the old one. (Simian foamy virus?) In other cases, the result is very severe disease for a limited number of people, after which the pathogen comes to a dead end. (Hendra, Ebola.) In still other cases, the pathogen achieves great and far-reaching success in its new host. It finds itself well enough suited to get a foothold; it makes itself still better suited by adapting. It evolves, it flourishes, it continues. The history of HIV is the story a leaping virus that might have come to a dead end but didn't.

Yes, HIV is a vivid example, I agreed. But is there any particular reason why other RNA viruses shouldn't have the same potential? For instance, Nipah?

"No reason at all. There's no reason at all," Epstein said. "A lot of what determines whether a pathogen becomes successful in a new host, I think, is odds. Chance, to a large degree." With their high rates of mutation, their high rates of replication, RNA viruses are very adaptable, he reminded me, and every spillover presents a new opportunity to adapt and take hold. We'll probably never know how often that occurs—how many animal viruses spill into people inconspicuously. Many of those viruses cause no disease, or they cause a new disease that—in some parts of the world, because health care is marginal—gets mistaken for an old disease. "The point being," he said, "that the more opportunity viruses have to jump hosts, the more opportunity they have to mutate when they encounter new immune systems." Their mutations are random but frequent, combining nucleotides in myriad new ways. "And, sooner or later, one of these viruses has the right combination to adapt to its new host."

This point about opportunity is a crucial idea, more subtle than it might seem. I had heard it from a few other disease scientists. It's crucial because it captures the randomness of the whole situation, without which we might romanticize the phenomena of emerging diseases, deluding ourselves that these new viruses attack humans with some sort of purposefulness. (Loose talk about "the revenge of the rain forest" is one form of such romanticizing. That's a nice metaphor, granted, but shouldn't be taken too seriously.) Epstein was talking, in an understated way, about the two distinct but interconnected dimensions of zoonotic transfer: ecology and evolution. Habitat disturbance, bushmeat hunting, the exposure of humans to unfamiliar viruses that lurk in animal hosts—that's ecology. Those things happen *between* humans and other kinds of organism, and are viewed in the moment. Rates of replication and mutation of an RNA virus, differential success for different strains of the virus, adaptation of the virus to a new host—that's evolution. It happens *within* a population of some organism, as the population responds to its environment over time. Among the most important things to remember about evolution—and about its primary mechanism, natural selection, as limned by Darwin and his successors—is that it doesn't have purposes. It only has results. To believe otherwise is to embrace a teleological fallacy that carries emotive appeal ("the revenge of the rain forest") but misleads. This is what Jon Epstein was getting at. Don't imagine that these viruses have a deliberate strategy, he said. Don't think that they bear some malign onus against humans. "It's all about opportunity." They don't come after us. In one way or another, we go to them.

But what *is* it about bats? I asked. Why do so many of these zoonotic viruses—or what seems like so many—spill over onto humans from the chiropteran order of mammals? Or is that the wrong question?

"It is the right question," he said. "But I don't think there's a good answer for it yet."

76

There may not be a good answer, but efforts have been made. I've put the same question—*why bats?*—to emerging-disease experts around the world. One of them was Charles H. Calisher, an eminent virologist recently retired as professor of microbiology at Colorado State University.

Calisher came out of the Georgetown School of Medicine with a PhD in microbiology in 1964. He made his bones doing classic lab-table virology, which meant growing live viruses, passaging them experimentally through mice and cell cultures, looking at them in electron micrographs, figuring out where to place them on the viral family tree—the kind of work that Karl Johnson had done on Machupo, and that traced back before Johnson to Frank Fenner and Macfarlane Burnet and others still earlier. Calisher's career included a long stretch at the CDC as well as academic appointments, during which he had focused on arthropod-borne viruses (aka arboviruses, such as West Nile, dengue, and La Crosse virus, all carried by mosquitoes) and rodent-borne viruses (notably the hantaviruses). As a scientist who studied viruses in their vectors and in their reservoirs for more than four decades, but with no particular attention to chiropterans, he too eventually found himself wanting to know: Why are so many of these new things emerging from *bats*?

Charlie Calisher is a smallish man with a dangerous twinkle, famed throughout the profession for his depth of knowledge, his caustic humor, his disdain for pomposity, his brusque manner, and (if you happen to get past those crusts) his big, affable heart. He insisted on buying me lunch, at a favorite Vietnamese restaurant in Fort Collins, before we got down to serious talk. He wore a fisherman's sweater, chinos, and hiking boots. After the meal I followed his red pickup truck back to a CSU laboratory compound, where he still had a few projects going. He pulled a flat-sided flask from an incubator, put it under a microscope, focused, and said, Look here: La Crosse virus. I saw monkey cells, in a culture medium the color of cherry Kool-Aid, under attack by something so tiny it could only be discerned by the damage it did. People around the world—doctors, veterinarians—send him tissue samples, Calisher explained, asking him to grow a virus from the stuff and identify it. Okay. That sort of thing has been his life's work, especially with regard to hantaviruses in rodents. And then came this little excursion into bats.

We repaired to his office, now almost empty as he eased into retirement, except for a desk, two chairs, a computer, and some boxes. He tilted back in his chair, set his boots on the desk, and began to talk: arboviruses, the CDC, hantaviruses in rodents, La Crosse virus, mosquitoes, and a congenial group called the Rocky

Mountain Virology Club. He ranged widely but, knowing my interest, circled back to a consequential chat he'd had with a colleague about six years earlier, soon after news broke that SARS, the new killer coronavirus, had been traced to a Chinese bat. The colleague was Kathryn V. Holmes, an expert on coronaviruses and their molecular structure, at the University of Colorado Health Sciences Center near Denver, just down the highway from Fort Collins. Charlie told me the story in his own vivid way, complete with dialogue:

"We oughta write a review paper about bats and their viruses," he said to Kay Holmes. "This bat coronavirus is really interesting."

She seemed intrigued but a little dubious. "What would we include?"

"Well, this and that, something else," Charlie said vaguely. The idea was still taking shape. "Maybe immunology."

"What do we know about immunology?"

Charlie: "I don't know shit about immunology. Let's ask Tony."

Tony Schountz, another professional friend, is an immunologist at the University of Northern Colorado, in Greeley, who does research on responses to hantaviruses in humans and mice. At that time Schountz, like Calisher, had never studied chiropterans. But he is a burly young guy, a former athlete, who had played college baseball as a catcher.

"Tony, what do you know about bats?"

Schountz thought Charlie meant Louisville Sluggers. "They're made of ash."

"Hello, Tony? I'm talkin' about *bats*." Wing-flapping gesture. As distinct from: DiMaggio gesture.

"Oh. Uh, nothing."

"You ever read anything about the immunology of bats?"

"No."

"Have you ever *seen* any papers on the immunology of bats?"

"No."

Neither had Charlie—nothing beyond the level of finding antibodies that confirmed infection. Nobody seemed to have addressed the deeper question of how chiropteran immune systems respond.

"So I said to Kay, 'Let's write a review paper,'" Charlie told me. "Tony said, 'Are you crazy? We don't *know* anything!'"

"Well, *she* doesn't know anything, *you* don't know anything, and *I* don't know anything. This is great. We don't have any biases."

"*Biases*?" said Schountz. "We don't have any *information*!"

"I said, 'Tony, that shouldn't hold us back.'"

Thus the workings of science. But Calisher and his two pals didn't plan to flaunt their ignorance. If we don't know anything in this or that area, he proposed, we'll just get somebody who *does*. They enlisted James E. Childs, an epidemiologist and rabies expert at the Yale School of Medicine (and an old friend of Charlie's from CDC days), and Hume Field, who by now was turning up everywhere. This five-member team, with their patchwork of expertise and their sublime lack of biases, then wrote a long, wide-ranging paper. Several journal editors voiced interest but wanted the manuscript cut; Charlie refused. It appeared finally, intact, in a more expansive journal, under the title "Bats: Important Reservoir Hosts of Emerging Viruses." It was a review, as Charlie had envisioned, meaning that the five authors made no claim of presenting original research; they simply summarized what had previously been done, gathered disparate results together (including unpublished data contributed by others), and sought to highlight some broader patterns. That much, it turned out, was a timely service. The paper offered a rich compendium of facts and ideas—and where facts were scarce, directive questions. Other disease scientists noticed. "All of a sudden," Charlie told me, "the phone's ringing off the hook." They met hundreds of requests for reprints, maybe thousands, sending their "Bats: Important Reservoir Hosts" to colleagues worldwide in the form of a PDF. Everybody wanted to know—everybody in that professional universe anyway—about these new viruses and their chiropteran hideouts. Yes, what *is* the deal with bats?

The paper made a handful of salient points, the first of which put the rest in perspective: Bats come in many, many forms. The order Chiroptera (the "hand-wing" creatures) encompasses 1,116 species, which amounts to 25 percent of all the recognized species of mammals. To say again: One in every four species of mammal is a bat. Such diversity might suggest that bats *don't* harbor more than their share of viruses; it could be, instead, that their viral burden is proportional to their share of all mammal diversity, and thus just *seems* surprisingly large. Maybe their virus-per-bat ratio is no higher than ratios among other mammals.

Then again, maybe it *is* higher. Calisher and company explored some reasons why that might be so.

Besides being diverse, bats are very abundant and very social. Many kinds roost in huge aggregations that can include millions of individuals at close quarters. They are also a very old lineage, having evolved to roughly their present form about 50 million years ago. Their ancientness provides scope for a long history of associations between viruses and bats, and those intimate associations may have contributed to viral diversity. When a bat lineage split into two new species, their passenger viruses may have split with them, yielding more kinds of virus as well as more kinds of bat. And the abundance of bats, as they gather to roost or to hibernate, may help viruses to persist in such populations, despite acquired immunity in many older individuals. Remember the concept of critical community size? Remember measles, circulating endemically in cities of five hundred thousand people or more? Bats probably meet the critical community size standard more consistently than most other mammals. Their communities are often huge and usually large, offering a steady supply of susceptible newborns to become infected and maintain the viral presence.

That scenario assumes a virus that infects each bat only briefly, leaving recovered individuals with lifelong immunity, as measles does in humans. An alternative scenario involves a virus capable of causing chronic, persistent infection, lasting months or even years within a single bat. If the infection can persist, then the long average lifespan of a bat becomes advantageous for the virus. Some of the smaller, insectivorous bats live twenty or twenty-five years. Such longevity, if the bat is infected and shedding virus, vastly increases the sum of opportunities over time for passing the virus to other bats. In the language of the mathematicians: R_0 increases with the lifespan of a persistently infected bat. And a bigger R_0, as you know, is always good for the pathogen.

Social intimacy helps too, and many kinds of bat seem to love crowding, at least when they hibernate or roost. Mexican free-tailed bats in Carlsbad Caverns, for instance, snuggle together at about three hundred individuals per square foot. Not even lab mice in an overloaded cage would tolerate that. If a virus can be passed by direct contact, bodily fluids, or tiny droplets sprayed through the air, crowding improves its chances. Under conditions like those in Carlsbad, Calisher's group noted, even rabies has been known to achieve airborne transmission.

Speaking of airborne: It's not insignificant that bats fly. An individual fruit bat may travel dozens of miles each night, searching for food, and hundreds of miles in a season as it moves among roosting sites. Some insectivorous bats migrate as much as eight hundred miles between their summer and winter roosts. Rodents don't make such journeys, and not many larger mammals do. Furthermore, bats move in three dimensions across the landscape, not just two; they fly high, they swoop low, they cruise in between, inhabiting a far greater volume of space than most animals. The breadth and the depth of their sheer presence are large. Does that increase the likelihood that they, or the viruses they carry, will come in contact with humans? Maybe.

Then there's bat immunology. Calisher's group could only touch judiciously on this topic, even with Tony Schountz as a coauthor, because little is known by anyone. Mainly they raised questions. Is it possible that the cold temperatures endured by hibernating bats suppress their immune responses, allowing viruses to persist in bat blood? Is it possible that antibodies, which would neutralize a virus, don't last as long in bats as in other mammals? What about the ancientness of the bat lineage? Did that lineage diverge from other mammals before the mammalian immune system had been well honed by evolution, reaching the level of effectiveness seen in rodents and primates? Do bats have a different "set point" for their immune responses, allowing a virus to replicate freely so long as it doesn't do the animal any harm?

Answering those questions, according to Calisher's group, would require new data derived from new work. And that work couldn't be done just with the sleek tools and methods of molecular genetics, comparing long sequences of nucleotide bases by way of computer software. They wrote:

> Emphasis, sometimes complete emphasis, on nucleotide sequence characterization rather than virus characterization has led us down a primrose path at the expense of having real viruses with which to work.

The paper was a collaborative effort but that sentence sounds like Charlie Calisher. What it means is: *Hello, people? We've gotta grow these bugs the old-fashioned way, we've gotta look at them in the flesh, if we're gonna understand how they operate.* And if we don't, the paper added, "we are simply waiting for the next disastrous zoonotic virus outbreak to occur."

77

Charlie Calisher and his coauthors, besides touching on broad principles, discussed a handful of bat-related viruses in detail: Nipah, Hendra, rabies and its close relatives (the lyssaviruses), SARS-CoV, and a couple of others. They mentioned Ebola and Marburg, though carefully omitting those two from the list of viruses for which bats had been proven to serve as reservoirs. "The natural reservoir hosts of these viruses have not yet been identified," they said about Marburg and Ebola—accurately, as of the time of publication. Their paper appeared in 2006. Fragments of Ebola RNA had been detected by then in some bats; antibodies against Ebola virus had been found in other bats. But that wasn't quite proof enough. Nobody had yet isolated any live filovirus from a bat, and the unsuccessful efforts to do so left Ebola and Marburg well hidden.

Then, in 2007, Marburg virus reappeared, this time among miners in Uganda. It was a small outbreak, affecting only four men, of whom one died, but it served as an opportunity to gain new insight into the virus, thanks in part to a quickly responsive multinational team. The four victims all worked at a site called Kitaka Cave, not far from Queen Elizabeth National Park, in the southwestern corner of Uganda. They dug galena, which is lead ore, plus a little bit of gold. The word "mine" caught the attention of some scientists within the CDC's Special Pathogens Branch, in Atlanta, because they already had reason to suspect that Marburg's reservoir, whatever it was, might be associated with cavelike environments. Several of the previous Marburg outbreaks included patients whose case histories involved visits to, or work in, caves or mines. So when the response team arrived at Kitaka Cave, in August 2007, they were ready to go underground.

This group included scientists from the CDC, the National Institute for Communicable Diseases in South Africa, and WHO in Geneva. The CDC sent Pierre Rollin and Jonathan Towner, whom we've met before, as well as Brian Amman and Serena Carroll. Bob Swanepoel and Alan Kemp of the NICD flew up from Johannesburg; Pierre Formenty arrived from WHO. All of them possessed extensive experience with Ebola and Marburg, gained variously through outbreak responses, lab research, and field studies. Amman was a mammalogist with a special affinity for bats. During a conversation at the CDC, he described to me what it was like to go to Kitaka Cave.

The cave served as roosting site for about a hundred thousand individuals of the Egyptian fruit bat (*Rousettus aegyptiacus*), a prime suspect as reservoir for Marburg. The team members, wearing Tyvek suits, rubber boots, goggles, respirators, gloves, and helmets, were shown to the shaft by miners, who as usual were clad only in shorts, T-shirts, and sandals. Guano covered the ground. The miners clapped their hands to scatter low-hanging bats as they went. The bats, panicked, came streaming out. These were sizable animals, each with a two-foot wingspan, not quite so large and hefty as the flying foxes of Asia but still daunting, especially with thousands swooshing at you in a narrow tunnel. Before he knew it, Amman had been conked in the face by a bat and suffered a cut over one eyebrow. Towner got hit too, Amman said. Fruit bats have long, sharp thumbnails. Later, because of the cut, Amman would get a postexposure shot against rabies, though Marburg was a more immediate concern. "Yeah," he thought, "this could be a really good place for transmission."

The cave had several shafts, Amman explained. The main shaft was about eight feet high. Because of all the mining activity along there, many of the bats had shifted their roosting preference "and went over to what we called the cobra shaft." That was a smaller shaft, branching off, which—

I interrupted him. "'Cobra' because there were *cobras*?"

"Yeah, there was a black forest cobra in there," he said.

Or maybe a couple. It was good dark habitat for snakes, with water and plenty of bats to eat. Anyway, the miners showed Amman and Towner into the cave, past another narrow shaft that led to a place called the Hole, a pit about ten feet deep accessed by shinnying down a pole, from the bottom of which came much of the ore. The two Americans were looking for the Hole but, following their guides, inadvertently passed that shaft by, continuing about two hundred meters along the main shaft to a chamber containing a body of brown, tepid water. Then the local fellows cleared out, leaving Towner and Amman to do a bit of exploring on their own. They dropped down beside the brown lake and found that the chamber branched into three shafts, each of which seemed blocked by standing water. Peering into those shafts, they could see many more bats. The humidity was high and the temperature maybe ten or fifteen degrees hotter than outside. Their goggles fogged up. Their respirators became soggy and wouldn't pass much oxygen. They were panting and sweating, zipped into their Tyvek suits, which felt like wearing a trash bag, and by now they were becoming "a little loopy," Amman recalled. One lakeside shaft seemed to curve back around, possibly connecting with the cobra shaft. They didn't know how deep the water might be, and the airspace above it was limited. Should they proceed? No, they decided, the increased risk wasn't worth the potential benefit. Formenty, their

WHO colleague, eventually found them down there and said, Hey guys, the Hole is back this way. They crawled out and retraced their path, "but by that time we were spent," Amman said. "We had to get out and cool off." It was only their first underground excursion at Kitaka. They would make several.

On a later day, the team investigated a grim, remote chamber they dubbed the Cage. It was where one of the four infected miners had been working just before he got sick. This time Amman, Formenty, and Alan Kemp of the NICD went to the far recesses of the cave. The Cage itself could only be entered by crawling through a low gap at the base of a wall—like sliding under a garage door that hasn't quite closed. Brian Amman is a large man, six foot three and 220 pounds, and for him the gap was a tight squeeze; his helmet got stuck and he had to pull it through separately. "You come out into this sort of blind room," he said, "and the first thing you see is just hundreds of these dead bats."

They were Egyptian fruit bats, the creature of interest, left in various stages of mummification and rot. Piles of dead and liquescent bats seemed a bad sign, potentially invalidating the hypothesis that Egyptian fruit bats might be a reservoir host of Marburg. If these bats had died in masses from the virus, then they couldn't also be its reservoir. Then again, they might have succumbed to earlier efforts by the locals to exterminate them with fire and smoke. Their cause of death was indeterminable without more evidence, and that's partly why the team was there. If these bats *had* died of Marburg, suspicion would shift elsewhere—to another bat, or maybe a rodent, or a tick, or a spider? Those other suspects might have to be investigated. Ticks, for instance: There were plenty of them in crevices near the bat roosts, waiting for a chance to drink some blood. Meanwhile, when Amman and Kemp stood up in the Cage, they realized that not every bat in there was dead. The room was aswirl with live ones, circling around their heads.

The two men went to work, collecting. They stuffed dead bats into bags. They caught a few live bats and bagged them too. Then, back down on their bellies, they squooched out through the low gap. "It was really unnerving," Amman told me. "I'd probably never do it again." One little accident, he said, a big rock rolls in the way, and that's it. You're trapped.

Wait a minute, lemme get this straight: You're in a cave in Uganda, surrounded by Marburg and rabies and black forest cobras, wading through a slurry of dead bats, getting hit in the face by live ones like Tippi Hedren in *The Birds*, and the walls are alive with thirsty ticks, and you can hardly breathe, and you can hardly see, and . . . you've got time to be *claustrophobic?*

"Uganda is not famous for its mine rescue teams," he said.

By the end of this fieldtrip, the scientists had collected about eight hundred bats for dissection and sampling, half of those belonging to *Rousettus aegyptiacus*. The CDC team, including Towner and Amman, returned to Kitaka Cave seven months later, in April 2008, catching and sampling two hundred more individuals of *R. aegyptiacus* to see if Marburg persisted in the population. If so, that would strongly suggest that this species was in fact a reservoir. During the second trip, they also marked and released more than a thousand bats, hoping that from later recaptures they could deduce the overall size of the population. Knowing the population size, as well as the prevalence of infection among their sampled bats, would indicate how many infected bats might be roosting in Kitaka at any one time. Towner and Amman used beaded collars (which seemed less discomfiting to the bats than the usual method of marking, leg bands), each collar coded with a number. The two scientists took some heat for this mark-recapture study; skeptical colleagues argued that it was wasted effort, given the vast size of the bat population and the odds against recapture. But, in Amman's words, "we kind of stuck to our guns," and they eventually released 1,329 tagged bats.

Less speculative, less controversial, were the samples of blood and tissue from dissected bats. Those went back to Atlanta, where Towner took part in the laboratory efforts to find traces of Marburg virus. One year later came a paper, authored by Towner, Amman, Rollin, and their WHO and NICD colleagues, announcing some important results. All the cave crawling, bat sampling, and lab work had yielded a dramatic breakthrough in the understanding of filoviruses, meaning both Marburg and Ebola. Not only did the team detect antibodies against Marburg (in thirteen of the roughly six hundred fruit bats sampled) and fragments of Marburg RNA (in thirty-one of the bats), but they also did something more difficult and compelling. Antibodies and RNA fragments, though significant, were just the same sorts of secondary evidence that had provisionally linked the Ebola virus to bats. This team had gone a step farther: They'd found live virus.

Working in one of the CDC's BSL-4 units, Towner and his co-workers had isolated viable, replicating Marburg virus from five different bats. Furthermore, the five strains of virus were genetically diverse, suggesting an extended history of viral presence and evolution within Egyptian fruit bats. Those data, plus the fragmentary RNA, constituted strong evidence that the Egyptian fruit bat is a reservoir—if not *the* reservoir—of Marburg virus. Based on the isolation work, it's definitely there in the bats. Based on the RNA fragments, it seems to infect about 5 percent of the bat population at a given time. Putting those numbers together with the overall population estimate of a hundred thousand bats at Kitaka, the team could say that about five thousand Marburg-infected bats flew out of the cave every night.

An interesting thought: five thousand infected bats passing overhead. Where were they going? How far to the fruiting trees? Whose livestock or little gardens got shat upon as they went? Jon Epstein's advice would have been apt: "Keep your mouth closed when you look up." And the Kitaka aggregation, Towner and his coauthors added, "is only one of many such cave populations throughout Africa."

Where else might Marburg virus be traveling on the wings of these bats? An answer to that arrived in the summer of 2008.

78

Astrid Joosten was a forty-one-year-old Dutch woman who, in June 2008, went to Uganda with her husband on an adventure vacation. It wasn't their first, but it would be more consequential than the others.

At home in Noord-Brabant (the same area, by coincidence, then being hard hit with Q fever), Joosten worked as a business analyst for an electrical company. Both she and her spouse, a financial manager, enjoyed escaping from the Netherlands on annual getaways to experience the landscapes and cultures of other countries, especially in Africa. In 2002 they had flown to Johannesburg and, stepping off the airplane, felt love at first sight. On later trips they visited Mozambique, Zambia, and Mali. The journey in 2008, booked through an adventure-travel outfitter, would allow them to see mountain gorillas in the southwestern highlands of the country as well as some other wildlife and cultures. They worked their way south toward Bwindi Impenetrable Forest, where the Ugandan gorillas reside. On one intervening day, the operators offered a side trip, an option, to a place called the Maramagambo Forest, where the chief attraction was a peculiar site that everyone knew as Python Cave. African rock pythons lived there, languid and content, grown large and fat on a diet of bats.

Joosten's husband, later her widower, was a fair-skinned man named Jaap Taal, a calm fellow with a shaved head and dark, roundish glasses. Most of the other travelers didn't fancy this offering, Jaap Taal told me, over a cup of coffee at a café in southwestern Montana. Never mind, for the moment, why he turned up there. Python Cave had been an add-on, he explained, price not included in their Uganda package. "But Astrid and I always said, maybe you come here only once in your life, and you have to do everything you can." They rode to Maramagambo Forest and then walked a mile or so, gradually ascending, to a small pond. Nearby, half-concealed by moss and other greenery, like a crocodile's eye barely surfaced, was a low dark opening. Joosten and Taal, with their guide and one other client, climbed down into the cave.

The footing was bad: rocky, uneven, slick with bat guano. The smell was bad too: fruity and sour. Think of a dreary barroom, closed and empty, with beer on the floor at 3 a.m. The cave seemed to have been carved by a creek, or at least to have channeled its waters, and part of the overhead rock had collapsed, leaving a floor of boulders and coarse rubble, a moonscape, coated with guano like a heavy layer of vanilla icing. The ceiling was thick with bats, big ones, many thousands of them, agitated and chittering at the presence of human intruders, shifting position, some dropping free to fly and then settling again. Astrid and Jaap kept their heads low and watched their step, trying not to slip, ready to put a hand down if needed. "I think that's how Astrid got infected," he told me. "I think she put her hand on a piece of rock, which contained droppings of a bat, which are infected. And so she had it on her hand." Maybe she touched her face an hour later, or put of piece of candy in her mouth, or something such, "and that's how I think the infection got in her."

Python Cave, in Maramagambo Forest, is just thirty miles west of Kitaka Cave. It too harbors Egyptian fruit bats. Thirty miles isn't far and individuals from the Kitaka aggregation are quite capable—as the CDC team's mark-recapture study would later prove—of finding their way to roost at Python.

No one had warned Joosten and Taal about the potential hazards of an African bat cave. They knew nothing of Marburg virus (though they had heard of Ebola). They only stayed in the cave about ten minutes. They saw a python, large and torpid. Then they left, continued their Uganda vacation, visited the mountain gorillas, did a boat trip, and flew back to Amsterdam. Thirteen days after the cave visit, home in Noord-Brabant, Astrid Joosten fell sick.

At first it seemed no worse than flu. Then her temperature went higher and higher. After a few days, she began suffering organ failure. Her doctors, knowing her case history, with recent time in Africa, suspected Lassa virus or maybe Marburg. Marburg, said Jaap, what's that? Astrid's brother looked it up on Wikipedia and told him: Marburg virus, it kills, could be bad trouble. The doctors moved her to a hospital in Leiden, where she could get better care and be isolated from other patients. There she developed a rash and conjunctivitis; she hemorrhaged. She was put into an induced coma, a move dictated by the need to dose her more aggressively with antiviral medicine. Before she lost consciousness, though not long before, Jaap went back into the isolation room, kissed his wife, and said to her, "Well, we'll see you in a few days." Blood samples, sent to a lab in Hamburg, confirmed the diagnosis: Marburg. She worsened. As her organs shut down, she lacked for oxygen to the brain, she suffered cerebral edema, and before long Astrid Joosten was declared brain-dead. "They kept her alive for a few more hours, until the family arrived," Jaap told me. "Then they pulled the plug out and she died within a few minutes."

The doctors, appalled by his recklessness in kissing her goodbye, had prepared an isolation room for Jaap himself, but that was never needed. "There's so much they don't know about Marburg and those other viral infections," he said to me. Then, still a venturesome traveler, he went off on a snow tour of Yellowstone National Park.

79

The news of Astrid Joosten's death carried far. She was the first person known to have left Africa with an active filovirus infection and died. The Swiss graduate student from Côte d'Ivoire, back in 1994, had recovered. Did any other person, apart from those two, ever pass through an international airport and depart the continent with Ebola or Marburg virus incubating in his or her body? No one of whom the experts were aware. Joosten's case proved that Marburg could travel in a human, though admittedly it didn't travel so well as SARS or influenza or HIV-1. Five thousand miles away, in Colorado, another woman heard the news with a shudder of recognition. She had visited Python Cave too.

Michelle Barnes is an energetic late-fortyish woman with blue eyes and auburn hair, one sibling among seven from an Irish Catholic family in Iowa. She's an avid rock climber and bicyclist, a camper and hiker, who has worked for Outward Bound in the past and now serves as an interim executive (stepping in when needed during transitions) and troubleshooter for nonprofit organizations. On the day I met her, at an office in downtown Boulder, she wore a red sweater and a scarf and looked healthy and professional. The auburn, she told me cheerily, came from a bottle. It approximates the original color, she said, but the original is gone. In early 2008 her hair started falling out; the rest went gray, "pretty much overnight." This was among the lesser effects of a mystery illness that had nearly killed her, during January that year, just after she'd returned from Uganda.

Her story paralleled the one Jaap Taal had told me about Astrid, with several key differences—the main difference being that Michelle Barnes was still alive. Another was that her case showed how hard it could be to get correctly diagnosed. Michelle and her husband, Rick Taylor, who runs a construction company, were entranced with Africa, like Jaap and Astrid. They too had made earlier trips, usually traveling to remote places on their own. And they too, this time, wanted to see mountain gorillas. So they booked with an adventure-travel outfitter, because those companies control permits for visiting the gorillas. Their itinerary had them progressing southward through the landscape attractions of western Uganda, again as Jaap and Astrid would later do, leaving the big apes down in Bwindi to be a crescendo near the end of the trip. One intermediate stop was Queen Elizabeth National Park, along the east shore of Lake Edward. It was a drier and flatter ecosystem, offering classic East African savanna full of lions and elephants and other big mammals, which converge on the water holes around dawn and dusk. Midday at Queen Elizabeth, blazing hot and bright, tends to be an off time for viewing wildlife. So on one of the days there, with about five hours to kill, the guide announced that they would go see a cave. Change of pace from the lions and elephants: pythons and bats.

Barnes and her group walked the same mile through Maramagambo Forest and entered the same cave, crossing an irregular floor of large rocks besmeared with guano, which made for poor footing. The walls were acrawl with large, hairy spiders, by her recollection. The ceiling was low and the roosting bats dangled down within two or three feet of a person's head. Some bats flew in and out, screeching as they went. The stench was ammoniac and horrible. You had to clamber across those slippery boulders. As a rock climber, Barnes said, she tends to be very conscious of where she places her hands. No, she didn't touch any guano. No, she was not bumped by a bat. Her party entered a short distance and found themselves on a sort of mezzanine, overlooking a lower level, with bats just above and two pythons below. Some of the other tourists left quickly. She and Rick lingered, trying to absorb the scene. "When again are we going to see pythons and bats in a cave?" she said to me, then caught herself, adding with caustic hindsight: "I can assure you, never."

After twenty minutes, they had seen enough. And that was it: no mishaps, nothing dramatic. "I definitely didn't touch a bat or knowingly touch guano." They hiked back to their vehicle, where the guide spread out a picnic lunch. Before eating, Barnes used some hand sanitizer that she had brought for such moments. By late afternoon they were back at Queen Elizabeth, in time for a sunset of watching the more conventionally appealing forms of African wildlife. It was Christmas evening, 2007.

They arrived home on New Year's Day. Michelle left again quickly for a postholiday visit with her parents in Iowa. So she was already in Sioux City, on January 4, when she woke up feeling like someone had driven a needle into her skull.

She was achy all over, feverish, and had this fierce, drilling headache. Suspecting that she might have been bitten by an insect, she asked her parents to check her scalp. "Of course, there was nothing. And then, as the day went on, I started developing a rash across my stomach." The rash spread. Besides the aches and pains, the exhaustion, the rash, she began to feel discombobulated. "Over the next forty-eight hours, I just went down really fast." She was still on malaria prophylaxis, from the trip, and to that she now added some Cipro and ibuprofen. No relief. But she toughed out the visit, flew back to Colorado, and stopped into an Urgent Care near her home in Golden, where they don't see a lot of Marburg virus disease. The doctor there took blood for testing, gave her painkillers, and sent her home. The blood sample got lost.

After that inconclusive consultation, plus two more with her regular doctor in the following two days, Michelle Barnes turned up at a hospital in suburban Denver. She was dehydrated; her white blood count was imperceptible; her kidneys and liver had begun shutting down. Once admitted, she faced a parade of doctors and a litany of questions. Among the first questions was: What have you been *doing* for the past four days? Most people seek help before multiple organ failure sets in. I've been sucking it up, Barnes answered. Her far-flung sisters, one of whom was a doctor in Alaska, converged on the hospital—which was gratifying to Michelle, but also alarming. Clearly, they had been given to understand that she might be going down. The doctor-sister, Melissa, played a key role in pressing Michelle's physicians for information and action. That's when an infectious disease specialist, Dr. Norman K. Fujita, joined the team. Fujita arranged for Michelle to be tested for leptospirosis, malaria, schistosomiasis, and other infections that might be contracted in Africa, such as Ebola and Marburg. All came back negative, including the test for Marburg.

Nobody knew what she had. But they could see her declining. The hospital doctors tried to stabilize her with hydration and antibiotics and oxygen, tried to ease her suffering with pain meds, while hoping her body would pass through the onslaught, whatever it was, and heal. The crisis must have arrived on the night of January 10 or 11, by Michelle's blurry recollection, when another of her sisters sat with her all night and showed signs of dire concern that Michelle was about to check out. One curious thing about that night, Barnes recalled, was that she'd been placed in a pediatrics ward. There was no room anymore in the ICU. "So, for whatever reason, they transferred me to pediatrics. I know because someone came around and gave me a teddy bear." Unlike Astrid Joosten in

Leiden, unlike Kelly Warfield at USAMRIID, Michelle Barnes was never put into an isolation unit. Sometimes her caregivers wore masks, as a precaution, and often they didn't. Gradually her body regained strength and her organs (all except her gall bladder, which had been surgically removed) began to recover. The teddy bear may have helped more than the antibiotics.

After twelve days, she left the hospital, still weak and anemic, still undiagnosed. In March she saw Norman Fujita on a follow-up visit and he had her serum tested again for Marburg. Again, negative. Three more months passed and Michelle, now gray-haired, lacking her old energy, suffering abdominal pain, unable to focus, got an email from a knowing friend—a journalist she and Rick had met during the Uganda trip—who had just seen a news article about which he thought Michelle should know. In the Netherlands, a woman had died of Marburg after a Uganda vacation during which she had visited a cave full of bats.

Barnes spent the next twenty-four hours googling up every article on the case she could find. By a small-world coincidence, she had lived in the Netherlands for three years herself, during the 1990s, so she could read the coverage in Dutch as well as in English. Early the following Monday morning, she was at Dr. Fujita's door. "I'm an emergency, I need to speak with you," she said. Fujita welcomed her in and listened to the new information. Beyond his polite demeanor, she felt, he must be rolling his eyes and thinking, *Great, another person who diagnoses herself from the Internet.* But he agreed to test her a third time for Marburg. That sample went to the CDC, as had the earlier ones, and again tested negative; but this time a lab technician, aware that the patient had visited a cave inhabited by Marburg-infected bats, cross-checked the third sample, and then the first sample also, using a more sensitive and specific assay. *Boing.*

The new results went to Fujita, who called Barnes with some left-handed congratulations: "You're now an honorary infectious disease doctor. You've self-diagnosed, and the Marburg test came back positive."

80

News of the Joosten case also reverberated at the CDC. Soon afterward, in August 2008, another team was dispatched to Uganda, this time including the veterinary microbiologist Tom Ksiazek, a veteran of field responses against zoonotic outbreaks, as well as Towner and Amman. Bob Swanepoel and Alan Kemp were again mustered from South Africa. "We got the call, 'Go investigate,'" Amman told me. Their mission now was to sample bats at Python Cave, where this Dutch woman (unnamed in the epidemiological traffic) had become infected. Her death, her case history, implied a change in the potential scope of the situation. That local Ugandans were dying of Marburg was a severe and sufficient concern—sufficient to bring a response team in haste from Atlanta and Johannesburg. But if tourists too were involved, tripping in and out of some lovely python-infested Marburg repository, in Tevas and hiking boots, blithe, unprotected, and then boarding their return flights to other continents, the place was not just a peril for Ugandan miners and their families. It was also an international threat.

The team converged at Entebbe and drove southwest. They walked the same trail that Joosten and Barnes and their husbands had walked, to the same opening amid the forest vegetation. Then, unlike the others, they donned their Tyvek pajamas, their rubber boots, their respirators, and their goggles. This time, with cobras in mind, they added snake chaps. Then they went in. Bats were everywhere overhead; guano was everywhere underfoot. In fact, the rain of guano seemed to come so continuously, Amman told me, that if you left something on the floor it would be covered within days. The pythons were indolent and shy, as well-fed snakes tend to be. One of them, by Amman's estimate, stretched about twenty feet long. The black forest cobras (yes, more of them here too) kept to the deeper recesses, away from heavy traffic. Towner was gazing at a python when Amman noticed something glittery on the floor.

At first glance it looked like a bleached vertebra, lying in the excremental glop. Amman picked the thing up.

It wasn't a vertebra. It was a string of aluminum beads with a number attached. More specifically, it was one of the beaded collars that he and Towner had placed on captured bats at Kitaka Cave, the *other* Marburg cave, three months earlier and thirty miles away. The code tag spoke one simple fact: Here was collar K-31, from the thirty-first animal they had released. "And of course, I just lost my mind," Amman told me. "I was, 'Yeah!' and jumping around. Jon and I were so excited." Amman's insane jubilance was in fact just the sane, giddy thrill that a scientist feels when two small bits of hard-won data click together and yield an epiphany. Towner got it and shared it. Picture two guys in a dark stone room, wearing headlamps, high-fiving in nitrile gloves.

Retrieving the collar at Python Cave vindicated, in a stroke, their mark-recapture study. "It confirmed my suspicions that these bats are moving," Amman said—and moving not only through the forest but from one roosting site to another. Travel of individual bats (such as K-31) between far-flung roosts (such as Kitaka and Python) implied circumstances whereby Marburg virus might ultimately be transmitted all across Africa, from one bat encampment to another. It suggested opportunities for infecting or reinfecting bat populations in sequence, like a string of blinking Christmas lights. It voided the comforting assumption that this virus is strictly localized. And it highlighted the complementary question: Why don't outbreaks of Marburg virus disease happen more often?

Marburg is only one instance to which that question applies. Why not more Hendra? Why not more Nipah? Why not more Ebola? Why not more SARS? If bats are so abundant and diverse and mobile, and zoonotic viruses so common within them, why don't those viruses spill into humans and take hold more frequently? Is there some mystical umbrella that protects us? Or is it fool's luck?

81

The ecological dynamics of the virus itself may be part of the reason that such diseases aren't constantly raining down. Yes, viruses *do* have ecological dynamics, just as do creatures that are more unambiguously alive. What I mean is: They're interconnected with other organisms at the scale of landscapes, not just at the scale of individual hosts and cells. A virus has geographical distribution. A virus can go extinct. The abundance, survival, and range of a virus all depend upon other organisms and what those do. That's viral ecology. In the case of Hendra, to take another instance, the changing ecology of the virus may partly account for its emergence as a cause of human disease.

This line of thought has been explored by an Australian scientist named Raina Plowright. Trained first as a veterinarian, Plowright worked on domestic animals and wildlife in New South Wales and overseas—Britain, Africa, Antarctica—before fetching up at the University of California, Davis, to do a master's degree in epidemiology and then a doctorate in the ecology of infectious diseases. She is one of this new breed of cross-trained disease specialists I've mentioned, veterinarian-ecologists who recognize the intimate connectedness of human health, wildlife health, livestock health, and the habitats we all share. For her doctoral fieldwork, Plowright returned to Australia to investigate the dynamics of Hendra virus within one of its reservoir hosts: the little red flying fox. She did some of her trapping and sampling in the Northern Territory, south of Darwin, amid the eucalyptus and melaleuca forests in and around Litchfield National Park. That's where I spoke with her, during one idle morning in 2006, as Cyclone Larry swept across northern Australia, drenching the land and raising the rivers and creeks. We had some time to kill before she went out once again and tried to catch bats amid the monsoonal flooding.

An interesting thing about Hendra, Plowright told me, is that it's one of four new viruses that emerged around the same time from this single group of bats, the pteropids. Soon after Hendra virus made its debut north of Brisbane, in 1994, there was Australian bat lyssavirus, appearing at two other sites along the Queensland coast, in 1996; then Menangle virus, emerging near Sydney, in 1997; and then Nipah virus, up in Malaysia, in September 1998. "For four viruses to emerge from one host genus within a short period of time is unprecedented," she said. "So we feel there's been some change in the ecology of *Pteropus* species that could precipitate disease emergence." Hume Field had helped identify such contributing factors in the case of Nipah virus among the pig farms of Malaysia. Now, eight years later, with Field on her committee of dissertation advisers, Plowright was looking for similar factors in the matter of Hendra. Changes in habitat, she knew, had affected population size, distributional patterns, and migratory behavior of Hendra reservoir hosts—not just the little red flying fox but also its congenerics, the black flying fox, the grey-headed, and the spectacled. Her task was to investigate how those changes had affected in turn the distribution, prevalence, and spillover likelihood of the virus.

Plowright's project, like much work in ecology these days, entailed a combination of data-gathering from the field and mathematical modeling by computer. The basic conceptual framework, she explained, "was developed by two guys in the 1920s, Kermack and McKendrick." She meant the *SIR* model (susceptible-infected-recovered), which I described earlier. Having alluded to the intellectual heritage, she began talking about susceptible individuals, infected individuals, and recovered individuals in a given bat population. If the population is isolated and insufficiently large, then the virus will move through it, infecting the susceptibles and leaving them recovered (and immune to reinfection), until there are virtually no susceptibles left. Then it will die out, just as measles dies out in an isolated human village. Eventually the virus will return, brought back to that population by a wayward, infected bat. This represents the same blinking-Christmas-light pattern that I invoked with regard to Marburg. The ecologists call it a *metapopulation*: a population of populations. The virus avoids extinction by infecting one relatively isolated population of bats after another. It dies out here, it arrives and infects there; it may not be permanently present in any one population but it's always somewhere. The lights blink off/on in their turns, never all lit, never all dark. If the bat populations are separated by distances great enough that those distances are seldom crossed, then the rate of reinfection is slow. The lights blink off and on languidly.

Now imagine one such bat population within the metapopulation. It has progressed through the *SIR* sequence, every individual infected, every one recovered, and the virus is gone. But not gone forever. As years pass, as the birth of new bats and the death of old ones raise back the proportion of susceptibles, the population regains its collective vulnerability to the virus. Greater isolation means greater elapsed time before the virus returns; greater elapsed time yields more newborn susceptibles; more susceptibles mean a richer potential for explosive infection. "So when you do introduce the virus again," Plowright said, describing the godlike role of the modeler, "you get a much bigger outbreak." This is where the Christmas-light metaphor fails to serve, because one light suddenly glows like a supernova among ordinary stars.

Plowright of course was working with numbers, not analogies. But her numbers reflected roughly this scenario. The relevance of such modeling to the facts on the ground is that Australian populations of flying foxes *have* become more isolated in recent decades. "The east coast of Australia used to be one big contiguous forest," she told me, "and so you had bat populations pretty evenly dispersed along the coastline." Their roosting aggregations, in the old days, were relatively mobile. Their food resources—mainly nectar and fruit—were diverse, seasonally variable, and scattered patchily throughout the forest. Each group of bats, comprising maybe a few hundred or a few thousand individuals, would fly out to a feeding site at night, return at daylight, and also migrate seasonally to put themselves closer to concentrations of food. With all the coming and going, individual bats would sometimes transfer from one group to another, carrying Hendra virus with them if they happened to be infected. There was a continual mixing and a continual reinfection of the smallish groups. This seems to have been the situation—for the little red flying fox, for the other flying foxes, and for Hendra virus—from time immemorial. Then things changed.

Habitat alteration was an ancient tradition in Australia, in the form of burning by Aboriginal people, but in recent decades land clearance has become a more drastic and mechanized trend, with less-reversible results, especially in Queensland. Vast areas of old forest have been cut, or chained down with bulldozers, to make way for cattle ranching and urban sprawl. People have planted orchards, established urban parks, landscaped their yards with blossoming trees, and created other unintended enticements amid the cities and suburbs. "So bats have decided that, as their native habitat is disappearing, as climate is becoming more variable, and their food source is becoming less diverse, it's easier to live in an urban area." They gather now in larger aggregations, traveling shorter distances to feed, living at closer proximity to humans (and to the horses that humans keep). Flying foxes in Sydney, flying foxes in Melbourne, flying foxes in Cairns. Flying foxes in the Moreton Bay fig trees shading a paddock on the north side of Brisbane.

I saw where Plowright was going and tried to frame the last bit in my own mind. So those large aggregations—comprising bats that are more sedentary, more urban, less needful of flying long distances in search of wild food—tend to reinfect one another less frequently? And in the interim they accumulate more susceptible individuals? So when the virus does arrive, the spread of new infections is more sudden and intense? The virus is more prevalent and abundant?

"Exactly. That's it," she said.

"And then a great likelihood of spillover into another species?" I wanted to leap toward that easy epiphany but Plowright, with many bats yet to catch, many data to assemble, many model parameters to explore, held me back. Five years after our conversation, with the PhD finished, now a respected voice on Hendra herself, she would present her work and ideas in an august journal, the *Proceedings of the Royal Society*. But for the moment, amid the rains and high waters of the Northern Territory, she spoke provisionally.

"This is a theory," she said.

82

Theories require testing, as Raina Plowright well knew. Science proceeds by observation and supposition and testing. Another such supposition pertains to ebolaviruses. If you've been paying close attention, you'll have noticed that just a few pages back I lumped Ebola virus, along with Hendra and Nipah and others, among viruses for which bats serve as reservoirs. So to clarify: That inclusion is tentative. It's a hypothesis awaiting assessment against further evidence. No one, as of this writing, has isolated any live ebolavirus from a bat—and virus isolation is still the gold standard for identifying a reservoir. That may happen soon; people are trying. Meanwhile the Ebola-in-bats hypothesis seems stronger since Jonathan Towner's team achieved their isolations of Marburg virus, so closely related, also from bats. And it has been strengthened further, at least a little, by another bit of data added to the ebolavirus dossier about the same time. This bit came in the form of a story about a little girl.

Eric Leroy, the Paris-trained virologist based in Franceville, Gabon, who had been chasing Ebola for more than a decade, led the team that reconstructed the girl's story. Their new evidence derived not from molecular virology but from old-fashioned epidemiological detective work—interviewing survivors, tracing contacts, discerning patterns. The context was an outbreak of Ebola virus that occurred in and around a village called Luebo, along the Lulua River, in a southern province of the Democratic Republic of the Congo. Between late May and November 2007, more than 260 people sickened with what seemed to be or (in some confirmed cases) definitely was Ebola virus. Most of them died. The lethality was 70 percent. Leroy and his colleagues arrived in October, as part of an international WHO response team in cooperation with the DRC's Ministry of Health. Leroy's study focused on the network of transmissions, which all seemed traceable to a certain fifty-five-year-old woman. She became known, in their report, as patient A. She wasn't necessarily the first human to get infected;

she was merely the first identified. This woman, elderly by Congo village standards, died after suffering high fever, vomiting, diarrhea, and hemorrhages. Eleven of her close contacts, mainly family, who helped care for her, sickened and died too. The outbreak spread onward from there.

Leroy and his group wondered how the woman herself had gotten infected. No one in her village showed symptoms before she did. So the investigators broadened their search to surrounding villages, of which there were quite a few, both along the river and in the forest nearby. From their interviews and their legwork, they learned that the villages were interconnected by footpaths, and that on Mondays the heavy traffic led to one particular village, Mombo Mounene 2, the site of a big weekly market. They also learned about an annual aggregation of migrating bats.

The bats generally arrived in April and May, stopping over amid a longer journey, finding roost sites and wild fruit trees on two islands in the river. In an average year, there might be thousands or tens of thousands of animals, according to what Leroy's group heard. In 2007, the migration was especially large. From their island roosts, the bats ranged the area. Sometimes they fed at a palm oil plantation along the river's north bank; the plantation was a leftover from colonial times, now abandoned and gone derelict, but still offering palm fruits in April on its remaining trees. Many or most of the animals were hammer-headed fruit bats (*Hypsignathus monstrosus*) and Franquet's epauletted fruit bats (*Epomops franqueti*), two of the three in which Leroy had earlier found Ebola antibodies. While roosting, the bats dangled thickly on tree branches. Local people, hungry for protein or a little extra cash, hunted them with guns. Hammer-headed bats, big and meaty, were especially prized. A single shotgun blast could bring down several dozen bats. Many of those animals ended up, freshly killed, raw and bloody, in the weekly market at Mombo Mounene 2, from which buyers carried them home for dinner.

One man who regularly walked from his own village to the market, and often bought bats, seems to have suffered a mild case of Ebola. The investigators eventually labeled him patient C. He wasn't a bat hunter himself; he was a retail consumer. During late May or early June, according to patient C's own recollection, he weathered some minor symptoms, mainly fever and headache. He recovered, but that wasn't the end of it. "Patient C was the father of a 4-year-old girl (patient B)," Leroy and his team later reported, "who suddenly fell ill on 12 June and died on 16 June 2007, having had vomiting, diarrhoea, and high fever." The little girl didn't hemorrhage, and she was never tested for Ebola, but it's the most plausible diagnosis.

How had she contracted it? Possibly she had shared in eating a fruit bat that carried the virus. What are the odds faced by bat-eaters? Hard to say; hard even to guess. If the hammer-headed bat *is* an Ebola reservoir, what's the prevalence of the virus within a given population? That's another unknown. Towner found 5 percent prevalence of Marburg in Egyptian fruit bats, meaning that one animal in twenty could be infected. Assuming a roughly similar prevalence in the hammer-headed bat, the little girl's family had been unlucky as well as hungry. They might have eaten nineteen other bats and gotten no exposure. Then again, if a bat meal was shared, why didn't the girl's mother and other family members get sick? Possibly her father, infected or besmeared after purchasing bats in the market, had carried the girl (common practice with small children thereabouts) along the footpath back to their village. The father, patient C, seems to have passed the virus to nobody else.

But his little daughter did pass it along. Her dead body was washed for burial, in accord with local traditions, by a close friend of the family. That friend was the fifty-five-year-old woman who became patient A.

"Thus, virus transmission may have occurred when patient A prepared the corpse for burial ceremony," Leroy's group wrote. "When interviewed, the two other preparers, the girl's mother and grandmother, reported they did not have direct contact with the corpse and they did not develop any clinical sign of infection in the four following weeks." Their role in the funerary washing was apparently observational. They didn't touch the dead body of their daughter and granddaughter. But patient A did, performing faithfully the service of a close family friend, after which she went back to her life—what was left of it. She resumed her social interactions, and 183 other people caught Ebola and died.

Leroy's team reconstructed this story and then, keen to extract meaning, asked themselves several questions. Why had the father infected his daughter but no one else? Maybe because he had a mild case, with a low level of virus in his body and not much leaking out. But if his case was mild, why was his daughter's so severe, killing her within four days? Maybe because, as a small child racked with vomiting and diarrhea, she had died of untreated dehydration. Why was there only one bat-to-human spillover event? Why was patient C unique, as the sole case linked directly to the reservoir? Well, maybe he wasn't. He was just the only one that came to notice. "In fact, it is highly likely that several other persons were infected by bats," Leroy's group wrote, "but the circumstances required for subsequent human-to-human transmission were not present." They were alluding to dead-end infections. A person sickens, suffers solitarily or with carefully distanced succor from wary family or friends (food and water left at the door of a hut), and dies. Is buried unceremoniously. Eric Leroy didn't know how many unfortunate people in the Luebo area may have eaten a bat, touched a bat, become infected with Ebola, succumbed to it, and been dropped into a hole, having infected no one else. Amid the horrific confusion of the outbreak, in those remote villages, the number of such dead-end cases might have been sizable.

This brought Leroy's team to the pivotal question. If the circumstances required for human-to-human transmission hadn't been met, what *were* those circumstances? Why hadn't the Luebo outbreak gone really big? Why hadn't the tinder ignited the logs? It had started in May, after all, and WHO didn't get there until October.

83

Human-to-human transmission is the crux. That capacity is what separates a bizarre, awful, localized, intermittent, and mysterious disease (such as Ebola) from a global pandemic. Remember the simple equation offered by Roy Anderson and Robert May for the dynamics of an unfolding epidemic?

$$R_0 = \beta N/(\alpha + b + v)$$

In that formulation, β represents the transmission rate. β is the letter beta, in case you're not a mathematician or a Greek. Here it's a multiplier in the single expression that stands as numerator of the fraction, a strong position. What that means is, when β changes muchly, R_0 changes muchly. And R_0, your good memory tells you, is the measure of whether an outbreak will take off.

In some zoonotic pathogens, efficient transmissibility among humans seems to be inherent from the start, a sort of accidental preadaptedness for spreading through the human population, despite a long history of residence within some other host. SARS-CoV had it, from the earliest days of its 2002–2003 emergence in Guangdong and Hong Kong. SARS-CoV *has* it, no matter where or why SARS-CoV may be hiding since then. Hendra virus does not have it. Hendra achieves fluent transmission among horses but not among humans. Of course, a pathogen may also *acquire* that capacity by mutation and adaptation within human hosts. Have you noticed the persistent, low-level buzz about avian influenza, the strain known as H5N1, among disease experts over the past fifteen years? That's because avian flu worries them deeply, though it hasn't caused many human fatalities. Swine flu comes and goes periodically in the human population (as it came and went during 2009), sometimes causing a bad pandemic and sometimes (as in 2009) not so bad as expected; but avian flu resides in a different category of menacing possibility. It worries the flu scientists because they know that H5N1 influenza is (1) extremely virulent in people, with a high lethality though a relatively low number of cases, and yet (2) poorly transmissible, so far, from human to human. It'll kill you if you catch it, very likely, but you're unlikely to catch it except by butchering an infected chicken. Most of us don't butcher our own chickens, and health officials all over the world have been working hard to assure that the chickens we handle—dead, disarticulated, wrapped in plastic or otherwise—have not been infected. But if H5N1 mutates or reassembles itself in just the right way, if it adapts for human-to-human transmission, then H5N1 could become the biggest and fastest killer disease since 1918.

How does a pathogen acquire such an adaptation? The process of genetic variation (by mutation or other means) is random. A game of craps. But an abundance of opportunity helps to increase a virus's likelihood of rolling its point—that is, chancing into a highly adaptive change. The more rolls before sevening out, the more opportunities to win. And there's Jon Epstein's word again: opportunity.

Back in Dhaka after my nights of bat catching with Epstein, I returned to the ICDDR,B for some further conversations, because I wanted to learn more about the capacity for human-to-human transmission in Nipah. I spoke with a handful of people from Steve Luby's program on infectious diseases. One was an American epidemiologist named Emily Gurley, who had spent several years of her youth as a diplomat's kid in Bangladesh and then returned as an adult to work in public health. Gurley is in her middle thirties, with curly brown hair, pale freckles, and blue eyes that widen when she discusses important details of disease sleuthing. She had helped investigate the outbreak in Faridpur District in 2004, the one with thirty-six identified case patients, of whom twenty-seven died. The most notable aspect of the Faridpur episode was that many of those people had evidently been infected by contact with a single person, a superspreader, who sat like a spider at the center of a web of transmissions.

This man was a religious leader, the venerated head of an unorthodox Islamic sect, an informal group that seems to have been nameless, with a small number of fervent followers in a village called Guholaxmipur and roundabouts. Unlike orthodox Muslims, the sect members declined to pray five times a day or to fast during Ramadan, and they sometimes sat up all night, men and women together, praying, smoking cigarettes (or stronger weed), and singing. Their ecstatic practices offended the conventionally pious believers around them, and so when the leader died of a brief, mysterious illness, and then his family and followers started dying too, neighbors attributed the deaths to *asmani bala*: a curse from above.

Okay, that was one possible explanation. Epidemiology would offer another.

The religious leader had already died and been buried, his grave made a shrine, the outbreak underway, by the time Gurley's group arrived. She and some colleagues drove out from Dhaka, in early April, in response to an urgent if belated call from the Faridpur civil surgeon, who alerted them that people were dying and the cause seemed to be Nipah. (The surgeon would have been at least roughly aware of what Nipah looked like from the outbreak in that neighboring district, Rajbari, just four months earlier.) As their car reached Guholaxmipur, Gurley told me, "it was very dramatic. We were met by a funeral procession coming out of the village, body wrapped in a white shroud. Which didn't bode well." People began carrying comatose relatives out of their homes, imploring the visitors for help. "There were a lot of people sick in that village." The doctors arranged for seventeen cases to be transferred to a district hospital in Faridpur city, where they were placed together in a separate small building apart from the main one—a makeshift isolation ward. This "ward" was a single large room. Gurley and her colleagues began taking specimens and histories. Some of the people showed severe respiratory symptoms. "There was one man," Gurley recalled, "who was sitting up talking to us, coughing, coughing, coughing—but gave us his whole illness history, and was dead the next morning."

"Were you wearing masks?"

"We were." They had N95 masks, simple and relatively cheap but effective against small particles, standard equipment in this sort of situation. If they had known what to expect in Faridpur, they might have wanted something better, but Gurley's chief regret was simply that they hadn't brought more N95s, enough for the local health-care staff as well as themselves. And then, because it was storm season, a heavy squall blew through town and knocked out the electricity. The lights went off, and the staff closed all the windows—"which is not what you want," Gurley said, laughing grimly. By morning, not just the coughing man but also two other patients in that crowded, stuffy room had died.

Gurley gathered the interview data and, as she started charting an epidemiological curve, realized that "everyone who was in that hospital ward had had very close contact with another person"—one in particular—"who died from this a couple weeks before." She meant the religious leader. This pattern was quite different from earlier Nipah outbreaks, in which most patients seemed to have gotten infected directly from some environmental source (sick livestock? treetops? the palm-sap hypothesis hadn't yet arisen), not from human contagion, and in which the symptoms had been mainly neurological, not respiratory. Gurley's group even doubted, for a while, that Nipah was the cause at Faridpur. But then samples shipped back to Atlanta tested positive for Nipah, at which point the CDC sent a small team of specialists to work alongside Gurley and her colleagues.

The investigation at Faridpur eventually yielded a new understanding of Nipah—as a disease in which person-to-person transmission could be far more important than supposed. Of the thirty-six cases, twenty-two were linked to the religious leader. Those people had gathered closely around him during his final illness. Presumably they had been infected by aerosolized virus, or touch, or spittle, or some other sort of direct transfer. Most cases among the other fourteen also seemed to reflect person-to-person transmission. A rickshaw driver in a nearby village, who worked seasonally as a date-palm-sap collector, fell ill and was nursed by his mother, his son, his aunt, and a neighbor; they all got sick too. The rickshaw driver's aunt received care from an in-law, a man from Guholaxmipur, who visited her in the hospital; that in-law was the religious leader. One of the sect followers, infected, his condition worsening, was helped to a hospital by another rickshaw driver; that driver fell ill, about ten days later, and died . . . and so on.

Nipah was passing horizontally through the community, like a rumor, not just down from the sky, like a divine curse or a dollop of bat poop. And its seeming ubiquity was confirmed by one other finding of the combined response team. This bit of data was especially spooky. The investigators took swabs from the wall of a hospital room in which one of the patients had been treated, five weeks earlier, and from the soiled frame of a bed in which that patient had lain. None of those surfaces had been cleaned in the meantime; bleach and labor were in short supply. Some swabs, both from the wall and the bed frame, tested positive for Nipah RNA. I'll repeat that: Fragments (at least) of Nipah virus, left from what the patient had spewed out, were still present after *five weeks*, invisibly decorating the room. To the sanitarian, such spewing represents contamination. To the virus: opportunity.

I spoke also with Rasheda Khan, a medical anthropologist just down the hall from Emily Gurley. Khan is a Bangladeshi with dark eyes and a severe, professional manner. Her job was to investigate the cultural and social factors that affect a disease event like the Faridpur outbreak. She had been there in Faridpur, interviewing villagers in their native tongue, Bangla, to collect testimony about behaviors and attitudes as well as to learn who got sick when. She talked about asmani bala ("a curse inflicted by Allah," was her translation, slightly more blunt than others I'd heard) and how that fateful idea may have dissuaded some victims from seeking hospital care. She helped me understand the sort of little interpersonal intimacies, characteristic of her country, that could be relevant to disease transmission. "In Bangladesh," she said, "physical contact is very common. We hug, we hold hands all the time." Even along the road, she said, you see men walking together, holding hands. Such physicality only increased, from a sense of concern, if a person were sick—and more still if the sick person were a venerated figure, like the sect leader in Guholaxmipur. This man was beloved by his followers and seen as close to God. People came as he lay on his deathbed to be favored with a last touch, or to whisper blessings in his ear, or to sponge his body, or to offer him a sip of water or milk or juice. "That is one of the customs here," Khan explained, "that you give water to the dying person's mouth." Many people came to his bedside, bent close, offered him water, she said, "and he was coughing all the time. And the fog was everywhere on people's—"

I think she was going to say "faces" but like a fool I interrupted her.

"The fog?"

"Yeah, the saliva," Khan said. "His coughing. So the spit was . . . people told us that he was coughing, and his coughing, the spit, on body, hands . . ." Eliding these thoughts, she left me to fill in the blanks, then mentioned that hand washing, unlike hand holding, is not common practice in Bangladesh. Unlucky followers and family members may have come away from their final audiences lightly glazed with the holy man's spittle—and then rubbed their eyes, taken food with those hands, or otherwise accepted the virus. You don't need date-palm sap if you've got that.

84

Over the course of three days I made several trips to the ICDDR,B, which occupies a complex of buildings behind a high wall in the Mohakhali neighborhood of Dhaka. In addition to the talks with Khan and Gurley, I spoke with some high administrators and some bright young researchers, who gave me a wide range of perspectives and insights on Nipah virus. But the most affecting moment occurred when my taxi through the crazed Dhaka traffic pulled up to the wrong gate of the compound, leaving me just disoriented enough to walk in the wrong door. This wasn't the sleek building that housed Steve Luby's infectious diseases program. This was the old Cholera Hospital itself.

A solicitous Bangladeshi man, who noticed me looking lost,

asked my destination and pointed me along, suggesting I simply cross through the hospital. A guard opened the next door and saluted me. No one asked for a badge. I found myself intruding through an open ward lined with dozens of beds. A few of those beds were empty, sheetless, showing a mattress of red or green vinyl with a bedpan hole in the middle: cold, practical, ready for the next case. Many other beds were filled with the thin, bony bodies of suffering patients, sorrowful brown-skinned people, alone or consoled quietly by relatives. Here came I, a white man with a briefcase, into this hangar of souls eagerly awaiting attention from a doctor. One woman caught my eyes, then whispered to her child, held beside her on the bed, and pointed at me. Out on the street such a gesture would suggest idle curiosity or maybe a prelude to begging, but here it surely indicated hope—deep hope, hope of deliverance, but misplaced. I averted my eyes and walked on, acutely aware that I had no skills, no knowledge, no training, no medicines that could be helpful to this woman and her child, more's the worse for me. Through further corridors, other doors, more saluting guards, I found my way to the next interview.

The Cholera Hospital was founded in 1962, as a clinical adjunct to an earlier Cholera Research Laboratory, both of which were eventually bundled into the ICDDR,B. The hospital provides free treatment to more than a hundred thousand patients each year, not only for cholera but also for blood dysentery and other diarrheal diseases. Most of its patients are children under the age of six. Eighty percent of those children arrive at the hospital malnourished. I can't tell you how many survive. I can't even tell you how many cholera cases occur annually when the flood season in Bangladesh brings infected waters up into villages and slums, because most cases go unreported and there is no systematic national tally. One authoritative guess: a million. What I can tell you is that Bangladesh, wondrous in so many ways, engaging and fascinating as well as horrifying to an affluent visitor, is an especially difficult country in which to be a poor citizen, either urban or rural, because if you're poor it's a difficult country in which to remain healthy. Thousands of people, young and old, die of cholera and other diarrheal diseases and pneumonia and tuberculosis and measles. Note that none of those afflictions is newly emergent, mysterious, or zoonotic. Together they dwarf the impact—at least so far—of Nipah virus encephalitis.

Why are zoonotic diseases important? I've been asked that question, and have asked it of others, more than a few times during my six years of chasing the subject. (One fellow, a respected historian I met at a conference, suggested that I forget about Ebola and write a book on asthma, which afflicts 22 million Americans. He happened to be asthmatic.) Given the global scorecard of morbidity and mortality caused by old-fashioned and nonzoonotic infectious diseases—such as cholera, typhoid, TB, rotavirus diarrhea, malaria (excepting *Plasmodium knowlesi*), not to mention chronic illnesses such as cancer and heart disease—why divert attention to these boutique infections, these anomalies, that spill out of bats or monkeys or who knows where to claim a few dozen or a few hundred people now and then? *Why?* Isn't it misguided to summon concern over a few scientifically intriguing diseases, some of them new but of relatively small impact, while boring old diseases continue to punish humanity? After my detour through the Cholera Hospital, after being pinioned by that mother's expectant stare, I found myself asking the same thing: Why obsess about zoonoses? In the larger balance of miseries, what makes anyone think they should be taken so seriously?

It's a fair question but there are good answers. Some of those answers are intricate and speculative. Some are subjective. Others are objective and blunt. The bluntest is this: AIDS.

VIII

THE CHIMP AND THE RIVER

85

There are many beginnings to what we think we know about the AIDS pandemic, most of which don't even address the subject of its origin in a single zoonotic spillover.

For instance: In autumn of 1980, a young immunologist named Michael Gottlieb, an assistant professor at the UCLA Medical Center, began noticing a strange pattern of infections among certain male patients. The patients, eventually five of them, were all active homosexuals and all suffering from pneumonia caused by a usually harmless fungus then known as *Pneumocystis carinii*. (Nowadays, after a name change, it's *Pneumocystis jirovecii*.) The stuff is ubiquitous; it floats around everywhere. Their immune systems should have been able to clear it. But their immune systems evidently weren't working, and this fungus filled their lungs. Each man also had another sort of fungal infection—oral candidiasis, meaning a mouthful of slimy *Candida* yeast, more often seen in newborn babies, diabetics, and people with compromised immune systems than in healthy adults. Blood tests, done on several of the patients, showed dramatic depletions of certain lymphocytes (white blood cells) that are crucial in regulating immune responses. Specifically, it was thymus-dependent lymphocytes (T cells, for short) that were "profoundly depressed" in number. Although Gottlieb noted some other symptoms, those three stood out: *Pneumocys-*

tis pneumonia, oral candidiasis, dearth of T cells. In mid-May of 1981, he and a colleague wrote a brief paper describing their observations. They didn't speculate about causes. They just saw the pattern as a befuddling, ominous trend and felt they should publish quickly. An editor at *The New England Journal of Medicine* was interested but his lead time would be at least three months.

So Gottlieb turned to the streamlined CDC newsletter, *Morbidity and Mortality Weekly Report.* His barebones text, less than two pages long, appeared in *MMWR* on June 5, 1981, under the dry title "*Pneumocystis* Pneumonia—Los Angeles." It was the first published medical alert about a syndrome that didn't yet have a name.

The second alert came a month later, again in the CDC newsletter. While Gottlieb was noticing *Pneumocystis* pneumonia and candidiasis, a New York dermatologist named Alvin E. Friedman-Kien spotted a parallel trend involving a different disease: Kaposi's sarcoma. A rare form of cancer, not usually too aggressive, Kaposi's sarcoma was known primarily as an affliction of middle-aged Mediterranean males—the sort of fellows you'd expect to find in an Athens café, drinking coffee and playing dominoes. This cancer often showed itself as purplish nodules in the skin. Within less than three years, Friedman-Kien and his network of colleagues had seen twenty-six cases of Kaposi's sarcoma in youngish homosexual men. Some of those patients also had *Pneumocystis* pneumonia. Eight of them died. Hmm. *Morbidity and Mortality Weekly Report* carried Friedman-Kien's communication on July 3, 1981.

Kaposi's sarcoma also figured prominently in a set of clinical observations made in Miami around the same time. The symptoms among this group of patients were similar; the cultural profile was different. These sick people, twenty of them, hospitalized between early 1980 and June 1982, were all Haitian immigrants. Most had arrived in the United States recently. By their own testimony during medical interviews they were all heterosexuals, with no history of homosexual activity. But their cluster of ailments resembled what Gottlieb had seen among gay men in Los Angeles and Friedman-Kien among gay men in New York: *Pneumocystis* pneumonia, candidiasis in the throat, plus other unusual infections, irregularities in lymphocyte counts, and aggressive Kaposi's sarcoma. Ten of the Haitians died. The team of doctors who published these observations saw a "syndrome" that seemed "strikingly similar to the syndrome of immunodeficiency described recently among American homosexuals." The early connection to Haitian heterosexuals would later come to seem like a false lead and be largely ignored in discussions of AIDS. It was hard to confirm, based on interview data, and harder still to construe. Calling attention to it even came to seem politically incorrect. Then, later still, its real significance would emerge from work at the level of molecular genetics.

Another perceived starting point was Gaëtan Dugas, the young Canadian flight attendant who became notorious as "Patient Zero." You've heard of him, probably, if you've heard much of anything about the dawning of AIDS. Dugas has been written about as the man who "carried the virus out of Africa and introduced it into the Western gay community." He wasn't. But he seems to have played an oversized and culpably heedless role as a transmitter during the 1970s and early 1980s. As a flight steward, with almost cost-free privileges of personal travel, he flew often between major cities in North America, joining in sybaritic play where he landed, notching up conquests, living the high life of a sexually voracious gay man at the height of the bathhouse era. He was handsome, sandy-haired, vain but charming, even "gorgeous" in some eyes. According to Randy Shilts, author of *And the Band Played On* (which includes much heroic research and a fair bit of presumptuous reimagining), Dugas himself reckoned that in the decade since becoming actively gay he had had at least twenty-five hundred sexual partners. Dugas paid a price for his appetite and his daring. He developed Kaposi's sarcoma, underwent chemotherapy for that, suffered from *Pneumocystis* pneumonia and other AIDS-related infections, and died of kidney failure at age thirty-one. During the brief stretch of years between his Kaposi's diagnosis and his final invalidism, Gaëtan Dugas didn't slow down. But he seems to have tipped, in his lonely despair, from hedonism to malice; he would have sex with a new acquaintance at the Eighth-and-Howard bathhouse in San Francisco, then turn up the lights—so Randy Shilts claimed—display his lesions, and say: "I've got gay cancer. I'm going to die and so are you."

In the same month as Dugas's death, March 1984, a team of epidemiologists from the CDC published a landmark study of the role of sexual contact in linking cases of what by then was called AIDS. The world had a label now but not an explanation. "Although the cause of AIDS is unknown," wrote the CDC team, whose lead author was David M. Auerbach, "it may be caused by an infectious agent that is transmissible from person to person in a manner analogous to hepatitis B infection." Hepatitis B is a blood-borne virus. It moves primarily by sexual contact, intravenous drug use with shared needles, or transfusion of blood products carrying the virus as a contaminant. It seemed like a template for understanding what otherwise was still a bewildering convergence of symptoms. "The existence of a cluster of AIDS cases linked by homosexual contact is consistent with an infectious-agent hypothesis," the CDC group added. Not a toxic chemical, not an accident of genetics, but some kind of bug, is what they meant.

Auerbach and his colleagues gathered information from nineteen AIDS cases in southern California, interviewing each patient or, if he was dead, his close companions. They spoke with another twenty-one patients in New York and other American cities, and from their forty case histories they created a graphic figure of forty interconnected disks, like a Tinkertoy structure, showing who had been linked sexually with whom. The patients' identities were coded by location and number, such as "SF 1," "LA 6," and "NY 19." At the center of the network, connected directly to eight disks and indirectly to all the rest, was a disk labeled "0." Although the researchers didn't name him, that patient was Gaëtan Dugas. Randy Shilts later transformed the somewhat bland "Patient 0," as mentioned in this paper, to the more resonant "Patient Zero" of his book. But what the word "Zero" belies, what the number "0" ignores, and what the central position of that one disk within the figure fails to acknowledge, is that Gaëtan Dugas didn't conceive the AIDS virus himself. Everything comes from somewhere, and he got it from someone else. Dugas himself was infected by some other human, presumably during a sexual encounter—and not in Africa, not in Haiti, somewhere closer to home. That was possible because, as evidence now shows, HIV-1 had already arrived in North America when Gaëtan Dugas was a virginal adolescent.

It had also arrived in Europe, though on that continent it hadn't yet gone far. A Danish doctor named Grethe Rask, who had been working in Africa, departed in 1977 from what was then Zaire and returned to Copenhagen for treatment of a condition that had been dragging her downward for several years. During her time in Zaire, Rask had first run a small hospital in a remote town in the north and then served as chief surgeon at a large Red Cross facility in the capital, Kinshasa. Somewhere along the way, possibly during a surgical procedure done without adequate protective supplies (such as latex gloves), she became infected with something for which no one at the time had a description or a name. She felt ill and fatigued. Drained by persistent diarrhea, she lost weight. Her lymph nodes swelled and stayed swollen. She told a friend: "I'd better go home to die." Back in Denmark, tests revealed a shortage of T cells. Her breath came with such difficulty that she depended on bottled oxygen. She struggled against staph infections. *Candida* fungus glazed her mouth. By the time Grethe Rask died, on December 12, 1977, her lungs were clogged with *Pneumocystis jirovecii*, and that seems to have been what killed her.

It shouldn't have, according to standard medical wisdom. *Pneumocystis* pneumonia wasn't normally a fatal condition. There had to be a broader explanation, and there was. Nine years later, a sample of Rask's blood serum tested positive for HIV-1.

All these unfortunate people—Grethe Rask, Gaëtan Dugas, the five men in Gottlieb's report from Los Angeles, the Kaposi's sarcoma patients known to Friedman-Kien, the Haitians in Miami, the cluster of thirty-nine (besides Dugas) identified in David Auerbach's study—were among the earliest recognized cases of what has retrospectively been identified as AIDS. But they weren't among the first victims. Not even close. Instead they represent midpoints in the course of the pandemic, marking the stage at which a slowly building, almost unnoticeable phenomenon suddenly rose to a crescendo. Again in the dry terms of the disease mathematicians, whose work is vitally applicable to the story of AIDS: R_0 for the virus in question had exceeded 1.0, by some margin, and the plague was on. But the real beginning of AIDS lay elsewhere, and more decades passed while a few scientists worked to discover it.

86

In the early years after its detection, the new illness was a shifting shape that carried several different names and acronyms. GRID was one, standing for Gay-Related Immune Deficiency. That proved too restricted as heterosexual patients began to turn up: needle-sharing addicts, hemophiliacs, other unlucky straights. Some doctors preferred ACIDS, for Acquired Community Immune Deficiency Syndrome. The word "community" was meant to signal that people acquired it *out there*, not in hospitals. A more precise if clumsier formulation, favored briefly by the CDC's *Morbidity and Mortality Weekly Report*, was "Kaposi's sarcoma and opportunistic infections in previously healthy persons," which didn't abbreviate neatly. KSOIPHP lacked punch. By September 1982, *MMWR* had switched its terminology to Acquired Immune Deficiency Syndrome (AIDS), and the rest of the world followed.

Naming the syndrome was the least of the early challenges. More urgent was to identify its cause. I just alluded to "the virus in question," but remember: No one knew, back when those reports from Gottlieb and Friedman-Kien began capturing attention, what sort of pathogen caused this combination of puzzling, lethal symptoms—nor even if there *was* a single pathogen. The virus idea arose as a plausible guess.

One scientist who made the guess was Luc Montagnier, then a little-known molecular biologist at the Institut Pasteur in Paris. Montagnier's research had focused mainly on cancer-causing viruses, especially the group known as retroviruses, some of which cause tumors in birds and mammals. Retroviruses are fiendish beasts, even more devious and persistent than the average virus. They take their name from the capacity to move backward (retro) against the usual expectations of how a creature translates its genes into working proteins. Instead of using RNA as a template for translating DNA into proteins, the retrovirus converts its RNA into DNA within a host cell; its viral DNA then penetrates the cell nucleus and gets itself integrated into the genome of the host cell, thereby guaranteeing replication of the virus whenever the host cell reproduces itself. Luc Montagnier had studied these things in animals—chickens, mice, primates—and wondered about the possibility of finding them in human tumors too. Another disquieting possibility about retroviruses was that the new disease showing up in America and Europe, AIDS, might be caused by one.

There was still no solid proof that AIDS was caused by a virus at all. But three kinds of evidence pointed that way, and Montagnier recalls them in his memoir, a book titled *Virus*. First, the incidence of AIDS among homosexuals linked by sexual interactions suggested that it was an infectious disease. Second, the incidence among intravenous drug users suggested a blood-borne infectious agent. Third, the cases among hemophiliacs implied a blood-borne agent that escaped detection in processed blood products such as clotting factor. So: It was infinitesimal, contagious, blood-borne. "AIDS could not be caused by a conventional bacterium, a fungus, or protozoan," Montagnier wrote, "since these kinds of germs are blocked by the filters through which the blood products necessary to the survival of hemophiliacs are passed. That left only a smaller organism: the agent responsible for AIDS thus could only be a virus."

Other evidence hinted that it might be a retrovirus. This was new ground, but then so was AIDS. The only known human retrovirus as of early 1981 was something called human T-cell leukemia virus (HTLV), recently discovered under the leadership of a smart, outgoing, highly regarded, and highly ambitious researcher named Robert Gallo, whose Laboratory of Tumor Cell Biology was part of the National Cancer Institute in Bethesda, Maryland.

HTLV, as its name implies, attacks T cells and can turn them cancerous. T cells are one of the three major types of lymphocyte of the immune system. (Later the acronym HTLV was recast to mean human T-lymphotropic virus, which is slightly more accurate.) A related retrovirus, feline leukemia virus, causes immune deficiency in cats. So a suspicion arose among cancer-virus researchers that the AIDS agent, destroying human immune systems by attacking their lymphocytes (in particular, a subcategory of T cells known as T-helper cells), might likewise be a retrovirus. Montagnier's group began looking for it.

Gallo's lab did too. And those two weren't alone. Other scientists at other laboratories around the world recognized that finding the cause of AIDS was the hottest, the most urgent, and potentially the most rewarding quest in medical research. By late spring of 1983, three teams working independently had each isolated a candidate virus, and in the May 20 issue of *Science*, two of those teams published announcements. Montagnier's group in Paris, screening cells from a thirty-three-year-old homosexual man who'd been suffering from lymphadenopathy (swollen lymph nodes), had found a new retrovirus, which they called LAV (for lymphadenopathy virus). Gallo's group came up with a new virus also, one that Gallo took for a close relative of the human T-cell leukemia viruses (by now there was a second, called HTLV-II, and the first had become HTLV-I) that he and his people had discovered. He called this newest bug HTLV-III, nesting it proprietarily into his menagerie. The French LAV and the Gallo HTLVs had at least one thing in common: They were indeed retroviruses. But within that family exists some rich and important diversity. An editorial in the same issue of *Science* trumpeted the Gallo and Montagnier papers with a misleading headline: HUMAN T-CELL LEUKEMIA VIRUS LINKED TO AIDS, despite the fact that Montagnier's LAV was *not* a human T-cell leukemia virus. Woops, mistaken identity. Montagnier knew better, but his *Science* paper seemed to blur the distinction, and the editorial occluded it entirely.

Then again, neither was Gallo's "HTLV-III" an HTLV, once it had been clearly seen and correctly classified. It turned out to be something nearly identical to Montagnier's LAV, of which Montagnier had given him a frozen sample. Montagnier had personally delivered that sample, carrying it on dry ice during a visit to Bethesda.

Confusion was thus sown early—confusion about what exactly had been discovered, who had discovered it, and when. That confusion, irrigated with competitive zeal, fertilized with accusation and denial, would grow rife for decades. There would be lawsuits. There would be fights over royalties from the patent on an AIDS blood-screening test that derived from virus grown in Gallo's lab but traceable to Montagnier's original isolate. (Contamination from one experiment to another, or from one batch of samples to another, is a familiar problem in lab work with viruses.) It wasn't a petty squabble. It was a big squabble, in which pettiness played no small part. What was ultimately at stake, besides money and ego and national pride, was not just advancing or retarding research toward an AIDS cure or vaccine but also the Nobel Prize in medicine, which eventually went to Luc Montagnier and his chief collaborator, Françoise Barré-Sinoussi.

Meanwhile the third team of researchers, led quietly by a man named Jay A. Levy in his lab at the University of California School of Medicine, in San Francisco, also found a candidate virus in 1983 but didn't publish until more than a year afterward. By summer of 1984, Levy noted, AIDS had affected "more than 4000 individuals in the world; in San Francisco, over 600 cases have been reported." Those numbers sounded alarmingly high at the time, though in retrospect, compared with 30 million deaths, they seem poignantly low. Levy's discovery was also a retrovirus. His group detected it in twenty-two AIDS patients and grew more than a half dozen isolates. Because the bug was an AIDS-associated retrovirus, Levy called it ARV. He suspected, correctly, that his ARV and Montagnier's LAV were simply variant samples of the same evolving virus. They were very similar but not *too* similar. "Our data cannot reflect a contamination of our cultures with LAV," he wrote, "since the original French isolate was never received in our laboratory." Harmless as that may sound, it was an implicit jab at Robert Gallo.

The details of this story, the near-simultaneous triple discovery and its aftermath, are intricate and contentious and seamy and technical, like a ratatouille of molecular biology and personal politics left out in the sun to ferment. They lead far afield from the subject of zoonotic disease. For our purposes here, the essential point is that a virus discovered in the early 1980s, in three different places under three different names, became persuasively implicated as the causal agent of AIDS. A distinguished committee of retrovirologists settled the naming issue in 1986. They decreed that the thing would be called HIV.

87

The next phase began, appropriately, with a veterinarian. Max Essex studied retroviruses in monkeys and cats.

Dr. Myron (Max) Essex, DVM, PhD, is not your ordinary small-animal vet. (Then again, this book is filled with extraordinary veterinarians who are keen scientists as well as caring animal doctors.) Essex is a professor in the Department of Cancer Biology at the Harvard School of Public Health. He worked on feline leukemia virus (FeLV), among other things, and cancer-causing viruses formed the broad frame of his interests. Having seen the effects of FeLV in wrecking the immune systems of cats, he suspected as early as 1982, along with Gallo and Montagnier, that the new human immune deficiency syndrome might be caused by a retrovirus.

Then something strange came to his notice, by way of a grad student named Phyllis Kanki. She was a veterinarian like him, but now working on a doctorate at the School of Public Health. Kanki grew up in Chicago, spent her adolescent summers doing zoo work, and then studied biology and chemistry on the way toward veterinary medicine and comparative pathology. During the summer of 1980, while still amid her DVM studies, she worked at the New England Regional Primate Research Center, which was part of Harvard but located out in Southborough, Massachusetts. There she saw a weird problem among the center's captive Asian macaques—some of them were dying of a mysterious immune dysfunction. Their T-helper lymphocyte counts were way down. They wasted away from diarrhea or succumbed to opportunistic infections, including *Pneumocystis jirovecii.* It sounded too much like AIDS. Kanki later brought this to the attention of Essex, her thesis adviser, and together with colleagues from Southborough, they started to look for what was killing those monkeys. Based on their knowledge of FeLV and other factors, they wondered whether it might be a retrovirus infection.

Taking blood samples from macaques, they did find a new retrovirus, and saw that it was closely related to the AIDS virus. Because this was 1985, they used Gallo's slightly misleading label (HTLV-III) for what would soon be renamed HIV. Their monkey virus would be renamed too and become, by analogy, simian immunodeficiency virus: SIV. The group published a pair of papers in *Science*, which had grown hungry for AIDS breakthroughs. This discovery, they wrote, could help illuminate the pathology of the disease, maybe even advance efforts to develop a vaccine, by providing an animal model for research. Only a single sentence at the end of one of the papers, a modest but pertinent comment dropped in like an afterthought, noted that SIV might also be a clue toward the *origin* of HIV.

It was. Phyllis Kanki performed the lab analysis of samples from the captive macaques and then made it her business to wonder whether the same virus might exist in the wild. Kanki and Essex looked at Asian macaques, testing blood samples from wild-caught animals. They found no trace of SIV. They tested other kinds of wild Asian monkey. Again, no SIV. This led them to surmise that the macaques at Southborough had picked up their SIV in captivity by exposure to animals of another species. It was a reasonable guess, given that the primate center at one point had a monkey playpen in its lobby, where Asian and African infant monkeys were sometimes allowed to mingle. But then which kind of African monkey was the reservoir? Where exactly had the virus come from? And how might it be related to the emergence of HIV?

"In 1985, the highest rates of HIV were reported in the U.S. and Europe," Essex and Kanki wrote later, "but disturbing reports from central Africa indicated that high rates of HIV infection and of AIDS prevailed there, at least in some urban centers." The focus of suspicion was shifting: not Asia, not Europe, not the United States, but *Africa* might be the point of origin. Central Africa also harbored a rich fauna of nonhuman primates. So the Harvard group got hold of blood from some wild-caught African simians, including chimpanzees, baboons, and African green monkeys. None of the chimps or the baboons showed any sign of SIV infection. Some of the African green monkeys did. It was an epiphany. More than two dozen of the monkeys carried antibodies to SIV, and Kanki grew isolates of live virus from seven. That finding too went straight into *Science,* and the search continued. Kanki and Essex eventually screened thousands of African green monkeys, caught in various regions of sub-Saharan Africa or held captive in research centers around the world. Depending on the population, between 30 and 70 percent of those animals tested SIV-positive.

But the monkeys weren't sick. They didn't seem to be suffering from immune deficiency. Unlike the Asian macaques, the African green monkeys "must have evolved mechanisms that kept a potentially lethal pathogen from causing disease," Essex and Kanki wrote. Maybe the virus had changed too. "Indeed, some SIV strains might also have evolved toward coexistence with their monkey hosts." The monkeys evolving toward greater resistance, the virus evolving toward lesser virulence—this sort of mutual adaptation would suggest that SIV had been in them a long time.

The new virus, SIV as found in African green monkeys, became the closest known relative of HIV. But it wasn't *that* close; many differences distinguished the two at the level of genetic coding. The resemblance, according to Essex and Kanki, was "not close enough to make it likely that SIV was an immediate precursor of HIV in people." More likely, those two viruses represented neighboring twigs on a single phylogenetic branch, separated by lots of evolutionary time and probably some extant intermediate forms. Where might the missing cousins be? "Perhaps, we thought, one could find such a virus—an intermediate between SIV and HIV—in human beings." They decided to look in West Africa.

With help from an international team of collaborators, Kanki and Essex gathered blood samples from Senegal and elsewhere. The samples arrived with coded labeling, for blind testing in the laboratory, so that Kanki herself didn't know their country of origin, nor even whether they derived from humans or monkeys. She screened them using tests for both SIV and HIV. Despite one possible misstep involving a lab contamination, her team found what they had thought they might: a virus intermediate between HIV and SIV. With the code unblinded, Kanki learned that the positive results came from Senegalese prostitutes. In retrospect it made sense. Prostitutes are at high risk for any sexually transmitted virus, including a new one recently spilled into humans. And the density of the rural human population in Senegal, where African green monkeys are native, makes monkey-human interactions (crop-raiding by monkeys, hunting by humans) relatively frequent.

Furthermore, the new bug from Senegalese prostitutes wasn't just halfway between HIV and SIV. It *more* closely resembled SIV strains from African green monkeys than it did the Montagnier-Gallo version of HIV. That was important but puzzling. Were there *two* distinct kinds of HIV?

Luc Montagnier now reenters the story. Having tussled with Gallo over the first HIV discovery, he converged more amicably with Essex and Kanki on this one. Using assay tools provided by the Harvard group, Montagnier and his colleagues screened the blood of a twenty-nine-year-old man from Guinea-Bissau, a tiny country, formerly a Portuguese colony, along the south border of Senegal. This man showed symptoms of AIDS (diarrhea, weight loss, swollen lymph nodes) but tested negative for HIV. He was hospitalized in Portugal, and his blood sample hand-delivered to Montagnier by a visiting Portuguese biologist. In Montagnier's lab, the man's serum again tested negative for antibodies to HIV. But from a culture of his white blood cells Montagnier's group isolated a new human retrovirus, which looked very similar to what Essex and Kanki had found. In another patient, hospitalized in Paris but originally from Cape Verde, an island nation off the west coast of Senegal, the French team found more virus of the same type. Montagnier called the new thing LAV-2. Eventually, when all parties embraced the label HIV instead, it would be HIV-2. The original became HIV-1.

The paths of discovery may be sinuous, the labels may seem many, and maybe you can't tell the players without a scorecard; but these details aren't trivial. The difference between HIV-2 and HIV-1 is the difference between a nasty little West African disease and a global pandemic.

88

During the late 1980s, as Kanki and Essex and other scientists studied HIV-2, a flurry of uncertainty arose about its provenance. Some challenged the idea that it was closely related to (and recently derived from) a retrovirus that infects African monkeys. An alternative view was that such a retrovirus had been present in the human lineage for as long as—or longer than—human time. Possibly it was already with us, a passenger riding the slow channels of evolution, when we diverged from our primate cousins. But that view left an unresolved conundrum: If the virus was an ancient parasite upon humans, unnoticed for millennia, how had it suddenly become so pathogenic?

Recent spillover seemed more likely. Still, the case against that idea got a boost in 1988, when a group of Japanese researchers sequenced the complete genome of SIV from an African green monkey. The animal came from Kenya. The nucleotide sequence of its retrovirus proved to be substantially different from the sequence for HIV-1, and different in roughly the same degree from HIV-2. So the monkey virus seemed no more closely related to the one human virus than to the other. That contradicted the notion that HIV-2 had lately emerged from an African green monkey. A commentary in the journal *Nature*, published to accompany the Japanese paper, celebrated this finding beneath a dogmatic headline: HUMAN AIDS VIRUS NOT FROM MONKEYS. But the headline was misleading to the point of falsity. *Not from monkeys*? Well, don't be so sure. It turned out that researchers were just looking at the wrong kind of monkey.

Confusion came from two sources. For starters, the label "African green monkey" is a little vague. It encompasses a diversity of forms, sometimes also known as savannah monkeys, that occupy adjacent geographical ranges spread out across sub-Saharan Africa, from Senegal in the west to Ethiopia in the east and down into South Africa. At one time those forms were considered a "superspecies" under the name *Cercopithecus aethiops*. Nowadays, their differences having been more acutely gauged, they are classified into six distinct species within the genus *Chlorocebus*. The "African green monkey" sampled by the Japanese team, because it was "of Kenyan origin," probably belonged to the species *Chlorocebus pygerythrus*. The species native to Senegal, on the other hand, is *Chlorocebus sabaeus*. Now that you've seen those two names you can forget them. The difference between one African green monkey and another is not what accounts for the genetic disjunction between SIV and HIV-2.

The trail backward from HIV-2 led to another monkey entirely: the sooty mangabey. This is not one of the six *Chlorocebus* species, not even close. It belongs to a different genus.

The sooty mangabey (*Cercocebus atys*) is a smoky-gray creature with a dark face and hands, white eyebrows, and flaring white muttonchops, not nearly so decorative as many monkeys on the continent but arresting in its way, like an elderly chimney sweep of dapper tonsorial habits. It lives in coastal West Africa, from Senegal to Ghana, favoring swamps and palm forests, where it eats fruit, nuts, seeds, leaves, shoots, and roots—an eclectic vegetarian—and spends much of its time on the ground, moving quadrupedally in search of fallen tidbits. Sometimes it ventures out of the bottomlands to raid farms and rice paddies. The sooty mangabey is hard to hunt within the swampy forests but, because of its terrestrial foraging habits and its taste for crops, easy to trap. Local people treat it as an annoying but edible sort of vermin. Sometimes also, if they're not too hungry, they adopt an orphan juvenile as a pet.

What brought the sooty mangabey to the attention of AIDS researchers was coincidence and an experiment on leprosy. It was an instance of the old scientific verity that sometimes you find much more than you're looking for.

Back in September 1979, scientists at a primate research center in New Iberia, Louisiana, south of Lafayette, noticed a leprosy-like infection in one of their captive monkeys. This seemed odd, because leprosy is a human disease caused by a bacterium (*Mycobacterium leprae*) not known to be transmissible from people to other primates. But here was a leprous monkey. The animal in question, a sooty mangabey, female, about five years old, had been imported from West Africa. The researchers called her Louise. Apart from her skin condition, Louise was healthy. She hadn't, so far as the records showed, yet been subjected to any experimental infection. They were using her in a study of diet and cholesterol. The New Iberia facility didn't happen to work on leprosy infections, so once Louise's condition had been recognized she was transferred to a place, also in Louisiana, that did: the Delta Regional Primate Research Center, north of Lake Pontchartrain. The researchers at Delta were glad to get her, for one very practical reason. If Louise had acquired her leprosy naturally, then (contrary to previous suppositions) the disease might be transmissible in populations of sooty mangabey. And if that were true, then the sooty mangabey could prove valuable as an experimental model for studies of human leprosy.

So the Delta team injected some infectious material from Louise into another sooty mangabey. This one was a male. Unlike Louise, he's nameless in the scientific record, remembered only by a code: A022. He became the first in a chain of experimentally infected monkeys that turned out to carry more than leprosy. The scientists at Delta had no idea, not at first, that A022 was SIV-positive.

The leprosy from Louise took hold easily in A022, which was notable, given that earlier attempts to infect monkeys with human leprosy had failed. Was this strain of *Mycobacterium leprae* a peculiarly monkey-adapted variant? If so, might it succeed in rhesus macaques too? That would be convenient for experimental purposes, because rhesus macaques were far cheaper and more available, in the medical-research chain of supply, than sooty mangabeys. So the Delta team injected four rhesus macaques with infectious gunk from A022. All four developed leprosy. For three of the four, that proved to be the least of their troubles. The unlucky three also developed simian AIDS. Suffering chronic diarrhea and weight loss, they wasted away and died.

Screening for virus, the researchers found SIV. How had their three macaques become SIV-positive? Evidently by way of the leprous inoculum from the sooty mangabey, A022. Was he unique? No. Tests of other sooty mangabeys at Delta revealed that the virus was "endemic" among them. Other investigators soon found it too, not just among captive sooty mangabeys but also in the wild. Yet the sooty mangabeys (native to Africa), unlike the rhesus macaques (native to Asia), showed no symptoms of simian AIDS. They were infected but healthy, which suggested that the virus had a long history in their kind. The same virus made the macaques sick, presumably because it was new to them.

The roster of simian immunodeficiency viruses was growing more crowded and complex. Now there were three known variants: one from African green monkeys, one from rhesus macaques (which they probably acquired in captivity), and one from sooty mangabeys. Needing a way to identify and distinguish them, someone hit upon the expedient of adding tiny subscripts to the acronym. Simian immunodeficiency virus as found in sooty mangabeys became SIV_{sm}. The other two were labeled SIV_{agm} (for African green monkeys) and SIV_{mac} (for Asian macaques). This little convention may seem esoteric, not to mention hard on the eyes, but it will be essential and luminous when I discuss the fateful significance of a variant that came to be known as SIV_{cpz}.

For now it's enough to note the upshot of the leprosy experiment in Louisiana. One scientist from the Delta team, a woman named

Michael Anne Murphey-Corb, collaborated with molecular biologists from other institutions to scrutinize the genomes of SIVs from sooty mangabeys and rhesus macaques, and to create a provisional family tree. Their work, published in 1989 with Vanessa M. Hirsch as first author, revealed that SIV_{sm} is closely related to HIV-2. So is SIV_{mac}. "These results suggest that SIV_{sm} has infected macaques in captivity and humans in West Africa," the group wrote, placing the onus of origination on sooty mangabeys, "and evolved as SIV_{mac} and HIV-2, respectively." In fact, those three strains were *very* similar, suggesting divergence fairly recently from a common ancestor.

"A plausible interpretation of these data," Hirsch and her coauthors added, to make the point plainly, "is that in the past 30–40 years SIV from a West African sooty mangabey (or closely related species) successfully infected a human and evolved as HIV-2." It was official: HIV-2 is a zoonosis.

89

But what about HIV-1? Where did the great killer come from? That larger mystery took somewhat longer to solve. The logical inference was that HIV-1 must be zoonotic in origin also. But what animal was its reservoir? When, where, and how had spillover occurred? Why had the consequences been so much more dire?

HIV-2 is both less transmissible and less virulent than HIV-1. The molecular bases for those fateful differences are still secrets embedded in the genomes, but the ecological and medical ramifications are clear and stark. HIV-2 is confined mostly to West African countries such as Senegal and Guinea-Bissau (the latter of which, during colonial times, was Portuguese Guinea), and to other areas connected socially and economically within the former Portuguese empire, including Portugal itself and southwestern India. People infected with HIV-2 tend to carry lower levels of virus in their blood, to infect fewer of their sexual contacts, and to suffer less severe or longer-delayed forms of immune deficiency. Many of them don't seem to progress to AIDS at all. And mothers who carry HIV-2 are less likely to pass it to their infants. The virus is bad, but not nearly so bad as it could be. HIV-1 provides the comparison. HIV-1 is the thing that afflicts tens of millions of people throughout the world. HIV-1 is the pandemic scourge. To understand how the AIDS catastrophe has happened to humanity, scientists had to trace HIV-1 to its source.

This takes us circling back to the city of Franceville, in southeastern Gabon, and its Centre International de Recherches Médicales (CIRMF), the same research institute at which Eric Leroy would later base his studies of Ebola. At the end of the 1980s, a young Belgian woman named Martine Peeters worked as a research assistant at CIRMF for a year or so, during the period between getting her diploma in tropical medicine and going on for a doctorate. The CIRMF facility maintained a compound of captive primates, including three dozen chimpanzees, and Peeters along with several associates was tasked with testing the captive animals for antibodies to HIV-1 and HIV-2. Almost all of the chimps tested negative—all except two. Both the exceptions were very young females, recently captured from the wild. Such baby chimps, like other orphan primates, are sometimes kept or sold off as pets after the killing and eating of their mothers. One of these animals, a two-year-old suffering from gunshot wounds, had been brought to CIRMF for medical treatment. She died of the wounds, but not before surrendering a blood sample. The other was an infant, maybe six months old, who survived. Blood serum from each of them reacted strongly when tested against HIV-1, less strongly when tested against HIV-2. That much was notable but slightly ambiguous. Antibody testing is an indirect gauge of infection, relatively convenient and quick, but imprecise. Greater precision comes with detecting fragments of viral RNA or, better still, isolating a virus—catching the thing in its wholeness and growing it in quantity—from which a confident identification can be made. Martine Peeters and her co-workers succeeded in isolating a virus from the baby chimp. Twenty years later, when I called on her at her office at an institute in southern France, Peeters remembered vividly how that virus showed up in a series of molecular tests.

"It was especially surprising," she said, "because it was so close to HIV-1."

Had there been any previous hints?

"Yes. At that time we knew already that HIV-2 most likely came from primates in West Africa," she said, alluding to the sooty mangabey work. "But there was no virus close to HIV-1 already detected in primates. And until now, it's still the only virus close to HIV-1." Her group had published a paper, in 1989, announcing the new virus and calling it $SIV_{cpz.}$ They did not crow about having found the reservoir of HIV-1. Their conclusion from the data was more modest: "It has been suggested that human AIDS retroviruses originated from monkeys in Africa. However, this study and other previous studies on SIV do not support this suggestion." Left implicit: Chimpanzees, not monkeys, might be the source of the pandemic bug.

By the time I met her, Martine Peeters was director of research at the Institut de Recherche pour le Développement, in Montpellier, a handsome old city just off the Mediterranean coast. She was a small, blonde woman in a black sweater and silver necklace, concise and judicious in conversation. What sort of response had met this discovery? I asked.

"HIV-2, people accepted it readily." They accepted, she meant, the notion of simian origins. "But HIV-1, people had more difficulties to accept it."

Why the resistance? "I don't know why," she said. "Maybe because we were young scientists."

The 1989 paper got little attention, which seems peculiar in retrospect, given the novelty and gravity of what it implied. In 1992 Peeters published another, describing a third case of SIV_{cpz}, this one in a captive chimpanzee that had been shipped to Brussels from Zaire. All three of her SIV-positive results had been in "wild-born" chimpanzees taken captive (as distinct from animals bred in captivity) but that still left a gap in the chain of evidence. What about chimps still *in* the wild?

With only such tools of molecular biology as available in the early 1990s, the screening of wild chimps was difficult (and unacceptable to most chimp researchers), because the diagnostic tests required blood sampling. Lack of evidence from wild populations, in turn, contributed to skepticism in the AIDS-research community about the link between HIV-1 and chimps. After all, if Asian macaques had become infected with HIV-2 in their cages, from contact with African monkeys, might not SIV-positive chimpanzees simply reflect cage-contact infections too? Another reason for skepticism was the fact that, by the end of the 1990s, roughly a thousand captive chimpanzees had been tested but, apart from Peeters's three, not a single one had yielded traces of SIV_{cpz}. These two factors—the absence of evidence from wild populations and the extreme rarity of SIV in captive chimps—left open the possibility that both HIV-1 and SIV_{cpz} derived directly from a common ancestral virus in some other primate. In other words, maybe those three lonely chimps had gotten their infections from some still-unidentified monkey, and maybe the same unidentified monkey had given HIV-1 to humans. With that possibility dangling, the origin of HIV-1 remained uncertain for much of the decade.

In the meantime, researchers investigated not just the source of HIV but also its diversity in humans, discovering three major lineages of HIV-1. "Groups" became the preferred term for these lineages. Each group was a cluster of strains that was genetically discrete from the other clusters; there was variation *within* each group, since HIV is always evolving, but the differences *between* groups were far larger. This pattern of groups had some dark implications that dawned on scientists only gradually and still haven't been absorbed in the popular understanding of AIDS. I'll get to them shortly, but first let's consider the pattern itself.

Group M was the most widespread and nefarious. The letter M stood for "main," because that group accounted for most of the HIV infections worldwide. Without HIV-1 group M, there was no global pandemic, no millions of deaths. Group O was the second to be delineated, its initial standing for "outlier," because it encompassed only a small number of viral isolates, mostly traceable to what seemed an outlier area relative to the hotspots of the pandemic: Gabon, Equatorial Guinea, and Cameroon, all in western Central Africa. By the time a third major group was discovered, in 1998, it seemed logical to label that one N, supposedly indicating "non-M/non-O" but also filling in the alphabetical sequence. (Years later, a fourth group would be identified and labeled P.) Group N was extremely rare; it had been found in just two people from Cameroon. The rarity of N and O put group M dramatically in relief. M was everywhere. Why had that particular lineage of virus, and not the other two (or three), spread so broadly and lethally around the planet?

Parallel research on HIV-2, the less virulent virus, also found distinct groups but even more of them. Their labeling came from the beginning of the alphabet rather than the middle, and by the year 2000 seven groups of HIV-2 were known: A, B, C, D, E, F, and G. (An eighth group, turning up later, became H.) Again, most of them were extremely rare—each represented, in fact, by a viral sample taken from only one person. Groups A and B *weren't* rare; they accounted for the majority of HIV-2 cases. Group A was more common than group B, especially in Guinea-Bissau and Europe. Group B was traceable mainly to countries on the eastern end of West Africa, such as Ghana and Côte d'Ivoire. Groups C through H, although tiny in total numbers, were significant in showing a range of diversity.

As the new century began, AIDS researchers pondered this roster of different viral lineages: seven groups of HIV-2 and three groups of HIV-1. The seven groups of HIV-2, distinct as they were from one another, all resembled SIV_{sm}, the virus endemic in sooty mangabeys. (So did the later addition, group H.) The three kinds of HIV-1 all resembled SIV_{cpz}, from chimps. (The eventual fourth kind, group P, is most closely related to SIV from gorillas.) Now here's the part that, as it percolates into your brain, should cause a shudder: Scientists think that each of those twelve groups (eight of HIV-2, four of HIV-1) reflects an independent instance of cross-species transmission. Twelve spillovers.

In other words, HIV hasn't happened to humanity just once. It has happened at least a dozen times—a dozen that we know of, and probably many more times in earlier history. Therefore it wasn't a highly improbable event. It wasn't a singular piece of vastly unlikely bad luck, striking humankind with devastating results—like a comet come knuckleballing across the infinitude of space to smack planet Earth and extinguish the dinosaurs. No. The arrival of HIV in human bloodstreams was, on the contrary, part of a small trend. Due to the nature of our interactions with African primates, it seems to occur pretty often.

90

Which raises a few large questions. If the spillover of SIV into humans has happened at least twelve times, why has the AIDS pandemic happened only once? And why did it happen when it did? Why didn't it happen decades or centuries earlier? Those questions entangle themselves with three others, more concrete, less speculative, to which I've already alluded: When, where, and how *did* the AIDS pandemic begin?

First let's consider when. We know from Michael Gottlieb's evidence that HIV had reached homosexual men in California by late 1980. We know from the case of Grethe Rask that it lurked in Zaire by 1977. We know that Gaëtan Dugas wasn't really Patient Zero. But if those people and places don't mark a real beginning point in time, what does? When did the fateful strain of virus, HIV-1 group M, enter the human population?

Two lines of evidence call attention to 1959.

In September of that year, a young print-shop worker in Manchester, England, died of what seemed to be immune-system failure. Because he spent a couple years in the Royal Navy before returning to his hometown and his job, this unfortunate man has been labeled "the Manchester sailor." His health went into decline after his naval hitch, which he served mainly but not entirely in

England. At least once he sailed as far as Gibraltar. Back in Manchester by November 1957, he wasted away, suffering some of the symptoms later associated with AIDS, including weight loss, fevers, a nagging cough, and opportunistic infections, including *Pneumocystis jirovecii*, but no underlying cause of death could be determined by the doctor who did the autopsy. That doctor preserved some small bits of kidney, bone marrow, spleen, and other tissues from the sailor—embedding them in paraffin, a routine method for fixing pathology samples—and reported the case in a medical journal. Thirty-one years later, in the era of AIDS, a virologist at the University of Manchester tested some of those archived samples and believed he found evidence that the sailor had been infected with HIV-1. If he was correct, then the Manchester sailor would be recognized retrospectively as the first case of AIDS ever documented in the medical literature.

But wait. Retesting of the same samples by a pair of scientists in New York, several years later, showed that the earlier HIV-positive result must have reflected a laboratory mistake. The bone marrow now tested negative. The kidney material again tested positive but in a way that rang alarms of doubt: HIV-1 evolves quickly, and the genetic sequence of virus from the kidney sample seemed far too modern. It looked more like a modern variant than like something that could have existed in 1959. That suggested contamination with some recent strain of the virus to account for the positive tests. Conclusion: The Manchester sailor may have died from immune-system failure but HIV probably wasn't the cause. His case merely illustrates how tricky it can be to make a retrospective diagnosis of AIDS, even with the presence of what seems to be good evidence.

Soon after that false lead from Manchester was debunked, another lead emerged in New York. By now it was 1998. A team of researchers including Tuofu Zhu, based at the Rockefeller University, obtained an archival specimen from Africa dating back to the same year as the sailor's, 1959. This time it wasn't tissues; it was a small tube of blood plasma, drawn from a Bantu man in what had been Léopoldville, capital of the Belgian Congo (nowadays

Kinshasa, capital of DRC) and stored for decades in a freezer. The man's name and his cause of death weren't reported. His sample had been screened during an earlier study, in 1986, along with 1,212 other plasmas—some archival, others new—from various locations in Africa. This man's was the only one that tested unambiguously positive for HIV. Tuofu Zhu and some colleagues probed further, working with what little remained of the original sample and using PCR to amplify fragments of the viral genome. Then they sequenced the fragments to assemble a genetic portrait of the Bantu man's virus. In their paper, published in February 1998, they called the sequence ZR59, referencing Zaire (as the country had long been known) and the year 1959. Comparative analysis showed that ZR59 was quite similar to both subtype B and subtype D (finer divisions within the HIV-1 group M lineage) but fell about halfway between, which suggested that it must closely resemble their common ancestor. In other words, ZR59 was a glimpse back in time, a genuinely old form of HIV-1, not a recent contamination. ZR59 proved that HIV-1 had been present—simmering, evolving, diversifying—in the population of Léopoldville by 1959. In fact it proved more. Further analysis of ZR59 and other sequences, led by Bette Korber of the Los Alamos National Laboratory, yielded a calculation that HIV-1 group M might have entered the human population around 1931.

For a decade, from the Zhu publication in 1998 until 2008, that landmark stood alone. ZR59 was the only known version of HIV-1 from a sample taken earlier than 1976. Then someone found another. This one became known as DRC60, and by now you can probably decode the label yourself: It came from the Democratic Republic of the Congo (same nation, latest name) and had been collected in 1960.

DRC60 was a biopsy specimen, a piece of lymph node snipped from a living woman. Like the Manchester sailor's bits of kidney and spleen, it had been locked away in a little pat of paraffin. Thus preserved, it needed no refrigeration, let alone freezing. It was as inert as a dead butterfly and less fragile. It could be stored and ignored on a dusty shelf—as it had been. After more than four decades, it emerged from a specimen cabinet at the University of Kinshasa and offered a new jolt of insight to AIDS researchers.

91

The University of Kinshasa sits on a hilltop near the edge of the city, reachable by an hour's taxi ride through the broken streets, the smoggy sprawl, the snarled traffic of vans and busses and pushcarts, past the street-side vendors of funerary wreaths, the cell-phone-recharge kiosks, the fruit markets, the meat markets, the open-air hardware stores, the tire-repair shops and cement brokers, the piles of sand and gravel and garbage, the awesome decrepitude of a postcolonial metropolis shaped by eight decades of Belgian opportunism, three decades of dictatorial misrule and egregious theft, and then a decade of war, but filled with 10 million striving people, some of whom are dangerous thugs (as in all cities) and most of whom are amiable, hopeful, and friendly. The university campus, on its hill, loosely called "the mountain," presents a relatively verdant and halcyon contrast to the city below. Students go there, climbing by foot from a crowded bus stop, to learn and to escape.

Professor Jean-Marie M. Kabongo is head of pathology in the university's Department of Anatomic Pathology. He's a small, natty man with a huge graying handlebar mustache and full muttonchops, making a forceful visual impression that's vitiated by his gentle manner. When I met him in his office, on the second floor of a building that overlooks a grassy concourse shaded with acacias, he pleaded imperfect knowledge of DRC60 and the patient from whom that specimen came. An old case, after all, going back long before his time. Yes, a woman, he believed. His memory was vague but he could check the records. He began taking notes as I questioned him and suggested I come back in a couple days, when he might be better prepared with answers. But then I asked about the room where DRC60 had been stored, and he brightened. Oh, of course, he said, I can show you *that*.

He fetched a key. He unlocked a blue door. Swinging it open, he welcomed me into a large sunlit laboratory with walls of white tile and two long, low tables down the middle. On one of the tables rested an old-fashioned folio ledger, with curling pages, like something from Chancery in the time of Dickens. On the far windowsill stood a row of beakers containing liquids in increments of color, beaker by beaker, from piss-yellow to vodka-clear. The yellowest, Professor Kabongo told me, was methanol. The clearest was xylol. We use these in preparing a tissue sample, he said. The point of such organic solvents is to extract the water; desiccation is prerequisite to fixing tissues for the long term. The methanol was darkened from processing many samples.

He showed me a small orange plastic basket, with a hinged lid, about the size and shape of a matchbook. This is a "cassette," Professor Kabongo explained. You take a lump of tissue from a lymph node or some other organ and enclose it in such a cassette; you soak the whole thing in the beaker of methanol; from the methanol, it goes through the intermediate baths in sequence; finally you dunk it in the xylol. Methanol draws out the water; xylol draws out the methanol, preparing your specimen for preservation in paraffin. And this device, Professor Kabongo said, indicating a large machine on one of the tables, delivers the paraffin. You take a leached tissue sample from its cassette, he explained, and, from that spigot, you dribble out a stream of warm, liquid paraffin. It cools on the sample like a pat of butter. Now you remove the cassette lid and label the base with an individual code—for instance, A90 or B71. That's your archival specimen, he said. "A" means that it came from an autopsy. "B" indicates a biopsy. So the paraffin-caked bit of lymph node that yielded DRC60 would have been labeled B-something. Each coded specimen gets recorded in the big ledger. Then the specimens go into storage.

Storage. Storage where? I asked.

At the far end of the lab was another doorway, this one hung with a blue curtain. Professor Kabongo pushed the curtain aside and I followed him into a specimen pantry, narrow and tight, lined with shelves and cabinets along one side. The shelves and cabinets contained thousands of dusty paraffin blocks and old microscope slides. The paraffin blocks were in stacks and cartons, some of the cartons dated, some not. It appeared to be organized chaos. A wooden stool awaited use by any curious, tireless soul wishing to rummage through the samples. Although I didn't plan to rummage, my tour had suddenly come to its crescendo. *Here?* Yes, just here, said the professor. This is where DRC60 sat for decades. He could have added, with local pride: before becoming a Rosetta stone in the study of AIDS.

92

From the pantry behind the blue curtain, that sample and hundreds of others had traveled a circuitous route, to Belgium and then the United States, ending up in the laboratory of a young biologist at the University of Arizona. Michael Worobey is a Canadian, originally from British Columbia, whose specialty is molecular phylogenetics. After his undergraduate work he went to Oxford on a Rhodes scholarship, which ordinarily means two years of mildly strenuous academic work plus lots of tea, sherry, tennis on grass, and genteel anglophilia before the "scholar" returns to professional school or a career. Worobey put Oxford to more serious use, staying on, finishing a doctorate and then a postdoc fellowship in evolutionary biology at the molecular level. From there he returned to North America in 2003, accepting an assistant professorship at Arizona and building himself a BSL-3 lab for work on the genomes of dangerous viruses. Several years later, it was Worobey who detected evidence of HIV in a certain Congolese biopsy specimen from 1960.

Worobey amplified fragments of the viral genome, pieced the

fragments together, recognized them as an early version of HIV-1, and named the sequence DRC60. Comparing his sequence with ZR59, the other earliest known strain, he reached a dramatic conclusion: that the AIDS virus has been present in humans for decades longer than anyone thought. The pandemic may have gotten its start with a spillover as early as 1908.

To appreciate Worobey's discovery and how it splashed down amid previous ideas, you'll need to know a little context. That context involved a heated dispute over just how HIV-1 entered the human population. The prevailing notion as of the early 1990s, based on what had been learned about HIV-2 and the sooty mangabey, among other factors, was that HIV-1 also came from an African primate, and that it had probably gotten into humans by way of two separate instances (for groups M and O, the ones then recognized) of butchering bushmeat. This became known as the cut-hunter hypothesis. In each instance, a man or a woman had presumably butchered the carcass of an SIV-positive primate and suffered exposure through an open wound—maybe a cut on the hand, or a scratch on the arm, or a raw spot on any skin surface that got smeared with the animal's blood. A wound on the back might have sufficed, if the carcass were draped over shoulders for carrying home. A wound in the mouth, if some of the meat were consumed raw. All that mattered was blood-to-blood contact. The cut-hunter hypothesis was speculative but plausible. It was parsimonious, requiring few complications and no unlikelihoods. It fit the known facts, though the known facts were fragmentary. And then in 1992 a contrary theory arose.

This one was heterodox and highly controversial: that HIV-1 first got into humans by way of a contaminated polio vaccine tested on a million unsuspecting Africans. The vaccine itself, by this theory, had been an unintended delivery system for AIDS. Someone, according to the theory, had monumentally goofed. Someone was culpable. Scientific hubris had overridden caution, with catastrophic results. The scariest thing about the polio-vaccine theory was that it *also* seemed plausible.

Viruses are subtle, as you've seen. They get in where they

shouldn't. Laboratory contaminations occur. Even viral or bacterial contamination of a vaccine at the production level—it has happened. Back in 1861, a group of Italian children vaccinated against smallpox, with material direct from a "vaccinal sore," came down with syphilis. Smallpox vaccine administered to kids in Camden, New Jersey, at the start of the twentieth century, seems to have been contaminated with tetanus bacillus, resulting in the death of nine vaccinated children from tetanus. Around the same time, a batch of diphtheria antitoxin prepared in St. Louis, using blood serum from a horse, also turned out to carry tetanus, which killed another seven children. Producers then began filtering vaccines, an effective precaution against bacterial contamination; but viruses passed through the filters. Formaldehyde was sometimes added to inactivate a target virus, and that supposedly killed unwanted viruses too, but the supposition wasn't always correct. As late as midcentury, some of the early batches of the Salk polio vaccine were contaminated with a virus known as SV40, endemic in rhesus macaques. SV40 in vaccine became a hot issue, several years later, when suspicions arose that this virus causes cancer.

Whether vaccine contamination happened with HIV-1, and far more consequentially, is another matter. That the vaccine in question had been given to Africans was not in dispute. Between 1957 and 1960, a Polish-born American researcher named Hilary Koprowski—a lesser-known competitor in the same vaccine-development race that engaged Salk and Sabin—arranged for his candidate vaccine to be widely administered in areas of the eastern Belgian Congo and adjacent colonial holdings. These were parts of what would eventually be DRC, Rwanda, and Burundi. Koprowski himself visited Stanleyville, in 1957, and made contacts who later oversaw the trials. Children and adults lined up trustingly, in places like the Ruzizi Valley north of Lake Tanganyika, to receive oral doses of liquid vaccine from a tablespoon or a squirting pipette. Spritz, you're good. Next! The numbers are uncertain. By one account, roughly seventy-five thousand kids were vaccinated just in Léopoldville. The heterodox theory argued two additional points about this enterprise: First, that Koprowski's vaccine was produced by growing the virus in chimpanzee kidney cells (rather than in monkey kidney cells, the standard technique); second, that at least some batches of that vaccine were produced from chimpanzee kidneys drawn from animals infected with SIV_{cpz}.

The result of that flawed vaccinating, certain people have argued, was iatrogenic infection (disease caused by medical treatment) of an unknown number of Central Africans with what later became recognized as HIV-1. By this notion, known for short as the OPV (oral polio vaccine) theory, a single reckless researcher had seeded the continent—and the world—with AIDS.

The OPV theory has been around and notorious since 1992, when a freelance journalist named Tom Curtis described it in a long article for *Rolling Stone.* Curtis's piece ran under the headline: THE ORIGIN OF AIDS: A STARTLING NEW THEORY ATTEMPTS TO ANSWER THE QUESTION, 'WAS IT AN ACT OF GOD OR AN ACT OF MAN?' Several other researchers had mooted the idea earlier, more obscurely, and one of them had put Tom Curtis onto the story. When Curtis started looking into it, some eminent scientists responded with defensive dismissals, which served only to suggest that maybe the theory did merit consideration. Curtis even drew a brusque comment from the head of research for WHO's Global Programme on AIDS, Dr. David Heymann: "The origin of the AIDS virus is of no importance to science today." He quoted another expert, William Haseltine of Harvard, as saying: "It's distracting, it's nonproductive, it's confusing to the public, and I think it's grossly misleading in terms of getting to the solution of the problem." After publication of the piece, lawyers for Hilary Koprowski filed a lawsuit against Curtis and *Rolling Stone,* charging defamation, and the magazine ran a "clarification," admitting that the OPV theory and Koprowski's role represented just an unsupported hypothesis. But as the dust settled at *Rolling Stone,* an English journalist named Edward Hooper took hold of the OPV theory as a personal obsession and an investigative crusade, giving it a second life.

Hooper spent years researching the subject with formidable tenaciousness (though not always critical good sense) and in 1999 made his case in a thousand-page book titled *The River: A Journey to the Source of HIV and AIDS*. Hooper's river was a metaphorical one: the flow of history, the stream of cause-and-effect, from a very small beginning to an ocean of consequences. In the book's prologue, he alluded to the quest by Victorian explorers for the source of the Nile. Does that river begin from Lake Victoria, pouring out at Ripon Falls, or is there another and more obscure source upstream from the lake? "The controversy surrounding the source of the Nile," Hooper wrote, "is strangely echoed by another controversy of a century and a half later, the long-running debate about the origins of AIDS." The Victorian explorers had been wrong about the Nile and, according to Hooper, so were the modern experts wrong about the starting point of the AIDS pandemic.

Hooper's book was massive, overwhelmingly detailed, seemingly reasonable, exhausting to plod through but mesmerizing in its claims, and successful at bringing the OPV theory to broader public attention. Some AIDS researchers (including Phyllis Kanki and Max Essex) had long been aware that vaccine contamination, with SIV from monkey cells, was at least a theoretical possibility; they had even conducted screening efforts on vaccine lines, and found no evidence of such a problem. Hooper, following Tom Curtis, raised the idea from a concern to an accusation. His vast river of information and his steamboat of argument didn't prove the essential thesis—that Koprowski's vaccine had been made from chimp cells contaminated with HIV. But his work did seem to raise the possibility that the vaccine *could* have been made from chimp cells that *might* have been contaminated.

The issue of possibility then gave way to the issue of fact. What had actually happened? Where was the evidence? At the urging of an eminent evolutionary biologist named William Hamilton, who believed that the OPV theory deserved investigation, the Royal Society convened a special meeting in September 2000 to discuss the subject within its broader context. Hamilton was a senior figure, liked and respected, whose early work in evolutionary theory helped inform Edward O. Wilson's *Sociobiology* and Richard Dawkins's *The Selfish Gene*. He swung the Royal Society into giving the OPV theory a fair hearing. Edward Hooper, though not a scientist himself, was invited to speak. Hilary Koprowski also came, as well as a roster of leading AIDS researchers. By the time that meeting convened, though, William Hamilton was dead.

He died suddenly in March 2000, of intestinal bleeding, after an attack of malaria contracted during a research trip to DRC. In his absence, his colleagues at the Royal Society discussed a wide range of matters related to the origins of HIV and AIDS. The OPV theory was just one topic among many, though implicitly it drove the agenda of the whole meeting. Did the available data from molecular biology and epidemiology tend to support, or to refute, the vaccine-contamination scenario? A corollary to that question was: When had HIV-1 first entered the human population? If the earliest infections occurred before 1957, those infections couldn't have resulted from Koprowski's OPV trials. Archival HIV-positives might be decisive.

This is the context that brought DRC60 out of Kinshasa. After the Royal Society meeting, a Belgian physician named Dirk Teuwen, who had taken part, recollected some references to early pathology work in the Congo that he had seen in archival reports of the colonial medical laboratories. Teuwen conceived the idea—and raised it with other attendees—that HIV-1 might be detected in some of the tissues preserved within those old paraffin blocks. He met skepticism; the others doubted that any useful traces of virus could have survived through the decades—decades of tropical heat, simple storage, administrative upheaval, and revolution. But Teuwen was stubborn. He enlisted an ally, a senior Congolese bacteriologist named Jean-Jacques Muyembe, and, with approval from the Ministry of Health, Muyembe started looking. He went up to the University of Kinshasa, rifled through the pantry behind the blue curtain, packed 813 paraffin-embedded specimens into an ordinary suitcase, and carried it with him on his next professional visit to Belgium. There he handed the trove to Dirk Teuwen. Teuwen, in accord with a prior agreement for collaborative study, sent the samples to Michael Worobey in Tucson.

These two lines of narrative fold back into each other. Worobey, as a grad student, knew both Bill Hamilton at Oxford and some of the disease biologists in Belgium. Impelled by his own interest in the origins of HIV, Worobey accompanied Hamilton to DRC on that last fatal fieldtrip. They went in January 2000, during the chaotic aftermath of the civil war, which had replaced President Mobutu Sese Seko with President Laurent Kabila. Hamilton wanted to collect fecal and urine samples from wild chimpanzees; those specimens, he hoped, might help confirm or refute the OPV theory. Worobey, for his part, put little stock in the OPV theory but wanted more data from which to chart the origin and evolution of HIV. It was a crazy time in DRC, more crazy than usual, because two rebel armies opposed to Laurent Kabila still controlled much of the eastern half of the country. Hamilton and Worobey flew into Kisangani (formerly Stanleyville), a regional capital along the upper Congo River, the same city where Koprowski had begun his vaccinating enterprise. Now it was occupied by Rwanda-backed forces on one riverbank and Uganda-backed forces on the other. Commercial airlines weren't flying, because of the war, so the two biologists shared a small, chartered plane with a diamond dealer. In Kisangani they paid their respects to the Rwanda-backed commander, whose ambit included most of the city, and as quickly as possible got out into the forest, where they would be safer among the leopards and snakes. They spent a month collecting fecal and urine samples from wild chimpanzees, with help from local guides, and by the time they left, Hamilton was sick.

Neither he nor Worobey knew *how* sick, but they caught the next exit flight they could, which took them to Rwanda. From there they bounced to Entebbe in Uganda, where Hamilton got a confirmed diagnosis of falciparum malaria and some treatment, then onward to Nairobi, and from Nairobi up to London Heathrow. By now Hamilton seemed past the worst of his illness; he was feeling much better. They had accomplished their mission and life was good. An American field biologist once expressed to me how he felt in such moments. "That's the name of the game: getting home with the data." This man's research too involved dangers—shipwreck, starvation, drowning, snakebite, though not malaria and AK rifles. "If you take too many risks, you don't get home," he said. "If you take too few, you don't get the data." Hamilton and Worobey got the data, got home, and then learned that the ice cooler containing their precious chimp specimens had gone astray in luggage handling somewhere between Nairobi and London.

I visited Michael Worobey in Tucson to hear about all this. "Everything was fine," he told me, "except we checked six bags, including the cooler that had samples, and five of our bags came through the carousel and the one with the samples disappeared." His friend Hamilton, feeling ill again the next morning, went to a hospital—and hemorrhaged catastrophically, perhaps due to anti-inflammatory drugs he'd been taking against the malarial fever. Worobey phoned and got the news from Hamilton's sister: *Who are you why are you calling Bill is in extremis.* Worobey meanwhile had been hassling by long-distance phone with a luggage handler in Nairobi, who assured him that the cooler had been found and would arrive on the next flight. What arrived was someone else's cooler, full of sandwiches. "So that was an extra bit of drama that unfolded as Bill was dying in the hospital," Worobey told me. The correct cooler arrived two days later but Hamilton was in no shape to celebrate. He went through a series of surgeries and transfusions and then, after weeks of struggle, he died.

The fecal samples from Congolese chimps, for which Hamilton had given his life, yielded no SIV-positives. A couple of urine samples registered in the borderline zone for antibodies. Those results weren't clear or dramatic enough to merit publication. Good data are where you find them, not always where you look. Several years later, when the human pathology samples from Kinshasa reached Tucson—those 813 little blocks of tissue in paraffin, the ones J. J. Muyembe had carried to Belgium in a suitcase—Michael Worobey was ready. He found DRC60 among them, and it told an unexpected story.

93

Screening paraffin-embedded hunks of old organ samples to find viral RNA isn't easy, not even for an expert. Those little blocks, Worobey said, turned out to be "some of the nastiest kinds of tissues to do molecular biology with." The problem wasn't forty-three years at room temperature in a dusty equatorial pantry. The problem was the chemicals used in fixing the tissues—the 1960 equivalent of the beakers of methanol and xylol that Professor Kabongo had shown me. Back in those days, pathologists favored something called Bouin's fixative, a potent little mixture containing mostly formalin and picric acid. It worked well for preserving the cellular structure of tissues, like salmon in aspic, so that samples could be sliced thin and examined under a microscope; but it was hell on the long molecules of life. It tended to break up DNA and RNA into tiny fragments, Worobey explained, and form new chemical bonds, leaving "sort of a big, tangled mess rather than a nice string of beads that you can do molecular biology on." Because the process was so laborious, he screened just 27 of the 813 tissue blocks from Kinshasa. Among those twenty-seven, he found one containing RNA fragments that unmistakably signaled HIV-1. Worobey persisted adeptly, untangling the mess and fitting the fragments to assemble the sequence of nucleotide bases he named DRC60.

That was the wet work. The dry work, done largely by computer, entailed base-by-base comparisons between DRC60 and ZR59. It also involved broader comparisons, placing those two within a family tree of known sequences of HIV-1 group M. The point of such comparisons was to see how much evolutionary divergence had occurred. How far had these strains of virus grown apart? Evolutionary divergence accumulates by mutation at the base-by-base level (other ways too, but those aren't relevant here), and among RNA viruses such as HIV, as I've explained, the mutation rate is relatively fast. Equally important, the average rate of HIV-1 mutation is known—or anyway, it can be carefully estimated from the study of many strains. That rate of mutation is considered the "molecular clock" for the virus. Every virus has its own rate, and therefore its own clock measuring the ticktock of change. The amount of difference between two viral strains can therefore reveal how much time has passed since they diverged from a common ancestor. Degree of difference factored against clock equals elapsed time. This is how molecular biologists calculate an important parameter they call TMRCA: time to most recent common ancestor.

Okay so far? You're doing great. Take a breath. Now those bits of understanding will boost us across a deep gulf of molecular arcana to an important scientific insight. Here we go.

Michael Worobey found that DRC60 and ZR59, sampled from people in Kinshasa during almost the same year, were *very* different. They both fell within the range of what was unmistakably HIV-1 group M; neither could be confused with group N or group O, nor with the chimp virus, SIV_{cpz}. But within M, they had diverged *far*. How far? Well, one section of genome differed by 12 percent between the two versions. And how different was that, measured in time? About fifty years' worth, Worobey figured. More precisely, he placed the most recent common ancestor of DRC60 and ZR59 in the year 1908, give or take a margin of error.

So the spillover had occurred by 1908? That's much earlier than anyone suspected, and therefore the sort of discovery that gets into an august journal such as *Nature*. Publishing in 2008, a century after the fact, with a list of coauthors that included Jean-Jacques Muyembe, Jean-Marie Kabongo, and Dirk Teuwen, Worobey wrote:

> Our estimation of divergence times, with an evolutionary timescale spanning several decades, together with the extensive genetic distance between DRC60 and ZR59 indicate that these viruses evolved from a common ancestor circulating in the African population near the beginning of the twentieth century.

To me he said: "This wasn't a new virus in humans."

Worobey's work directly refuted the OPV hypothesis. If HIV-1 existed in humans as early as 1908, then obviously it hadn't been introduced via vaccine trials beginning in 1958. Clarity on that point was valuable—but it was only part of Worobey's contribution. Placing the crucial spillover in time represented a big step toward understanding how the AIDS pandemic may have started and grown.

94

Placing the spillover in *space* was equally important, and achieved by a different laboratory. Beatrice Hahn is somewhat older than Worobey and had begun her work on the origin of AIDS long before he found DRC60.

Born in Germany, Hahn got a medical degree in Munich, then came to the United States in 1982 and spent three years as a postdoc in Robert Gallo's lab, studying retroviruses. She moved next to the University of Alabama at Birmingham, where she became Professor of Medicine and Microbiology and codirector of a center for AIDS research, with a group of bright postdocs and grad students working under her aegis. (She remained at Alabama from 1985 to 2011, a period encompassing most of the work described here, and then joined the Perelman School of Medicine at the University of Pennsylvania, in Philadelphia.) The broader purpose of Hahn's various projects, and the goal she shares with Worobey, is to understand the evolutionary history of HIV-1 and its relatives and antecedents. The fittest label for that sort of research is the one Worobey mentioned when I asked him to describe his field: molecular phylogenetics. A molecular phylogeneticist scrutinizes the nucleotide sequences in the DNA or RNA of different organisms, comparing and contrasting, for the same reason a paleontologist scrutinizes fragments of petrified bone from extinct giant saurians—to learn the shape of lineages and the story of evolutionary descent. But for Beatrice Hahn especially, as a medical doctor, there's an additional purpose: to detect how the genes of HIV-1 function in causing disease, toward the prospects of better treatment, prevention, and maybe even a cure.

Some very interesting papers have come out of Hahn's laboratory in the past two decades, many of them published with a junior researcher as first author and Hahn in the mentorship position, last. That was the case in 1999, when Feng Gao produced a phylogenetic study of SIV_{cpz} and its relationship to HIV-1. At the time there were only three known strains of SIV_{cpz}, all drawn from captive chimps, with Gao's paper adding a fourth. The work appeared in *Nature*, highlighted by a commentary calling it "the most persuasive evidence yet that HIV-1 came to humans from the chimpanzee, *Pan troglodytes*." In fact, Gao and his colleagues did more than trace HIV-1 to the chimp; their analysis of viral strains linked it to individuals of a particular subspecies known as the central chimpanzee (*Pan troglodytes troglodytes*), whose SIV had spilled over to become HIV-1 group M. That chimpanzee lives only in western Central Africa, north of the Congo River and west of the Oubangui. So the Gao study effectively identified both the reservoir and also the geographical area from which AIDS must have arisen. It was a huge discovery, as reflected in the headline of *Nature*'s commentary: FROM *PAN* TO PANDEMIC. Feng Gao at the time was a postdoc in Hahn's lab.

But because Gao based his genetic comparisons (as Martine Peeters had done earlier) on viruses drawn from captive chimps, the soupçon of uncertainty about infection among wild chimpanzees remained, at least for a few more years. Then, in 2002, Mario L. Santiago led a list of coauthors announcing in *Science* their discovery of SIV_{cpz} in the wild. Santiago was a PhD student of Beatrice Hahn's.

The most significant aspect of Santiago's work, for which he got his richly deserved doctorate, was that on the way toward detecting SIV in a single wild chimpanzee (just one animal, of fifty-eight tested), he invented methods by which such detections could be made. The methods were "noninvasive," meaning that a researcher didn't need to capture a chimp and draw its blood. The researcher needed only to follow animals through the forest, get under them when they pissed (or, better still, send a field assistant into that yellow shower), collect samples in little tubes, and then screen the samples for antibodies. Turns out that urine can be almost as telling as blood.

"That was a breakthrough," Hahn told me, during a talk at her lab in Birmingham. "We weren't sure it would work." But Santiago took the risk, cooked up the techniques, and it did work. The very first sample of SIV-positive urine from a wild chimpanzee came from the world's most famous community of chimps: the ones at Gombe National Park, in Tanzania, where Jane Goodall had done her historic field study, beginning back in 1960. That sample didn't match quite so closely with HIV-1 as Feng Gao's had done, and it came from an individual of a different subspecies, the eastern chimp (*Pan troglodytes schweinfurthii*). But it was SIV_{cpz} nonetheless.

The advantage of sampling at Gombe, Hahn told me, was that those chimps didn't run away. They were truly wild but, after four decades of study by Goodall and her successors, well habituated to human presence. For use elsewhere, the urine-screening method wasn't practical. "Because, you know, nonhabituated chimps don't stay close enough so you can catch their pee." You could collect their poop from the forest floor, of course, but fecal samples were useless unless preserved somehow; fresh feces contain an abundance of proteases, digestive enzymes, which would destroy the evidence of viral presence long before you got to your laboratory. These are the constraints within which a molecular biologist studying wild animals labors: the relative availability and other parameters of blood, shit, and piss.

Another of Hahn's young wizards, Brandon F. Keele, soon solved the problem of fecal sample decay. He did it by tinkering with a liquid stabilizer called RNAlater, a commercial product made by a company in Austin, Texas, for preserving nucleic acids in tissue samples. The nice thing about RNAlater is that its name is so literally descriptive: The stuff allows you to retrieve RNA from a sample . . . later. If it worked with RNA in tissues, Keele reasoned, maybe it could work also with antibodies in feces. And indeed it did, after he and his colleagues untangled the chemical complications of getting those antibodies released from the fixative. This technique vastly enlarged the scope of screening that was possible on wild chimpanzees. Field assistants could collect hundreds of fecal samples, scooping each into a little tube of RNAlater, and those samples—stored without refrigeration, transported to a distant laboratory—would yield their secrets later. "If we find the antibodies, we know that chimps are infected," Hahn told me. "And then we can home in on those we know are infected, and try to get the viruses out." Antibody screening is easy and quick. Performing PCR amplification and the other requisite steps to probe for fragments of viral RNA is far more laborious. The new methods allowed Hahn and her group to look first at a large number of specimens and then work more concertedly on a select few. They could separate the Shinola from the shit.

And they could expand their field surveying beyond Gombe. They could turn their attention back to the central chimpanzee, the animal whose SIV_{cpz} most closely matched HIV-1. Working now with Martine Peeters of Montpellier, plus some contacts in Africa, they collected 446 samples of chimpanzee dung from various forest sites in the south and southeast of Cameroon, after which Brandon Keele led the laboratory analysis. DNA testing showed that almost all the samples came from central chimpanzees (though a couple dozen derived from chimps belonging to a different subspecies, *P. t. vellerosus*, which range just north of a major river). Keele then looked for evidence of virus. The samples yielded two surprising results.

95

To hear about those surprises, I visited Brandon Keele, who by this time had finished his postdoc with Hahn and gone off to a research position at a branch of the National Cancer Institute, in Frederick, Maryland. He was still studying viral phylogenetics and AIDS, as head of a unit devoted to viral evolution. His new office and lab were on the grounds of Fort Detrick, inside the same fence as USAMRIID, where Kelly Warfield had worked on Ebola and, after her accident, spent three weeks in the Slammer. This time, since I was entering without an escort, soldiers at the guardhouse searched the underside of my rental car for a bomb before letting me pass. Keele, waiting to flag me down outside the door of his building, wore a blue dress shirt, jeans, his black hair moussed back, and a two-day stubble. He is a tall young man, extremely polite, raised and educated in Utah. We sat in his small office and looked at a map of Cameroon.

The first surprise to emerge from the fecal samples was the high prevalence of SIV_{cpz} in some communities of Cameroonian chimps. Two that scored highest, Keele said, were at sites labeled Mambele (near a crossroads by that name) and Lobeke (within a national park). Whereas all other sampling of chimps suggested that SIV infection was rare, the sampling in southeastern Cameroon showed prevalence rates up to 35 percent. But even there, the prevalence was "spotty," Keele said. "We can sample hundreds of chimps at a site and find nothing." But go just a little farther east, cross a certain river, sample again, and the prevalence spikes upward. That was unexpected. The rates were especially high in the farthest southeastern corner of the country, where two rivers converge, forming a wedge-shaped national boundary. This wedge of Cameroon appears to jab down into the Republic of the Congo, its neighbor to the southeast. The wedge was a hotspot for SIV_{cpz}.

The second surprise came once he extracted viral fragments from the samples, amplified those fragments, sequenced them, and fed the genetic sequences into a program that would compare these new strains with many other known strains of SIV and HIV. The program expressed its comparisons in the form of a most-probable phylogeny—a family tree. Keele recalled watching the results for a certain chimp, an individual labeled LB7, whose feces had been collected at Lobeke. "We were just shocked," he said. "I mean, I had ten people around my computer, all waiting to see what that sequence looked like." What it looked like was the AIDS virus.

When his computer delivered its latest tree, LB7's isolate of

SIV_{cpz} showed up as a twig amid the same little branch that held all known human strains of HIV-1 group M. (In scientific lingo, it fell within the same *clade.*) It was at that point "the closest thing" to a match, Keele told me, ever found in a wild chimp. "And then we find more, right? The more we dig, the more we find." The other close matches came from that same little area: southeastern Cameroon. A chilling, historic epiphany, at which Keele and his colleagues were thrilled. "You can't make this stuff up, as Beatrice would say. It's too good." Their joy lasted about ten seconds, after which everyone became hungry for more samples and more results. Your celebration is always provisional, Keele told me, until you've written the paper and gotten that congratulatory note of acceptance from the editors of *Science.*

Keele and the group now sequenced entire genomes (not just fragments) from four samples, all collected in the same area, and on those sequences ran their genetic analyses again. Again they found the new SIV_{cpz} shockingly similar to HIV-1 group M. The similarity was so close as to leave almost no chance that any other variant, yet undiscovered, could be much closer. Hahn's lab had located the geographical origin of the pandemic.

96

So much for *where* as well as *when.* AIDS began with a spillover from one chimp to one human, in southeastern Cameroon, no later than 1908 (give or take a margin of error), and grew slowly but inexorably from there. That leaves our third question: *how?*

The Keele paper appeared in *Science*, on July 28, 2006, under the title "Chimpanzee Reservoirs of Pandemic and Nonpandemic HIV-1." Brandon Keele was first author, with the usual list of coauthors, including Mario Santiago, Martine Peeters, several partners from Cameroon, and last again, Beatrice H. Hahn. The data were fascinating, the conclusions were judicious, the language was careful and tight. Near the end, though, the authors let supposition fly:

> We show here that the $SIV_{cpz\textit{Ptt}}$ strain that gave rise to HIV-1 group M belonged to a viral lineage that persists today in *P. t. troglodytes* apes in southeastern Cameroon. That virus was probably transmitted locally. From there it appears to have made its way via the Sangha River (or other tributaries) south to the Congo River and on to Kinshasa where the group M pandemic was probably spawned.

But the phrase "transmitted locally" was opaque. What mechanism, what circumstances? How did those crucial events occur and proceed?

Hahn herself, along with three coauthors, had addressed that back in 2000, when she first argued the idea that AIDS is a zoonosis: "In humans, direct exposure to animal blood and secretions as a result of hunting, butchering, or other activities (such as consumption of uncooked contaminated meat) provides a plausible explanation for the transmission." She was alluding to the cut-hunter hypothesis. More recently she addressed it again: "The likeliest route of chimpanzee-to-human transmission would have been through exposure to infected blood and body fluids during the butchery of bushmeat." A man kills a chimpanzee and dresses it out, hacks it up, in the course of which he suffers blood-to-blood contact through a cut on his hand. SIV_{cpz} passes across the species boundary, from chimp to human, and taking hold in the new host becomes HIV-1. This event is unknowable in its particulars but it's plausible, and it fits the established facts. Some variant of the cut-hunter scenario, occurring in a forest of southeastern Cameroon around 1908, would account not just for the Keele data but also for the timeline of Michael Worobey. But then what? One man in southeastern Cameroon is infected.

"If the spillover occurred there," I asked Hahn, "how was it that the epidemic began in Kinshasa?"

"Well, there are lots of rivers going down from that region to Kinshasa," she said. "And the speculation, the hypothesis, is that is how the virus traveled—in people, not in apes. It wasn't the apes that got into the canoe for a little visit of Kinshasa. It was the people who carried the virus down, most likely." Sure, she acknowledged, there was a slim chance that someone might have brought a live chimp, captive, infected, all the way down from the Cameroonian wedge—"but I think it is highly unlikely." More likely the virus traveled in humans.

Sexual contacts in the villages kept the chain of infection alive, though barely, by this line of speculation, and the disease didn't explode as a notable outbreak—not for a long while. When someone died of immune deficiency, the death may have seemed unremarkable amid all other sources of mortality. Life was hard, life was perilous, life expectancy was short even apart from the new disease, and many of those earliest HIV-positive people may have succumbed to other causes before their immune systems failed. There was no epidemic. But the chain of infection sustained itself. R_0 remained greater than 1.0. The virus seems to have traveled just as people traveled in those days: mainly by river. It made its way out of southeastern Cameroon along the headwaters of the Sangha, then down the Sangha to the Congo, then down the Congo to Brazzaville and Léopoldville, the two colonial towns on either side of what then was still known as the Stanley Pool. "Once it got into an urban population," Hahn said, "it had an opportunity to spread."

But still it moved slowly, like a locomotive just leaving the station. Léopoldville contained fewer than ten thousand people in 1908, and Brazzaville was even smaller. Sexual mores and the fluidity of interactions were unlike what prevailed in the boondocks, but not yet so unlike as they would become. R_0 for the virus must have continued to hover around 1.0. Time passed and more people drifted into the towns, drawn by the prospect of working for wages or selling their goods. Habits and opportunities changed. Women came as well as men, though not so many of them, and among those who did, more than a few entered the sex trade.

By 1914, Brazzaville contained about six thousand people and was "a hard mission field," according to one Swedish missionary, where "hundreds of women from upper Congo are professional prostitutes." The male population included French civil servants, soldiers, traders, and laborers, and they probably outnumbered females by a sizable margin, due to colonial policies that discouraged married men, coming there to work, from bringing their families. That gender imbalance heightened the demand for commercial sex. But the format for bought favors, in those early years, was generally different from what the word "prostitute" might suggest—grindingly efficient, wham-bam encounters with a long succession of strangers. Instead there were single women, known as *ndumbas* in Lingala and *femmes libres* in French, "free women" as distinct from wives or daughters, who would provide their clients with a suite of services, ranging from conversation to sex to washing clothes and cooking. One such ndumba might have just two or three male friends who returned on a regular basis and kept her solvent. Another variant was the *ménagère*, a "housekeeper" who lived with a white colonial official and did more than keep house. Commercial arrangements, yes, but these didn't represent the sort of prodigiously interconnected promiscuity that could cause a sexually transmitted virus to spread widely.

Across the pool in Léopoldville, meanwhile, the disparity of genders was even worse. This town was essentially a labor camp, controlled by its Belgian administrators, inhospitable to families, where the male-female ratio in 1910 was ten to one. Travel through the countryside and entry into Léopoldville was restricted, especially for adult females, though some women managed to get false documents or evade the police. If you were a restless, imaginative girl in one of the villages, poorly fed and poorly treated, to be a ndumba in Léopoldville could well have seemed enticing. Here too, though, even with ten horny men for each woman, commercial sex didn't happen in brothels or by streetwalking. Free women had their special friends, their clients, maybe several contemporaneously, but there was no dizzying permutation of multiple sexual contacts, not yet. One expert has called this "a low-risk type of prostitution," with regard to the prospects of HIV transmission.

Léopoldville also supported a lively market in smoked fish. Ivory, rubber, and slaves were traded there, for export, with profits going mainly to white concessionaires, well into the colonial era. Although a deep canyon and a set of forbidding cataracts stood between the Stanley Pool and the river's mouth, isolating both cities from the Atlantic, a portage railway built in 1898 breached that isolation, bringing more goods and commerce, which brought more people, and in 1920 Léopoldville replaced a downriver town as capital of the Belgian Congo. By 1940, its population had edged up to forty-nine thousand. Then the demographic curve steepened. Between 1940 and independence, which came in 1960, the city grew by almost an order of magnitude, to about four hundred thousand people. Léopoldville became Kinshasa, a twentieth-century African metropolis, where life was very different from what passed in a Cameroonian village. The tenfold population increase, along with the concomitant changes in social relations, might go a long way to explain why HIV "suddenly" took off. By 1959, the ZR59 carrier was infected, and a year later in the same city the carrier of DRC60 too. By that time the virus had proliferated to such a degree, mutating and diversifying, that DRC60 and ZR59 represented quite different strains. R_0 now must have well exceeded 1.0, and the new disease spread—through the two cities and eventually beyond. "You know," Hahn said, "a virus was at the right place at the right time."

When I read Keele's presentation of the chimp data and the analysis, in early 2007, my jaw dropped like a pound of ham. These folks had located Ground Zero, if not Patient Zero. And when I looked at the map—Figure 1 in Keele's paper, showing the Cameroonian wedge and its surroundings—I saw places I knew. A village where I had slept. A river I had ascended in a motor pirogue. Turns out that, during my journeys with Mike Fay across the Congo basin, seven years earlier, besides footslogging through Ebola country we had also passed very near the cradle of AIDS. After talking with Beatrice Hahn, I thought it might be illuminating to go back.

97

We rode east from Douala in a beat-up but sturdy Toyota truck, leaving at dawn, getting ahead of the crush, our gear stashed under tarps in the pickup's bed. Moïse Tchuialeu was my driver, Neville Mbah my Cameroonian fixer, and Max Mviri, from the Republic of the Congo, was along to handle things when we reentered his country in the course of the crazy loop I had planned. Max and I had flown up from Brazzaville the night before. We were a genial foursome, eager to move after the hassles of preparation, rolling past the closed shops and the billboards to the city's eastern fringe, where traffic thickened in a haze of blue diesel exhaust and the outlier markets were already open for business, selling everything from pineapples to phone minutes. Highway N3 would take us straight to Yaoundé, Cameroon's capital, and then another big two-lane onward from there.

During a stop in Yaoundé, around midday, I met with a man named Ofir Drori, head of an unusual activist group called LAGA (the Last Great Ape Organization) that helps government agencies in Central Africa to enforce their wildlife-protection laws. I wanted to see Drori because I knew that LAGA was especially engaged on the problem of apes being killed for bushmeat. I found him to be a lean Israeli expat with dark, alert eyes and a patchy goatee. Wearing a black shirt, black jeans, a black ponytail, and an earring, he looked like a rock musician or, at least, a hip New York waiter. But he seemed to be a serious fellow. He had come to Africa as an adventure-seeking eighteen-year-old, Drori told me, and gotten involved with human-rights work in Nigeria, then moved to Cameroon, did a little gorilla journalism (or was it *guerrilla* journalism?), and became a passionate antipoaching organizer. He founded LAGA, he said, because enforcement of Cameroon's antipoaching laws had been terrible, nonexistent, for years. The group now provides technical support to investigations, raids, and arrests. Subsistence hunting for duikers and other abundant, unprotected kinds of animal is legal in Cameroon, but apes, elephants, lions, and a few others are protected by law—and increasingly by actual enforcement. Perpetrators are finally getting busted, even doing time, for dealing in ape flesh and other contraband wildlife products. Drori gave me a LAGA newsletter describing the efforts to stem poaching of chimps and gorillas, and he warned me against the myth that ape hunting is a problem because local people are hungry. The reality, he said, is that local people eat duikers or rats or squirrels or monkeys—if they eat meat at all—whereas the fancy stuff, the illicit delicacies, the chimpanzee body parts, the gobs of elephant flesh, the hippopotamus steaks, get siphoned away by upscale demand from the cities, where premium prices justify the risks of poaching and illegal transport. "What brings the money are the protected species," he said. "Things that are rare." It sounded like the Era of Wild Flavor back in southern China.

Drori's newsletter mentioned a raid against a hidden storage room, at a train station, that served at least three different dealers; the room contained six refrigerators and its seized contraband included a chimpanzee hand. Another bust, against a dealer whose car held fifty kilos of marijuana plus a young chimp with a bullet wound, suggested diversified wholesale commerce. And if chimp meat travels toward money, chimp viruses presumably do too. "If you're thinking about infection," he said, knowing that I was, "don't just think of villages." Any chimpanzee killed in the southeastern corner of the country, including an SIV-positive individual, might easily end up here in Yaoundé, being sold for meat in a back alley or served through a very discreet restaurant.

We left the city in early afternoon, headed eastward again, moving against a stream of log trucks hammering toward us in the opposite lane, each one burdened to capacity with a load of just five or six gigantic stems. Somewhere out there, in that sparsely populated corner of the country, old-growth forests were being sheared. Around sundown we reached a town called Abong Mbang and stopped at the best local hotel, which meant running water and a lightbulb. Early the next day, an hour beyond Abong Mbang, the blacktop ended though the log trucks kept coming, now on a ribbon of rusty red clay. The temperature climbed toward midday equatorial heat and, wherever we encountered a little rain shower, the road steamed in red. Elsewhere the landscape was so dry that powdery red clay dust rose on the gusts from passing vehicles, coating trees along the roadside like bloody frost. We hit a police checkpoint and endured a routine but annoying shakedown, which Neville handled with aplomb, making two phone calls to influential contacts, refusing to pay the expected bribe, and yet somehow recovering our passports after only an hour. This guy is good, I thought. The road narrowed further, to a band of arsenical red barely wider than a log truck, leaving us hugging the shoulder when we encountered one, and the forest thickened on both sides. Around noon we crossed the Kadéï River, greenish brown and slow, meandering southeast, a reminder that we were now at the headwaters of the Congo basin. The villages through which we passed became smaller and looked progressively more spare and poor, with few gardens, little livestock, almost nothing for sale except bananas, mangoes, or a bowl of white manioc chips set out forlornly on an untended stand. Occasionally a goat or a chicken scampered out of our way. In addition to the log trucks, we now met flatbeds loaded with milled lumber, and I remembered hearing how such trucks sometimes carried a concealed stash of bushmeat, rumbling toward the black markets of Yaoundé and Douala. (A photographer and activist named Karl Ammann documented that tactic with a photo, taken at a road junction here in southeastern Cameroon, of a driver unloading chimpanzee arms and legs from the engine compartment of his log truck. The photo appeared in a book by Dale Peterson, titled *Eating Apes,* in which Peterson estimated that the human population of the Congo basin consumes roughly 5 million metric tons of bushmeat each year. Much of that wild meat—though no one knows just how much—travels out of the forest as contraband cargo on log trucks.) Apart from the trucks, today on this stretch of red clay, there was almost no traffic. By late afternoon we reached Yokadouma, a town of several thousand. The name translates as "Fallen Elephant," presumably marking the site of a memorable kill.

We found a local office of the World Wildlife Fund and, inside, two earnest Cameroonian employees named Zacharie Dongmo and Hanson Njiforti. Zacharie showed me a digital map plotting the distribution of chimpanzee nests in this southeastern corner of the country, which includes three national parks—Boumba Bek, Nki, and Lobeke. A chimpanzee nest is simply a small platform of interwoven branches, often in the fork of a smallish tree, which provides just enough support for the ape to sleep comfortably. Each individual makes one each night, though a mother will share hers with an infant. Tallying such nests, which remain intact for weeks after a one-night use, is how biologists estimate chimpanzee populations. The pattern on Zacharie's map was clear: a high density of nests (and therefore of chimpanzees) within the parks, a low density outside the parks, and none at all in areas adjacent to the roads leading to Yokadouma. Logging and bushmeat were the reasons. Logging operations bring roads and workers and firearms into the depths of the forest; dead wildlife consequently travels out. Zacharie and Hanson explained it as an informal, ad hoc form of commerce. "Most of the illegal trade is man-to-man," Hanson said. "A poacher meets you and says, I have meat." But it's also woman-to-man, he added: Much of the trading is done by "Buy 'em–Sell 'ems," women who travel between villages as petty traders, dealing openly in cloth, or spices, or other staples, and covertly in bushmeat. Such a woman buys directly from the hunter, often paying in bullets or shotgun shells, and sells to whomever she can. Commerce is relatively fluid; many of these women have cell phones. And there are all sorts of tricks, Hanson said, for getting the meat out. It could be tucked into a truckload of cocoa pods, for instance, a cash crop from this region. The police and the wildlife wardens get tips, and they can stop a truck and search it, but at some risk to themselves. If you stop a truck and demand it be unloaded, and then there's no illegal cargo, Hanson said, "the guy can sue you. The information has to be very good." That's why Ofir Drori's network has proved itself useful.

Most of the poachers, Zacharie added, are Kakaos, a tribe from the north with a strong propensity toward bushmeat. Many of them have drifted down here to the southeast, drawn by marital connections or opportunity in the bush. The local Baka tribe, on the other hand, has traditional strictures against eating apes, which are deemed too close to human. In fact, Zacharie reckoned, there was probably less eating of apes down here than in some other sectors of the country—apart from the totemic consumption of ape parts by Bakwele people in connection with a certain initiation ceremony for adolescent boys. And that offhanded comment from Zacharie was the first I'd heard of a Bakwele ritual known as *beka*.

We lingered in Yokadouma for two nights and a day, long enough for me to walk the dirt streets, admire the concrete statue of an elephant gracing the town's central roundabout, photograph a piteous pangolin about to be butchered for meat, and meet a fellow who told me more about beka. This man, whose name I'll omit, wrote a small report on the subject, which his organization declined to publish. He gave me a copy. Yes, he said, the Bakwele people here in the southeast use chimp and gorilla meat in their beka ceremony. They especially favor the arms. As a result, he said, "chimps are becoming more and more scarce." So scarce that gorilla arms are now often used as a substitute.

His report described a typical beka initiation, complete with slaughtered sheep and chickens, the neck of a tortoise (because it resembles a penis), and "virgin lasses" in attendance through a long prelude that culminates at four in the morning. The boy to be initiated is dressed in leaves and given drugs to keep him awake. Drums beat through the night until, before dawn, the boy is led into a special area of forest, where he's obliged to confront two chimpanzees. Some of what follows seems to be symbolic enactment, some of it blood-real. "A gong is sounded," according to a Bakwele chief who informed my source, "a voice calls out from the forest, and two chimpanzees respond. The male chimpanzee comes out first and touches the boy's head. The female chimpanzee emerges minutes after and the boy is expected to kill it." At dawn the boy bathes, then stays awake until late afternoon, pacing and expectant, at which point the circumciser comes at him with a homemade knife. "I nursed my wound for 45 days after," one initiate said. But now he was a man, no longer a boy. The unpublished report added:

> Until recently, the Bakweles have been using chimps for this ritual. They claim two chimps could be used for circumcision of as many as 36 people. They amputate the arms of the chimps. This part of the animal is eaten by elders of the village. Of late, however, due to the scarcity of chimps, Bakweles go for gorillas.

Eight gorilla arms had lately been seized when a poacher fled from game rangers, leaving the meat behind in a bag. The arms were intended for an impending beka. "We cannot do without these animals," the Bakwele chief complained, "if we must perform this important traditional rite."

It's no condescension against Bakwele culture to note that butchering chimpanzees to eat their arms, as part of an ancient and bloody ritual, could be a very good way to acquire SIV_{cpz}. Then again, in a landscape as lean and severe as southeastern Cameroon in 1908, beka might have been superfluous. Sheer hunger could account for the original spillover just as well.

98

Thirty miles farther south, at a crossroads known as Mambele Junction, with a central roundabout defined by three truck tires piled up like coins, we dined by kerosene lantern at a small cantina, eating smoked fish (at least, I hoped it was smoked fish) in peanut sauce and drinking warm Muntzig beer. This happened to be the place where Karl Ammann saw chimpanzee arms stashed under the hood of a log truck. It was also one of the locations featured in Brandon Keele's paper on the chimpanzee origins of HIV-1. Chimp fecal samples from hereabouts had shown high prevalence of the virus in its most fateful form. Somewhere very nearby was Ground Zero of the AIDS pandemic.

After dinner, my compadres and I stepped back outside and admired the sky. Although this was Saturday night, the lights of

Mambele Junction didn't amount to much and despite their dim glow we could see not just the Big Dipper, Orion's Belt, and the Southern Cross but even the Milky Way, arcing overhead like a great smear of glitter. You know you're in the boonies when the galaxy itself is visible downtown.

Two days later, at a modest building nearby that served as headquarters for Lobeke National Park, I met with the park's *conservateur*, its director, a handsomely bald man named Albert Munga, dressed in a floral shirt and (unmatched) floral pants. He sat aloof at his desk for several minutes, shuffling papers, before deigning to notice me, and then for a while longer he seemed cool to my questions about chimpanzees. The office was heavily air-conditioned; everything about it was cool. But after half an hour Mr. Munga warmed, loosened, and began to share some of his data and his concerns. The park's population of great apes (chimps and gorillas combined) had fallen abruptly since 2002, he told me: from about sixty-three hundred animals to about twenty-seven hundred. Commercial poachers were the problem, and by his account they came mainly across the eastern boundary of the park, the Sangha River, which happens also to be the southeastern border of Cameroon. Beyond the Sangha lie the Central African Republic and, slightly farther south, the Republic of the Congo, two countries that have known insurgency and war in the last couple decades. Those political conflicts brought military weapons (especially Kalashnikov rifles) into the region, vastly increasing the difficulty of protecting animals. Bands of well-armed poachers come across the river, mow down elephants and anything else they see, whack out the ivory and the elephant meat, lop off the heads and limbs of the apes, take the smaller creatures whole, and escape back across the water. Or else they move their booty downriver by boat. "There is a huge bushmeat traffic on the Sangha," Munga told me, "and the destination is Ouesso." The town of Ouesso, a river port of some twenty-eight thousand people, just over the border in Congo, is a major trading nexus on the upper Sangha. By no coincidence, it was my destination too.

Just outside Mr. Munga's office, I paused in the corridor to look at a wall poster with lurid illustrations and a warning in French: *LA DIARRHEA ROUGE TUE!* The red diarrhea kills. At first glance I thought that referred to Ebola, but no. "*Grands Singes et VIH/SIDA,*" read the finer print. SIDA is the French acronym for AIDS, and VIH likewise is HIV. The cartoonish but unfunny drawings depicted a stark parable about the connection between simian bushmeat and *la diarrhea rouge.* I lingered long enough for the oddness to strike me. Throughout the rest of the world you see AIDS-education materials crying out: *Practice safe sex! Wear a condom! Don't reuse needles!* Here the message was: *Don't eat apes!*

We drove onward, along a dirt track between walls of green, still farther into Cameroon's southeastern wedge. The country's southern border out here is formed by the Ngoko River, a tributary flowing east to its junction with the Sangha. The Ngoko, according to local lore, is one of the deepest rivers in Africa, but if so there must be a steep wrinkle of rock underneath, because it's only about eighty yards wide. We reached it around midday at a town called Moloundou, a scruffy place spread over small hills above the river. From any good point of vantage in Moloundou, the Republic of the Congo was easily visible across the water—close enough that, in the quiet of evening, we could hear the chainsaws of illegal loggers at work in the darkness over there. These log poachers would fell trees directly into the water and tangle them into rafts, I was told, and then float the rafts down to Ouesso, where a mill operator would pay cash, no questions asked. Ouesso again: the outlaw entrepôt. There was no government presence, no law, no timber concessionaires defending their interests, on that side—so said scuttlebutt on this side, anyway. We had reached the frontier zone, which was still a bit wild and woolly.

Early the next morning we walked up to the market and watched sellers setting out their goods in neat piles and rows: local peanuts and pumpkinseeds and red palm nuts, garlic and onions, manioc tubers, plantains, giant snails and smoke-blackened fish, hocks of meat. I hung back discreetly from the meat counters, leaving Neville and Max to investigate what was available. Mostly it was smoked duiker; no sign of ape meat being sold aboveboard; and even pangolin, a seller told Neville, was out of season. I hadn't expected different. Anything so valuable as a chimpanzee carcass would change hands in private, probably by prior arrangement, and not be slabbed out at a public market.

Downstream from Moloundou, the last Cameroonian outpost on the Ngoko River is Kika, a logging town with a big mill that provides jobs and lodging for hundreds of men and their families, plus a dirt airstrip for the convenience of its managerial elite. There was no direct riverside road (why would there be? the river *is* a road) so we circled back inland to get there. Arriving in Kika, we reported promptly to the police station, a small shack near the river that served also as immigration post, where an officer named Ekeme Justin roused himself, pulled on his yellow T-shirt, and performed the necessary formalities for me and Max: stamping our passports *sortie de Cameroon.* We would exit the country here. Officer Justin, upon receipt of a fee for his stamp work, became our great friend and host, offering us tent space there beside the police post and help in finding a boat. He went off to town with Neville, the all-purpose fixer, and by sunset they had arranged charter of a thirty-foot wooden pirogue, with an outboard, capable of getting Max and me to Ouesso.

I was up at five the next morning, packing my tent, eager to turn the corner on this big loop and head back into Congo. Then we waited through a heavy morning rain. Finally came our boatman, a languid young man named Sylvain in a green tracksuit and flip-flops, to mount his outboard and bail the pirogue. We loaded, covered our gear with a tarp against the lingering drizzle, and after warm goodbyes to the faithful Neville and Moïse, also Officer Justin, we launched, catching a strong current on the Ngoko. We pointed ourselves downriver. For me, of course, this journey was all about the cut-hunter hypothesis. I wanted to see the route HIV-1 had traveled from its source and imagine the nature of its passage.

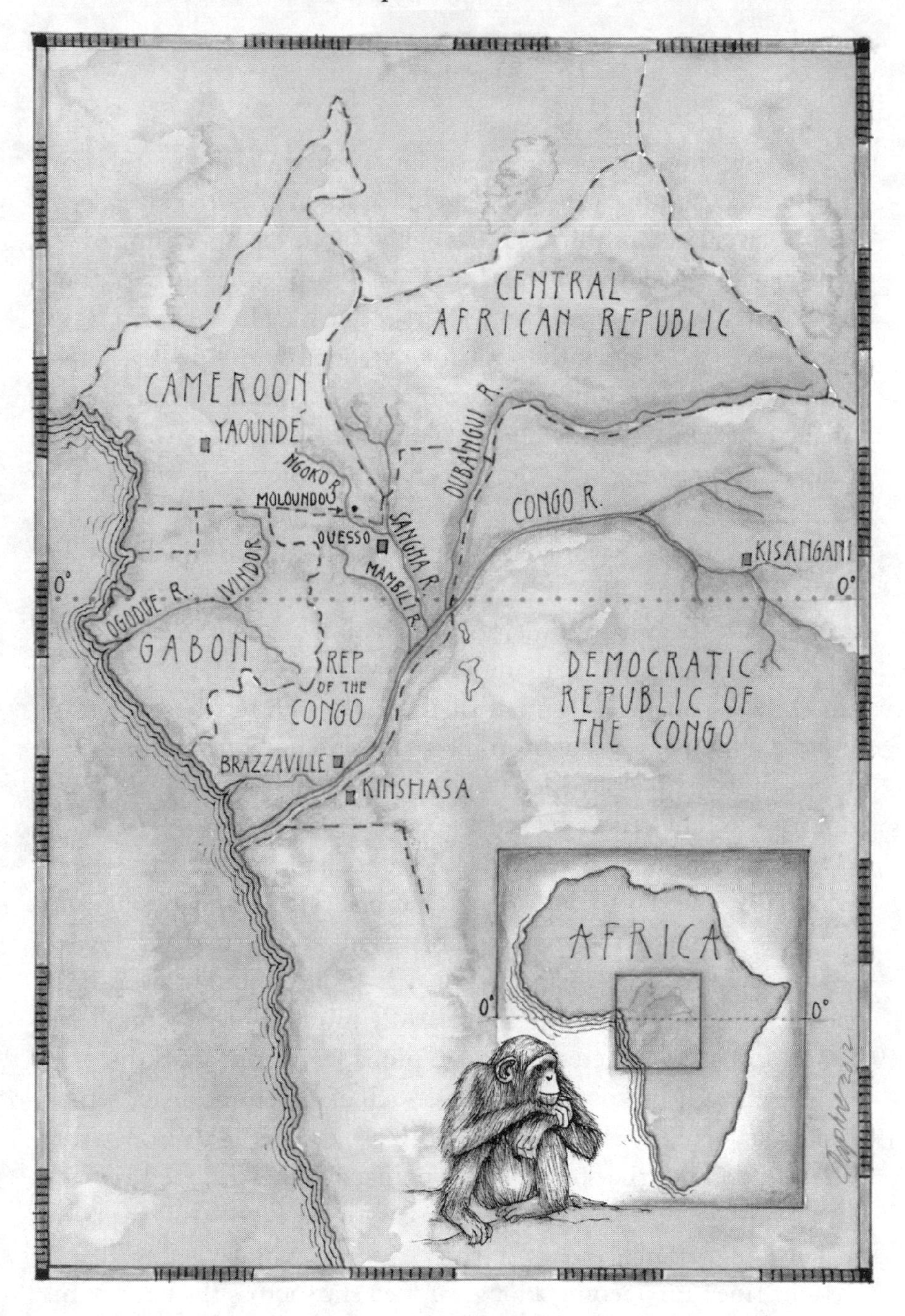
CENTRAL
AFRICAN REPUBLIC
CAMEROON
YAOUNDÉ
NGOKO R.
MOLOUNDOU
OUESSO
SANGHA R.
OUBANGUI R.
CONGO R.
KISANGANI
0°
0°
OGOOUE R.
IVINDO R.
MAMBILI R.
GABON
REP
OF THE
CONGO
DEMOCRATIC
REPUBLIC OF
THE CONGO
BRAZZAVILLE
KINSHASA
AFRICA
0°
0°

99

Let's give him due stature: not just a cut hunter but the Cut Hunter. Assuming he lived hereabouts in the first decade of the twentieth century, he probably captured his chimpanzee with a snare made from a forest vine, or in some other form of trap, and then killed the animal with a spear. He may have been a Baka man, living independently with his extended family in the forest or functioning as a sort of serf under the "protection" of a Bantu village chief. But probably he wasn't, given what I'd heard of Baka scruples about eating ape. More likely he was Bantu, possibly of the Mpiemu or the Kako or one of the other ethnic groups inhabiting the upper Sangha River basin. Or he may have been a Bakwele, involved in the practice of beka. There's no way of establishing his identity, nor even his ethnicity, but this remote southeastern corner of what was then Germany's Kamerun colony offered plenty of candidates. I imagine the man thrilled and a bit terrified when he found a chimpanzee caught in his snare. He had proved himself a successful hunter, a provider, a proficient member of his little community—and he wasn't yet cut.

The chimp too, tethered by a foot or a hand, would have been terrified as the man approached, but also angry and strong and dangerous. Maybe the man killed it without getting hurt; if so, he was lucky. Maybe there was an ugly fight; he might even have been pummeled by the chimp, or badly bitten. But he won. Then he would have butchered his prey, probably on the spot (discarding the entrails but not the organs, such as heart and liver, which were much valued) and probably with a machete or an iron knife. At some point during the process, perhaps as he struggled to hack through the chimp's sternum or disarticulate an arm from its socket, the man injured himself.

I imagine him opening a long, sudden slice across the back of his left hand, into the muscular web between thumb and forefinger, his flesh smiling out pink and raw almost before he saw the damage or felt it, because his blade was so sharp. And then immediately his wound bled. By a lag of some seconds, it also hurt. The Cut Hunter kept working. He'd been cut before and it was an annoyance that barely dimmed his excitement over the prize. His blood flowed out and mingled with the chimp's, the chimp's flowed in and mingled with his, so that he couldn't quite tell which was which. He was up to his elbows in gore. He wiped his hand. Blood leaked again into his cut, dribbled again into it from the chimp, and again he wiped. He had no way of knowing—no language of words or thoughts by which to conceive—that this animal was SIV-positive. The idea didn't exist in 1908.

The chimpanzee's virus entered his bloodstream. He got a sizable dose. The virus, finding his blood to be not such a different environment from the blood of a chimp, took hold. *Okay, I can live here.* It did what a retrovirus does: penetrated cells, converted its RNA genome into double-stranded DNA, then penetrated further, into the cells' nuclei, and inserted itself as DNA in the DNA genome of those host cells. Its primary targets were T cells of the immune system. A certain protein receptor (CD4) on the surface of those cells, in the Cut Hunter, was not very different from the equivalent receptor (another CD4) on the T cells of the butchered chimpanzee. The virus attached, entered the human cells, and made itself at home. Once integrated into the cellular genome, it was there for good. It was part of the program. It could proliferate in two ways: by cell replication (each time an infected T cell copied itself, the retroviral genome was copied also) and by activating its little subgenome to print off new virions, which then escaped from the T cell and floated off to attack other cells. The Cut Hunter was now infected, though apart from a slash on the hand he felt fine.

Forget about Gaëtan Dugas. This man was Patient Zero.

Maybe he carried the chimp carcass, or parts of it, back to his village in triumph—as the boys of Mayibout 2 later carried an Ebola-filled chimp carcass back to theirs. Maybe, if he was Baka, he delivered the whole thing to his Bantu overlord. He didn't want to eat it anyway. If he was Bantu himself, his family and friends feasted. Or maybe the chimp was a windfall from which he could afford to take special profit. If the season had been bounteous, with some duikers or monkeys killed, some forest fruits and tubers to eat, a good crop of manioc, so that his family wasn't starving, he may have lugged his chimpanzee to a market, like the one in Moloundou, and traded for cash or some valuable item, such as a better machete. In that case, the meat would have been parceled out retail and many people may have eaten bits of it, either roasted or smoked or dried. But because of how the virus generally achieves transmission (blood-to-blood or sexually) and how it doesn't (through the gastrointestinal tract), quite possibly none of those people received an infectious dose of virus, unless by contact of raw meat with an open cut on the hand or a sore in the mouth. A person might swallow plenty of HIV-1 particles but, if those virions are greeted by stomach acids and not blood, they would likely fail to establish themselves and replicate. Let's suppose that fifteen different consumers partook of the chimp meat and that they all remained fine. HIV-negative. Lucky folks. Let's suppose that only the Cut Hunter became infected directly from the chimp.

Time passed. The virus abided and replicated within him. His infectiousness rose high during the first six months, as virions in multitude bloomed in his blood; then the viremia declined some as his body mounted an early immune response, while it still could, and leveled off, for a period of time. He noticed no effects. He passed the virus to his wife, eventually also to one of the four other women with whom he had sex. He suffered no immune deficiency—not yet. He was a robust, active fellow who continued to hunt in the forest. He fathered a child. He drank palm wine and laughed with his friends. And then a year later, let's say, he died violently in the course of an elephant hunt, an activity even more perilous than butchering chimpanzees. He was one of seven men, all armed with spears, and the wounded elephant chose him. He took a tusk through the stomach, momentarily pinning him to the ground. You could see the tusk hole in the dirt afterwards, as though a bloody stake had been driven in and pulled. Of the men who scooped him up, the women who prepared him for burial, none had an open cut and so they were spared infection. His son was born HIV-negative.

The Cut Hunter's widow found a new man. That man was circumcised, free of genital sores, and lucky; he didn't become infected. The other woman who had been infected by the Cut Hunter took several partners. She infected one. This fellow was a local chief, with two wives and occasional access to young village daughters; he infected both wives and one of the girls. The chief's wives remained faithful to him (by constraint of circumstance if not by choice), infecting no one. The infected girl eventually had her own husband. And so, onward. You get the idea. Although sexual transmission of the virus occurred less efficiently from female to male, and not all so efficiently from male to female, it was just efficient enough. After several years, a handful of people had acquired the virus. And then still more, in time, but not many. Social life was constrained by small population size, absence of opportunity, and to some degree mores. The virus survived with an R_0 barely above 1.0. It passed to a second village, in the course of neighborly interactions, and then a third, but it didn't proliferate quickly in any of them. No one detected a wave of inexplicable deaths. It smoldered as an endemic infection at low prevalence in the populace of that little wedge of terrain, between the Ngoko River and the upper Sangha, where life tended to be short and hard. People died young from all manner of mishaps and afflictions. If a young man, HIV-positive, was killed in a fight, no one knew anything about his blood status except that it had been spilled. If a young woman, HIV-positive, died of smallpox during a local outbreak, likewise she left no unusual story.

In some cases, during those early years, an infected person may have lived long enough to suffer immune failure. Then there were plenty of ready bugs, in the forest, in the village, to kill him or her. That wouldn't have seemed remarkable either. People died of malaria. People died of tuberculosis. People died of pneumonia. People died of nameless fever. It was routine. Some of those people might have recovered, had their immune systems been capable, but no one noticed a new disease. Or if someone did notice, the report hasn't survived. This thing remained invisible.

Meanwhile the virus itself may have adapted, at least slightly, to its new host. It mutated often. Natural selection was at work. Given a marginal increase in its capacity to replicate within human cells, leading to increased levels of viremia, its efficiency of transmission may have increased too. By now it was what we would call HIV-1 group M. A human-infecting pathogen, rare, peculiar, confined to southeastern Cameroon. Maybe a decade went by. The bug survived. Spillovers of SIV_{cpz} into humans had almost certainly occurred in the past (plenty of chimps were butchered, plenty of hunters were cut) and resulted in previous chains of infection, but those chains had been localized and short. The smoldering outbreak had always come to a cold end. This time it didn't. Before such burnout could occur, another person entered the situation—also hypothetical but fitted to the facts—whom I'll call the Voyager.

The Voyager wasn't a hunter. Not an expert and dedicated one, anyway. He had other skills. By my imagining, he was a fisherman. He lived not in a forest clearing like the one at Mambele but in a fishing village along the Ngoko River. I picture him as a river boy from childhood; he knew the water; he knew boats. He owned a canoe, a good one, sturdy and long, made from a mahogany log with his own hands, and he spent his days in it. He was a young man with no wife, no children, and just a bit of an appetite for adventure. He had fallen away from his natal community at an early age, becoming a loner, because his father died and the village came to despise his mother, suspecting her of sorcery based on a piece of bad luck and a grudge. He took this as a deep personal bruise; he despised the villagers in return, screw them, and went his own way. It suited him to be alone. He was not an observant Bakwele. He never got circumcised.

The Voyager ate fish. He ate little else, in fact, besides fish and bananas—and sometimes manioc, which he didn't plant or process himself but which was easily acquired in trade for fish. He liked the taste and he loved the idea of fish, and there was always enough. He knew where to find fish, how to catch them, their varied types and names. He drank the river. That was enough. He didn't make palm wine or buy it. He was self-sufficient and contained within his small world.

He provided fish to his mother and her two younger children, as I see him, a loyal son though an alienated neighbor. His mother still lived at the fringe of the old village. His surplus catch he dried on racks, or in wet season smoked over a fire, at his solitary riverbank camp. Occasionally he made considerable journeys, paddling miles upstream or drifting downstream, to sell a boatload of fish in one of the market villages. In this way, he had tasted the empowerment of dealing for cash. Brass rods were the prevailing currency, or cowrie shells, and sometimes he may even have seen deutschmarks. He bought some steel hooks and one spool of manufactured line, which had come all the way from Marseille. The line was disappointing. The hooks were excellent. Once he had floated downstream as far as the confluence with the Sangha, a much larger river, powerful, twice as wide as the Ngoko, and had ridden its current for a day—a heady and fearful experience. On the right bank he had seen a town, which he knew to be Ouesso, vast and notorious; he gave it a wide berth, holding himself at mid-river until he was past. At day's end he stopped and slept on the bank; the next day he reversed, having tested himself enough. It took him four days of anxious effort to paddle back up, hugging the bank (except again at Ouesso), climbing through eddies, but the Voyager made it, relieved when he regained his own world, the little Ngoko River, and swollen with new confidence by the time he beached at his camp. This might have occurred, let's say, in the long dry season of 1916.

On another occasion, he paddled upstream as far as Ngbala, a river town some miles above Moloundou. It was during his return from that journey, as I posit, that he stopped at Moloundou and there, in his boat, where it was tied for the night in a shaded cove just below town, had sex with a woman.

She wasn't his first but she was different from village girls. She was a river trader herself, a Buy 'em–Sell 'em, several years older than he was and considerably more experienced. She traveled up and down the Ngoko and the Sangha, making a living with her wits and her wares and sometimes her body. The Voyager didn't know her name. Never heard it. She was outgoing and flirtatious, almost pretty. He didn't think much about pretty. She wore a print dress of bright calico, manufactured, not local raffia. She must have liked him, or at least liked his performance, because she returned to his boat in the shadows the next night and they coupled again, three times. She seemed healthy; she laughed merrily and she was strong. He considered himself lucky that night—lucky to have met her, to have impressed her, to have gotten at no cost what other men paid for. But he wasn't lucky. He had a small open wound on his penis, barely more than a scratch, where he'd been caught by a thorny vine while stepping ashore from a river bath. No one can know, not even in this imagined scenario, whether the lack of circumcision was crucial to his susceptibility, or the little thorn wound, or neither. He gave the woman some smoked fish. She gave him the virus.

It was no act of malice or irresponsibility on her part. Despite swollen and aching armpits, she had no idea she was carrying it herself.

100

River travel through tropical jungle is uncommonly soothing and hypnotic. You watch the walls of greenery slide by and, unless the channel is narrow enough for tsetse flies to notice your passing and come out from the shores, you suffer almost none of the discomforts. Because the riverbanks represent forest edges, admitting the full blast of sunlight, as closed canopy does not, the vegetation is especially tangled and rife: trees draped with vines, understory impenetrable, thick as an old velvet curtain at the Shubert Theater. It presents an illusion that the forest itself, its interior, might be as dense as a sponge. But to a river traveler that density is immaterial because you have your own open route down the middle. If you've walked the forest, which is difficult though not sponge-thick, river journeying is an escape from impediments that feels almost akin to flight.

For a while after leaving Kika, we favored the Congo side, riding a strong channel. Sylvain knew his preferred line. His assistant, a Baka man named Jolo, handled the outboard while Sylvain supervised, signaling directions from the bow. The pirogue was large and steady enough that Max and I could sit on the gunnels. Immediately we passed a small police post on the right bank, a Congolese counterpart to the Cameroonian one at Kika, and fortunately no one flagged us to stop. Every such checkpoint in Congo is an occasion for passport-stamping and minor shakedowns, and you want to avoid them when you can. Then we puttered past a few villages, widely spaced, each just a cluster of wattle-and-daub houses sited on a high bank to escape inundation in wet season. The houses were topped with thatch and surrounded by banana trees, an oil palm or two, children in rag dresses and shorts. The kids stood transfixed as we passed. How many hours to our destination? I asked Sylvain. Depends, he said. Ordinarily he would stop in villages along the way for trading or passengers, delaying long enough to enter Ouesso by darkness so as to escape notice by the immigration police. Not long after that explanation he did stop, guiding us ashore at a village on the Congo bank, to which he delivered a large plastic tarp and from which, on departure, we gained a passenger.

It was my charter but I didn't mind. She was a young woman carrying two bags, an umbrella, a purse, and a pot of lunch. She wore an orange-and-green dress and a bandana kerchief. I might have guessed if I hadn't been told: She was a Buy 'em–Sell 'em. Her name was Vivian. She lived down in Ouesso and would be glad for the ride home. She was lively and plump, confident enough to be traveling the river alone, trading in rice, pasta, cooking oil, and other staples. Sylvain liked to give her a lift because she was his sister—a statement of status that could be taken literally or not. She might have been his girlfriend or his cousin. Beyond this, I didn't learn much from Vivian except that her niche still exists, the Buy 'em–Sell 'em role, offering independent-spirited women a form of autonomy not easily found within village life, or even town life, and that the river still functions as a conduit of economic and social fluidity. Mostly she seemed a charming throwback and, though this might be unfair to her, put me in mind of women that the Voyager might have met almost a century earlier. She was a potential intermediary.

When the rain returned, Max and I and Sylvain and Vivian hunkered beneath our tarp, heads down but peeking out, while Jolo the Baka stolidly motored us on. We passed a solitary fisherman in his canoe, pulling a net. We passed another village from which children stared. Then the rain died again and the storm breeze fell off; the gentle chop disappeared, leaving the river as flat and brown as a cooled café au lait. Mangroves reached out from the banks like groping octopuses. I noticed a few egrets but no kingfishers. In midafternoon we approached the confluence with the Sangha. Along the left bank, the land fell gradually lower and then tapered, sinking away into the water. The Sangha River gripped us, swung us around, and I turned to watch that southeastern wedge of Cameroon recede to a vanishing point.

The air warmed slightly with an upstream breeze. We passed a large, wooded island. We passed a man standing upright in his dugout, paddling carefully. And then in the distance ahead, through haze, I saw white buildings. White buildings meant bricks and whitewash and governmental presence in something larger than a village: Ouesso.

Within half an hour we landed at the Ouesso waterfront, with its concrete ramp and wall, where an officer from the immigration police and a gaggle of tip-hungry, scuttering porters awaited. Stepping ashore, we had reentered the Republic of the Congo. We completed the immigration formalities in French and then Max dealt with the bag-grabbing porters in Lingala. Sylvain and Jolo and Vivian melted away. Max was a shier, less forceful fellow than Neville but conscientious and earnest, and now it was his turn to be my fixer. He made some inquiries here along the waterfront and soon had good news: that the big boat, the cargo-and-passenger barge known as *le bateau*, would be departing tomorrow for Brazzaville, many miles and days further downriver. I wanted us to be on it.

We found a hotel, Max and I, and in the morning walked to the Ouesso market, which was centered in a squat, pagoda-shaped building of red brick just blocks from the river. The pagoda was big and stylish and old, with a concrete floor and a circular hall beneath three tiers of corrugated metal roof, dating back at least to colonial times. The market had far outgrown it, sprawling into a warren of wood-frame stalls and counters with narrow lanes between, covering much of a city block. Business was brisk.

A study of bushmeat traffic in and around Ouesso, done in the mid-1990s by two expat researchers and a Congolese assistant, had found about 12,600 pounds of wild harvest passing through this market each week. That total included only mammals, not fish or crocodiles. Duikers accounted for much of it and primates were second, though most of the primate meat was monkey, not ape. Eighteen gorillas and four chimps were butchered and sold during the four-month study. The carcasses arrived by truck and by dugout canoe. As the biggest town in northern Congo, with no beef cattle to be seen, Ouesso was draining large critters out of the forest for many miles around.

Max and I snooped up and down the market aisles, stepping around mud holes, dodging low metal roofs, browsing as we had done in Moloundou. Because this was Ouesso, the merchandise was far more abundant and diverse: bolts of colorful cloth, athletic bags, linens, kerosene lanterns, African Barbie dolls, hair falls, DVDs, flashlights, umbrellas, thermoses, peanut butter in bulk, powdered fufu in piles, mushrooms in buckets, dried shrimp, wild fruits from the forest, freshly fried beignets, blocks of bouillon, salt by the scoop, blocks of soap, medicines, bins of beans, pineapples and safety pins and potatoes. On one counter a woman hacked at live catfish with a machete. Just across from her, another woman offered a selection of dead monkeys. The monkey seller was a large middle-aged lady, her hair in cornrows, wearing a brown butcher's apron over her paisley dress. Genial and direct, she slapped a smoked monkey down proudly in front of me and named her price. Its face was tiny and contorted, its eyes closed, its lips dried back to reveal a deathly smile of teeth. Split up the belly and splayed flat, it was roughly the size and shape of a hubcap. *Six mille francs*, she said. Beside the first monkey she tossed down another, in case I was particular. *Six mille* for that one too. She was talking in CFA, the weak Central African currency. Her six thousand francs amounted to US $13, and was negotiable, but I passed. She also had a smoked porcupine, five duikers, and another simian, this one so freshly killed that its fur was still glossy and I could recognize it as a greater spot-nosed monkey. That's a premium item, Max said, it'll go fast. Nearby, gobbets of smoked pork from a red river hog were priced at three thousand francs per kilo. All these animals could be hunted legally (though not with snares) and traded openly in Congo. There was no sign of apes. If you want chimpanzee or gorilla meat in Ouesso it can still be had, no doubt, but you've got to make private arrangements.

Our trip downriver on the bateau suffered complications and delays so that, four days later, Max and I were back in Ouesso. Revisiting the market, we passed again through the pagoda, down the narrow aisles between stalls, along the counters piled with catfish and monkeys and duikers, smoked and fresh. This time I noticed a wheelbarrow full of smallish crocodiles and saw one croc being whacked apart on a plank. You could locate the meat section from anywhere in the market maze, I realized, by that sound—the steady *thunk-thunk!* of machetes. And then we came again to the brown-aproned lady, who remembered me. "You've returned," she said in French. "Why don't you buy something?" This time she plunked down a little duiker, more as a challenge than as an offering: *Are you a shopper or a voyeur?* I prefer chicken, I said lamely. Or smoked fish. Unsurprised by the pusillanimity of the white man, she smiled and shrugged. Then, as a flyer, I said: But if you had *chimpanzee* . . . She ignored me.

Or *elephant*, Max added. Now she laughed noncommittally and turned back to her real customers.

101

The idea of Ouesso and its market served as a crucial enticement to get the Voyager, as I imagine him, on his way. That's where the wildcat notion of his wildcat journey began: Ouesso. He hadn't intended on going farther. A trip down to Ouesso and back (he *had* meant to come back, though life unfolded otherwise) would be ambitious and risky enough. But even before the idea of Ouesso, there was the dizzying happenstance of the tusks. If it was Ouesso that pulled him, it was the tusks that pushed him.

He had never gone looking for ivory. It came by accident. One day he was upriver on the Ngoko, working his net at the mouth of a feeder stream that drained from the Congo side. It was dry season—near the end of the long dry season, early March. The river was low and slow and warm, which was why he thought the freshening flow of the feeder stream might attract fish. As it happened, not many. The catch there scarcely repaid his effort. So in midafternoon he decided to walk inland, back-following this little stream into the forest, looking for pools where small fish might be trapped and vulnerable. He fought his way along the mud banks for almost half a mile, through the thorn vines, over the cobble of roots, finding few pools and no fish. It was frustrating but not surprising. He paused for breath, dipped up a handful of water to drink, and frowned ahead, deciding whether to continue. That's when he noticed a large gray mound in the stream bottom about forty yards on. To you or to me it would have looked like a granite boulder. But there are no granite boulders in northern Congo or southeastern Cameroon, and the Voyager had never seen one. He knew immediately what it was: an elephant. His heartbeat surged and his first instinct was to run.

Instead he stared. His legs didn't go. He lingered, unsure why. He sensed terror in the scene somewhere, but the terror wasn't his. Then he realized what seemed wrong—the elephant was down, and not in a position that might suggest sleep. Its face lay smashed into the mud, its trunk sideways, its hip canted up. He approached carefully. He noticed the purplish red holes along its lower sides and belly. Protruding from one of those holes was a Baka spear. He could see the awful way the beast had collapsed down over its left shoulder, its front leg on that side bent out at a ruinous angle. By the time he had crept within ten yards, he knew that it was dead.

A sizable male, middle-aged, with good ivory. Left to die alone in a stream bottom and rot. Quickly the Voyager made some deductions. Probably it had been killed by a hunting party of Baka men—but not quite killed, just mortally wounded. It had broken away, escaped, and to do that, presumably, it would have had to kill one or two of the Baka who surrounded it. The others must have lost heart for the chase. Maybe this had occurred on the north side of the river. Maybe the elephant, wounded and desperate, had swum across. But if the Baka took up the trail, got themselves over here, and reappeared now—that could be bad for him. Finding the Voyager with their costly trophy, the Baka might fill *him* with purple spear holes. So he worked fast. He whaled into the elephant's face with his machete, hacking through flesh and gristle, opening an ugly maw that no longer looked elephantine but like something else, something exploded and ogrish, and within half an hour he had twisted both tusks free. They surrendered with ripping noises, like any tooth drawn from its jaw.

He shaved the tusks free of tissue, then rubbed them with sandy mud and rinsed them white in the stream. Held in his hands, each one seemed huge. Bounteous. Maybe fifteen kilos. He had never experienced such an armload of wealth. He could only handle one at a time. He examined each in turn, passing his hand down the smooth white curve to the point. Then he gathered up both and staggered back to his canoe, crouching and dodging through the vines, and dropped them into the bilge with his few fish. Untied the boat quickly, caught the current, headed downstream. Having rounded one bend, he began to ease, his heart slowing back to normal.

What had just happened? He had stumbled upon half a fortune and stolen it, that's what. Claimed it, rather. Now what?

Back at his camp, the Voyager cached the tusks hastily beneath leaves and branches in a recess beside a fallen tree. Midway through the first night he woke, suddenly aware that his hiding place was inadequate, stupidly so, and he waited out the darkness impatiently. At daylight he rose, scraped away the coals and embers and ash from his campfire—his hearth site of several years' custom—and dug a pit on that spot, cracking through the layer of baked earth with his machete, slapping deep slices into the clay beneath. He went down four feet. He shaped a deep, narrow slot. He wrapped the two tusks in *ngoungou* leaves for protection and nestled them at the bottom of the trench. Then he refilled it, leveled the ground carefully, spread the old ashes back where they'd been, replaced the charred logs, and lit a new fire. Now his treasure was safe, maybe, for a while. And he could think about what to do.

There were no easy answers. There was opportunity and there was risk. He was not a man who hunted elephant, and everyone who knew him knew that. He was not supposed to have tusks. If he took them to Moloundou the agents of the French concessionaires, greedy for ivory, leaching it from the forest by all manner of compulsion and threat, would simply impound them. He might even be punished. Others would try to steal them, or to trade for them while cheating him of their value. He thought through the scenarios. He wasn't a cunning man but he was tough and stubborn.

Six months passed. He continued to live as before: fishing the river, drying fish at his camp, spending his days alone, making infrequent stops at Ngbala or Moloundou for trading. There was one man in Moloundou, a merchant, not a local Bantu and not a concessionaire's agent but a half-Portuguese outsider with connections, notoriously clever, known to deal discreetly in elephant meat and ivory. One day during a transaction over fish, salt, and fufu, the Voyager asked this merchant about the price for tusks. *It was just a question!* The merchant looked at him slyly and mentioned a number. The number seemed high but not very high, and the Voyager's face may have flickered with disappointment. He said nothing more.

Two nights later, the Voyager returned from upriver and found his camp wrecked. The half-Portuguese merchant had spoken with someone, and that someone had gone straight to rob him.

His hut had been ripped apart, his drying racks broken. His few possessions—his second net, some tin pots, a camp knife, a shirt, his raffia mat, and the rest—had been scattered disdainfully. His little tin box had been broken open and the fishhooks and tobacco dumped out. Dried fish lay on the ground, willfully trodden upon. There were signs of digging here and there—beside the fallen log, in the floor of his hut, a couple other places too. Desultory, petulant searching. The Voyager's campfire had been scattered, logs and ashes kicked away. His breath caught when he saw that. But the dirt beneath the ashes hadn't been disturbed. They hadn't found what they had come for.

So he turned his mind toward Ouesso. He waited out the night in his ruined camp, beside a fire burning low, with his machete in hand. At dawn he excavated his tusks and, leaving them leaf-wrapped and dirty, without pausing to savor their cool precious weight, put them into his canoe. He covered the tusks with dried fish, of which he had plenty, and smoked fish, of which he had just a bit, and then covered the fish with more ngoungou leaves in neat bundles, as though he were taking them to market. Ngoungou leaves had their value as wrapping, but it was minimal; a pathetic, countryman's product, and therefore plausible. Atop the leaves he placed his mat. He pushed off, paddled out, and let himself be swung downriver on the Ngoko, putting Moloundou behind him. He paddled steadily for hours, reached the Sangha, turned downstream there, and continued straight to Ouesso.

Half a mile below the town, he found an eddy and pulled his boat up into the forest. There was no landing, no trail, no camp, no sign of human presence—which was good. The next day he concealed the canoe beneath leafy branches and bushwhacked northwest until he struck the outer lanes of Ouesso. He walked straight to the market by following other people. He had never seen such a concentration of humans and, once he was amid the crowd, his heart began thumping as it had when he stood over the dead elephant. But no one hurt him; no one even looked at him twice, despite the fact that his clothes were shabby and he carried a machete. He saw other men in dirty clothes, a few, and one or two of them carried machetes also. He began to relax.

The market, sheltered in a huge round building with a metal roof, was wondrous. You could buy meat, you could buy fish, you could buy colorful clothing and dried manioc and vegetables and fishnets and things he had never seen. The Voyager had no money of any sort, not francs, not brass rods, but he wandered among the goods as though he might want something. He admired the duikers and the monkeys. He picked up a gorilla hand, while the seller woman watched him closely, and set it back down. The people spoke Lingala. He exchanged a few words with a man selling fish. The Voyager was more cautious than he had been in Moloundou. Do you buy smoked fish if I have some? he asked. Maybe, when I see it, the man said. The Voyager took note of another man nearby, behind a plank table upon which sat large chunks of elephant meat, smoky and gray. A man who sold elephant meat might also deal in ivory. The Voyager memorized that man's face but didn't speak with him. He would do it tomorrow.

He walked back out of town and into the forest, satisfied by his judicious preliminary excursion, and when he emerged through the undergrowth to his riverbank hiding spot, he was horrified to see the cut branches cast aside and someone bent over his boat. Horrified and enraged: at himself for his redoubled stupidity, at the world, and especially at the man coveting his tusks. The Voyager raised his machete, ran forward, and struck before the interloper had half turned around, splitting the man's skull like a dry coconut. That made a sickening, fateful sound. The man fell hard. Where his head had broken open, pink brains showed and blood surged around the pinkness, and then stopped.

It was scarcely midafternoon of the Voyager's first day in Ouesso and he had killed someone. What sort of hellish place was this?

His next shock came when he rolled the dead man over. It wasn't a man's face; it was a boy's. Smooth skin, baby cheeks, long jaw, barely old enough for initiation. The Voyager had been fooled by height. He had killed a tall youngster, a gangly boy who had dared to stoop over his canoe. A boy from the town, with relatives who would miss him. This wasn't good.

The Voyager stood for a moment, exhausted and pained, calculating his situation. Then again he moved quickly. He dragged the boy's body to the river. Splashing into the shallows, stumbling, he pulled it offshore just enough to be sure of current, released it, and watched it drift away. The body floated low in the water but it floated. Back on the bank, he rifled down into his canoe and confirmed that the tusks were still there. They were. He gripped each individually at the tip, assuring himself: one, two. He peeled back the leaf wrap and looked. Yes, ivory, two tusks. He dragged his canoe to the water, climbed in, and began paddling downstream. Within fifty yards he caught up with the boy's body and passed it by. He did not glance back toward Ouesso.

Now he was launched, untethered, no going back. For three weeks he journeyed downstream. Or maybe four weeks; he didn't keep a tally of the days. He had his canoe and his tusks, his machete, his fishing line and hooks, little else. His immediate purpose was to stay alive, day by day. His driving goal was to recoup a life from the ivory. He resumed fishing as he went, trolling with his line, seldom stopping except for the night. He ate what he caught, saving the dried and smoked fish for contingency. He was on the water again every morning by full light. He passed another town, avoiding it along the far bank, and paddled through a stretch where the river meandered slowly amid swamplands. He could see it was taking him generally south. There were adventures and mishaps and some further narrow escapes along the way. Maybe you can imagine them as well as I can. There was the encounter with the men on the log raft, drifting downriver, to whom he sold fish and by whom he was warned about the Bobangi, an imperious people controlling trade and passage at the mouth of the Sangha. He didn't know what that meant, the mouth of the Sangha; he pictured this river going on forever. There was the ambush by the crocodile, another hateful moment, but he had been lucky that morning. It was a nasty animal, not large, barely six feet, presumptuous and stupid to attack a human, and he'd had his revenge. He ate the belly meat and tail of the crocodile for six days afterward. He had never eaten chicken so to him it tasted like fish. He placed the crocodile's severed head into a column of driver ants and they cleaned it of flesh within an afternoon. Now the sun-bleached skull rode atop the other cargo in his canoe, toothy and grinning, like a totem. He reached the mouth of the Sangha and tried to elude the Bobangi, running midriver at night and laying up by day. But he couldn't stay with his treasures every moment. He left the boat unguarded once, for only a short time, to gather fruit beneath a *mobei* tree, and so there was his standoff with the solitary Bobangi man whom he found, as he had found the Tall Boy, committing an outrage: looking into his canoe. Unlike the Tall Boy, this man heard him and turned around.

The man had gray hair at his temples and his left eye was milky blue. His right eye was normal. He was old but not too old to be dangerous; his body appeared still strong. He carried a small iron knife, but no machete, and a little packet in animal hide strung around his neck. He looked like a magus or a sorcerer. He had unwrapped the Voyager's ivory. The Voyager knew that there were many other Bobangi on the river, maybe even some within earshot. The Voyager felt trapped. He remembered the sickening sound of his machete on the Tall Boy's head. He decided, very quickly, upon a desperate compromise. He addressed the blue-eyed man in Lingala, not sure whether a Bobangi would understand.

I give you one tusk, the Voyager said.

No sign of response.

I give you one tusk, he repeated, speaking very clearly. *You deliver it to your chief. Or . . . you don't.*

He waited, letting the blue-eyed man ponder.

One tusk, he said. He held up a finger. *Or I fight you and I kill you for two.*

It seemed a long delay. The Voyager began wishing he had simply cracked the man's skull, at least tried to, whatever the consequences. Then the blue-eyed man turned back to the Voyager's canoe. He rummaged, shoved away leaves, and lifted out one tusk. He stroked it, testing the smooth cool surface, and appeared satisfied. The Voyager watched him; willed him on his way. *All right. Take it. Go.* But then, no, the man stooped again. He picked up a single smoked fish. He gaped back at the Voyager with an expression of shameless, bemused defiance. The blue eye twitched—or was that a wink? He took the tusk and the fish and he departed.

That night the Voyager passed onward through Bobangi territory, slipping by their big village near the mouth of the Sangha, where this river debouched into another, unimaginably huge: the Congo. He was astonished when daylight revealed the extent of its braiding channels, islands, and strong currents. It was like a bundle of rivers, not just one. He paddled harder than ever now, but also more carefully, learning wariness of the eddy lines that could knock a canoe sideways, the whirlpools that could suck it under. He kept a distance between himself and other canoes. When he saw men on a raft, he paddled within shouting distance, offered to sell fish, sought information. Once he encountered a steamboat, like a great house proceeding upriver under power, with a machine inside thumping stupidly, passengers and bundled cargo on the deck. It was a strange sight. But the Voyager had seen other strange sights—the spilled brains of a boy, the Ouesso market, a blue-eyed Bobangi thief—and by now felt almost inured to astonishment. The boatman, he could see, was a white man. The Voyager hugged the opposite shore.

The river continued south. He entered the territory of the Tio, a more tractable people than the Bobangi—eager for trade but not demanding monopoly, according to what the Voyager heard. Maybe the Tio were humbler because the river was now so vast. No one could imagine himself owning such a river. No tribe, even. Here the Voyager saw dozens of other boats. It was a new universe. Many canoes, several more steamboats, people hollering and trading from one boat to another. The maze of channels and the traffic, plus the increasing distance from Ouesso, gave a sense of jumble and anonymity and security that allowed the Voyager to travel by daylight, which was fortunate in these formidable waters. He sold fresh fish to Tio boatmen and swapped fish for manioc. He chatted. *Yes, I've come from the upper river, very far.* But he didn't say which river. He didn't mention ivory. He gathered intelligence without revealing much. He was tired.

He had an intermediate goal now, between the daily purpose of survival and the dream of due reward for his troubles. He had a destination: a place called Brazzaville. It was a large town, downriver, some days ahead. It sat on the right, beside a great pool. He would know it when he saw it—so he'd been told. Another big town sat on the left bank, across the pool, but that one was owned by the Belgians. *Who are the Belgians?* he asked. *Are they a tribe like the Bobangi?* Worse. Yes, he heard, Brazzaville was a good market for fish or whatever you had.

And so the Voyager arrived. He rounded a last bend, came to a great pool where the river seemed as wide as it was long, put a large island to his left as advised, and saw white buildings on the right bank, some of them twice as tall as a house, taller even than the circular market hall at Ouesso. He paddled toward the white buildings. Coming near, he held himself some distance out, drifting, observing, until he was well past the docks and the big boats and the bustle of workingmen, and then beached his canoe in a quieter place. Several children gaped, as children do, but no one else noticed him. People were busy and no adults diverted their attention to the sight of a strong young Bakwele coming ashore in tattered clothes with a crocodile skull, a single fine tusk, and half a boatload of rotten fish.

He stepped out of the water and stood alone. No one greeted him.

No one knew what he had done. No one compared him to Lewis and Clark. No one hailed him as the Marco Polo of the upper Congo basin. No one knew that he was Huck Finn and Jim, John Wesley Powell on the Colorado, Teddy Roosevelt on the River of Doubt, Frank Borman circling the moon in *Apollo 8*, and Dr. Richard Kimble at large. No one knew.

The Voyager walked into town and sold his tusk the first afternoon, receiving 120 brass rods, which was a good price, he thought, but also somehow anticlimactic and unsatisfactory. For his crocodile skull, at the benign whim of the ivory buyer, he received another ten brass rods. He bought some palm wine, got drunk, found that experience not to his liking, and never did it again. The rest of his money he saved, or rather set aside, spending it slowly and variously until it was gone. He had arrived.

He found lodging in Poto-Poto, a neighborhood east of the city center, full of others from the upper river, and he found work on the waterfront. He made friends. He settled in. Urban life suited him. He became something of a colorful figure, confident, charming in his river-man way, with stories to tell. No one viewed him as the pariah son of a sorceress. No one guessed that he had ever been a surly young loner. No one knew his real name because he had invented another. And the other thing no one knew, not even he, was that he had brought a new element, a new circumstance, to Brazzaville. A virus, in his blood. More specifically: He had brought HIV-1 group M.

Seven and eight and nine years later, near the end of his life, the Voyager would tell some of his stories to friends, acquaintances, and a few of the women with whom he had relationships, transient or longer: about the Dead Elephant, the Half-Portuguese Merchant, the Tall Boy, the Crocodile, and the Blue-Eyed Bobangi. In his telling, the Tall Boy became an adult and the Crocodile was very large, a leviathan. No one doubted his word. They knew he had come down the river and it must have been perilous. The crocodile skull wasn't there to belie him. During those years he slept with thirteen women, all of whom were *femmes libres* to one degree or another. One of those, a young Tio girl who had recently arrived in Brazzaville from upriver, and who found that she fancied him more than she did her freedom, became his wife. Eventually he infected her with the virus. He also infected one other, a rather more professional woman who lived in a small house in the Bacongo neighborhood, west of town, where he visited her on an intermittent basis when his wife was pregnant. The other eleven women had only fleeting sexual contacts with him and were luckier. They remained HIV-negative. The Voyager's personal lifetime R_0 was therefore precisely 2.0. People liked him and were sorry when he fell ill.

The Bacongo girlfriend was vivacious and pretty and ambitious for wider horizons, so she crossed the pool to Léopoldville, where she had a successful career, though not a long one.

102

If the virus reached Léopoldville in 1920 or so, that still leaves a gap of four decades to the time of ZR59 and DRC60, those earliest archival HIV sequences. What happened during the interim? We don't know, but available evidence allows a rough sketch of the outlines of possibility.

The virus lurked in the city. It replicated within individuals. It passed from one person to another by sexual contact, and possibly also by the reuse of needles and syringes for treatment of well-known diseases such as trypanosomiasis. (More on that possibility, below.) Whatever its means of transmission, presumably HIV caused immune deficiency, eventually death, among most or all people infected—except those who died early from other causes. But it didn't yet assert itself conspicuously enough to be recognized as a distinct new phenomenon.

It may also have proliferated slowly in Brazzaville, across the pool, helped along there too by changing sexual mores and programs of therapeutic injection. It may have lingered in villages of southeastern Cameroon or elsewhere in the upper Sangha basin.

And wherever it was, but definitely in Léopoldville, it continued to mutate. The wide divergence between ZR59 and DRC60 tells us that. It continued to evolve.

Studying the evolutionary history of HIV-1 is more than an idle exercise. The point is to understand how one strain of the virus (group M) made itself so deadly and widespread among humans. Such understanding, in turn, may lead toward better measures to control the devastation of AIDS, possibly by way of a vaccine, more likely by way of improved treatments. That's why scientists such as Beatrice Hahn, Michael Worobey, and their colleagues explore the molecular phylogenetics of HIV-1, HIV-2, and the various SIVs. One issue they address is whether the virus became virulent before, or only after, its spillover from chimpanzees. To state the question more plainly: Does SIV_{cpz} kill chimps, or is it only an innocuous passenger? Answering that one could reveal something important about how human bodies respond to HIV-1.

For a while after the discovery of SIV_{cpz}, the prevailing impression was that it's harmless in chimpanzees, an ancient infection that may once have caused symptoms but no longer does. This impression was bolstered by the fact that, in the earlier years of AIDS research, more than a hundred captive chimpanzees were experimentally infected with HIV-1 and none showed immune system failure. When a single lab chimpanzee *did* progress to AIDS (ten years after experimental infection with three different strains of HIV-1), its case was remarkable enough to merit a six-page paper in the *Journal of Virology*. The researchers implied that this was good news, finally offering hope that chimpanzees do represent a relevant experimental model (that is, a sufficiently analogous test subject) for studying human AIDS. There was even a report, based on genetic analysis of captive animals in the Netherlands, suggesting that chimpanzees had "survived their own AIDS-like pandemic" more than 2 million years ago. They emerged from the experience, according to this line of thought, with genetic adaptations that render them resistant to the effects of the virus. They still carry it but apparently don't get sick. That notion, to repeat, was founded on captive chimpanzees. As for SIV-positive chimps in the wild, no one knew whether they suffer immunodeficiency. It was a difficult question to research.

These suppositions and guesses jibed with available information about other variants of the virus in other primates. SIV is highly diverse and broadly distributed, found as a naturally occurring infection in members of more than forty different species of African monkey and ape. (But it seems to be unique to that continent. Although some Asian primates have acquired the virus in captivity, it hasn't shown up among wild monkeys in either Asia or South America.) Most of those SIV-carrying African simians are monkeys. Each kind of monkey harbors its own distinct type of SIV, such as SIV_{gsn} in the greater spot-nosed monkey, SIV_{ver} in the vervet, SIV_{rcm} in the red-capped mangabey, and so forth. Based on evidence presently available, none of those SIVs seems to cause immunodeficiency in its natural host. A close evolutionary kinship between two kinds of simian, such as L'Hoest's monkey and the sun-tailed monkey, both classified in the genus *Cercopithecus*, is sometimes paralleled by a close similarity between their respective SIVs. Those deep taxonomic alignments, plus the absence of noticeable disease, led researchers to suspect that African monkeys have carried their SIV infections for a very long time—probably millions of years. That length of time would allow divergence among the viruses and mutual accommodation between each type of virus and its host.

The same two-part hypothesis applied also to chimps: that their virus, SIV_{cpz}, is (1) an ancient infection that now (2) causes no harm. But for chimps those were just tenuous assumptions. Then new evidence and analyses addressed them, and both parts turned out to be wrong.

The first premise, that SIV_{cpz} has lurked within chimpanzees for a very long time, began to look doubtful in 2003. That's when another team of researchers (led by Paul Sharp and Elizabeth Bailes of the University of Nottingham, and including again both Beatrice Hahn and Martine Peeters) noticed that SIV_{cpz} seems to be a hybrid virus. The Nottingham group reached that conclusion by comparing the genome of SIV_{cpz} with the genomes of several monkey SIVs. They found that one major section of the chimp virus's genome matches closely to a section of SIV_{rcm}. Another major section closely resembles a section in SIV_{gsn}. In plain words: The chimp virus contains genetic material from the virus of red-capped mangabeys and also genetic material from the virus of greater spot-nosed monkeys. How did it happen? By recombination—that is, genetic mixing. A chimpanzee infected with both monkey viruses must have served as the mixing bowl in which two viruses traded genes. And *when* did it happen? Possibly just hundreds of years ago, rather than thousands or tens of thousands.

How did a single chimpanzee become infected with two monkey viruses? Presumably that occurred through predation, or through the combined circumstances of predation (bringing aboard one virus) plus sexual transmission (bringing aboard a second), followed by a chance rearrangement of genes between one virus and the other during viral replication. Chimpanzees are omnivores who love an occasional taste of meat. They kill monkeys, rip them apart, fight over the pieces or share out gobbets and joints; then they eat the flesh, red and raw. It doesn't happen often, just whenever the opportunity and the hankering arise. Such gore fests must sometimes involve blood-to-blood contact. Chimpanzees, even without the use of machetes, suffer wounds on their hands and in their mouths. Bloody meat plus an open sore equals exposure. What the Nottingham group suggested was another chimpanzee version of the cut-hunter hypothesis—except in this case the cut hunter was the chimp.

103

So the very existence of SIV_{cpz} is relatively recent. It has no ancient association with chimpanzees. And now, based on a study published in 2009, part two of the two-part hypothesis has also been cast into doubt. The virus is not so harmless in its chimpanzee host. Evidence from the chimps of Gombe—Jane Goodall's study population, known and beloved around the world—suggests that SIV_{cpz} causes simian AIDS.

I've mentioned already that the first wild chimp to test SIV-positive was at Gombe. What I didn't say, but will here, is that SIV-positive status among Gombe's chimpanzees correlates strongly with failing health and early death. Again it was Beatrice Hahn and her group who made the discovery.

Having found SIV_{cpz} in captive chimps, Hahn wanted to look for it in the wild. But she and her team of young molecular biologists knew little about sampling chimpanzees in an African forest. What do you do, go out and dart one? Knock the ape out with ketamine, take blood, wake him up, and send him on his way? (That's what Billy Karesh had been equipped to do, with gorillas, during our eight-day stakeout at Moba Bai in the Republic of the Congo. But the protocols for well-studied and habituated chimpanzee populations are very different.) Egads, no! said field primatologists, horrified at the prospect of any such invasive violation of their sensitive, trusting subjects. It was a new realm for Hahn, with a new set of concerns and methods, to which she quickly became attuned. At a scientific meeting that brought primate researchers together with virologists, she met Richard Wrangham, of Harvard, highly respected for his work on the behavioral ecology and evolution of apes. Wrangham has for many years led a study of chimpanzees at Kibale National Park, in western Uganda; before that, four decades ago, he did his own PhD fieldwork at Gombe. He responded enthusiastically to Hahn's idea of screening wild chimps, and ultimately it was Wrangham, she recalled, "who convinced Jane that we were okay to work with." But before any such work began at Gombe, they looked at the chimps of Kibale, Wrangham's own research site. Crucial help came from a Wrangham grad student named Martin Muller, who in 1998 had collected urine samples for a study of testosterone, aggression, and stress. Mario Santiago, of Hahn's lab, cooked up the requisite tools for detecting SIV_{cpz} antibodies in a few milliliters of piss, and Martin Muller supplied some frozen samples from his collections at Kibale. For this part of the story, I went to Albuquerque and talked with Muller, now an associate professor of anthropology at the University of New Mexico.

The Kibale samples all tested negative for SIV. "We were slightly disappointed," Muller recalled. "That was because, at the time, the conventional wisdom was that this didn't have any negative impacts on chimps." Meanwhile, though, he was getting some interesting results in the hormone study and wanted to broaden his data. He and Wrangham agreed that it might be instructive to sample a few other chimp populations for comparison. That led Muller down to Gombe, in August 2000, with his urine-collecting bottles and all the cumbersome equipment necessary to keep samples frozen. He stayed only a couple weeks, training Tanzanian field assistants to continue the collecting, and brought away just a few samples himself. Back home in the United States, he emailed Hahn to ask whether she would like six tubes of frozen Gombe urine, to which she replied: "YES, YES, YES." He sent them with coded labels, standard procedure, so Hahn had no way of knowing whose was whose. Two of the six tested positive for SIV antibodies. Breaking the code, Muller informed her that both samples came from a chimp named Gimble, a twenty-three-year-old male.

Gimble was a well-known member of one of the famed Gombe families; his mother had been Melissa, a successful matriarch, and his brothers included Goblin, who rose to be the community's alpha male and lived to age forty. Gimble's life and career would be different—and shorter.

Soon after getting the results on Gimble, Beatrice Hahn wrote a long email to Jane Goodall, explaining the context and the implications. Goodall herself had trained as an ethologist (she earned a PhD at Cambridge), not as a molecular biologist, and the realm of western blot analysis for antibodies was as alien to her as field sampling had been to Hahn. Goodall's work on chimpanzees began back in July 1960, at what was then the Gombe Stream Game Reserve, on the east shore of Lake Tanganyika, and which later became Gombe National Park. She established the Gombe Stream Research Center in 1965, based in a small concrete building near the lake, and continued her study of chimps in the hilly forest for another twenty-one years. In 1986 Goodall published an imposing scientific opus, *The Chimpanzees of Gombe*, and then ended her own career as a field scientist because, appalled by the treatment of chimpanzees in medical labs and other captive situations around the world, she felt obliged to become an activist. The study of Gombe's chimps went ahead in her absence, thanks to well-trained Tanzanian field assistants and later generations of scientists, adding decades of data and precious continuity to what Goodall had started. She remained closely connected to Gombe and its chimps, both personally and through the programs of her Jane Goodall Institute, but she wasn't often present at the old research camp, apart from stolen interludes of rest and reinvigoration. Instead she traveled the world, roughly three hundred days a year, lecturing, lobbying, meeting with media people and schoolchildren, delivering her inspirational message. Hahn understood the intensity of Goodall's protectiveness toward chimps in general, toward Gombe's chimps in particular, and of her wariness toward anything that might put them in more jeopardy of exploitation, especially in the name of medical science. At the end of the long email, Hahn wrote:

> Let me finish by saying that finding SIV_{cpz} in the Gombe community is a virologist's DREAM-COME-TRUE. Given the wealth of behavioural and observational data that you and your colleagues have collected over decades, it is the IDEAL setting to study the natural history, transmission patterns and pathogenicity (or lack thereof) of natural SIV_{cpz} infection in wild chimpanzees. Moreover, all this can be done entirely non-invasively. AND there certainly are funding opportunities for such a unique study. So, the virologist's dream-come-true does not have to be the primatologist's nightmare, although I am sure it will take some time before I can convince you of that.

Eventually she did convince Goodall, but not before another nightmarish discovery emerged from the work.

Earlier in her email, Hahn had written: "With respect to the chimpanzees, it is probably safe to say that SIV infection will NOT cause them to develop immunodeficiency or AIDS." On that point, she would prove herself wrong.

104

Jane Goodall described her own concerns when I caught up with her during one of her stopovers. We knew each other from previous adventures—among chimps in the Congo, among black-footed ferrets in South Dakota, over single-malt scotch in Montana—but this was a chance to sit down quietly at a hotel in Arlington, Virginia, during a paralyzing snowstorm, and talk about Gombe. The fiftieth anniversary of her own chimp study was approaching, and I had been assigned by *National Geographic* to write about it. After we discussed her childhood influences, her dream of becoming a naturalist in Africa, her mentor Louis Leakey, her early days in the field, and her time as a PhD student at Cambridge, she herself mentioned genetics and virology. At that point I turned the conversation to SIV.

"I was really, really apprehensive about Beatrice Hahn's research," Jane volunteered. "We were, a lot of us, really nervous about the result of what might happen if she found HIV/AIDS." She had met Hahn, talked with her, and was reassured by the force of Hahn's concern for the chimps' welfare. "But still. I still have this unease because, even though she cares, once these results are out, as they are now, other people can use them in different ways." For instance? What sort of dangers, I asked, did Jane have in mind? "That this would start a whole new flurry of research on captive chimps in medical labs." The news of chimps with AIDS, she feared, would sound like a promising opportunity to learn more about AIDS in humans, never mind the chimps.

What about the impact of the virus at Gombe itself? We both knew that Hahn *had* found something resembling AIDS, and by now Gimble was dead. What about the prospect that other members of the Gombe community might die of immune failure? "Yeah, exactly," Jane said. "That's a very scary thought."

As scary as it was, though, she realized from the start of her conversations with Hahn that such a finding could be taken two ways. On the one hand, Jane said, there was a possible consolation: If people heard that wild chimps carry an AIDS-causing virus, they might stop hunting and butchering and eating them. "Because they'll be afraid. That was one side of it. Then the other side of it was, well, people will say, 'All these creatures are really dangerous for us, so let's kill them all.' It could have gone either way." Jane is a perspicacious woman. She has the aura of a secular saint but is actually quite human, grounded, savvy, and capable of ambivalence. As things have transpired so far, she noted, neither of the extreme outcomes has occurred.

Briefly we discussed Hahn's noninvasive sampling methodology:

Urine might contain antibodies, and feces could yield viral RNA. Jane allowed that that part was reassuring, not having to knock out chimps and jab them with needles. "Don't need blood," she said. "Just need a bit of poo." Amazing what they can do from a bit of poo, I agreed.

So she had given her consent for Hahn's study, and the work proceeded. At the end of November 2000, Hahn's lab in Alabama received the first batch of material, which included three fecal samples from poor Gimble. Hahn's grad student Mario Santiago did the screening, and again all three of Gimble's samples tested positive. Santiago then amplified a viral RNA fragment and sequenced it, confirming that Gimble's virus was indeed SIV_{cpz}. It seemed to be a new strain, distinct enough from other known strains that it might be unique to East Africa. This was significant on several counts. Yes, the chimps of Gombe were infected. No, they couldn't be source animals for the human pandemic. The variants of SIV found by Martine Peeters in western Africa (this was before Hahn's own findings from Cameroon) more closely matched HIV-1 group M than the Gombe virus did.

In mid-December, another email from Hahn's computer went out to Richard Wrangham, Jane Goodall, Martin Muller, and others. Under the subject line GOOD NEWS AT LAST, Hahn described the findings from Gimble and the position of his strain on the SIV family tree. Then, with her characteristic penchant for uppercase exuberance, she wrote: "THIS IS A HOME RUN!"

105

That was just the beginning. For nine years the study continued. Fieldworkers at Gombe collected fecal samples from ninety-four different chimpanzees, each of which was known by name and, in most cases, by its individual character and family history.

Beatrice Hahn's people did the analyses, finding that seventeen of those ninety-four chimps were SIV-positive. As time passed, some chimps died. Others disappeared in the forest and were presumed dead when they failed to reappear. Death is often a private matter for wild creatures, including chimpanzees, especially when it comes upon them by slow and painful degrees. They tend to go absent from the social group, if there is a social group, and meet the end alone. Gimble last showed himself to trackers on January 23, 2007. His body was never found.

Back in Birmingham, there was turnover of a different sort, as grad students and postdocs cycled through Hahn's lab. Mario Santiago departed, heading off for the next stage of his career, and Brandon Keele arrived. Samples continued coming from Gombe, in occasional batches, and those samples were analyzed—a slow and laborious process. Much of the work fell to Keele, though even for him this was "a backburner project." Keele described to me, during my visit with him at Fort Detrick, the moment of recognition that occurred near the end of his postdoc period, bringing that project to the front burner.

"I was trying to leave and finish up. I said to myself, 'I wonder what's happening with these chimps?'" He was aware that the number of known SIV-positives at Gombe had increased as the sampling stretched on, and that there was evidence of vertical transmission (mother to infant) as well as sexual transmission accounting for new infections. He thought the study might yield an interesting, undramatic paper about how a harmless virus spreads through a population. "And then we started compiling the data," he told me. That meant bringing in a dimension of behavioral observations from the field. So he contacted collaborators at the Jane Goodall Institute's research headquarters in Minnesota and, asking about one individual after another, heard a drumbeat of unsettling news.

"Oh, no, that chimp is dead."

"No, that chimp is dead. He died in 2006."

"No, that chimp is dead."

Keele recalled asking himself: "What the hell is going on?" Part of the answer, revealed when he saw an updated mortality list, was that a wave of untimely deaths had been sweeping through SIV-positive members of the Gombe population.

He and the team at Hahn's lab had lately written an abstract for a talk he planned to give at a meeting, which would lead in time to a journal publication. The draft abstract, by Keele's recollection, contained a sentence such as: "It doesn't really seem that there is a death hazard to infection in these chimps." They had sent the draft to their partners at Gombe, who responded quickly with news of seven additional chimpanzee deaths, about which Keele hadn't even known. He scrapped the abstract, thought again about what he was doing, and began working more closely with Gombe and Minnesota to assemble a more complete set of data. Then they would see where it led.

Around the same time, spring of 2008, Keele also heard about some unusual pathology results on tissues from one dead Gombe chimp. The chimp was known as Yolanda, a twenty-four-year-old female. She sickened in November 2007, of an unknown ailment, and came down from the mountains to languish near the research center. People tried to feed her, but Yolanda didn't eat. She sat in the rain amid thick vegetation, weakened and miserable, and then died. They put her body in a freezer. Two months later, it was thawed for necropsy.

The necropsy was performed by Jane Raphael, a Tanzanian veterinarian working at the Gombe Stream Research Center and specially trained for the task. Not knowing whether Yolanda had been SIV-positive or not, Raphael took the stipulated precautions. She wore a full Tyvek suit, two layers of gloves, an N95 respirator mask, a face shield, and rubber boots. She split open Yolanda's belly, cut through the ribs, and spread them wide to see what she could see.

"The main problem was in the abdominal cavity," Raphael told me, two years later, as we sat in her small office just up from the shore of Lake Tanganyika. "There was something like abdominal peritonitis. The intestines were very much adhered together." Raphael, a quiet woman, wearing a neat cornrow hairdo and a flowered print dress, chose her words carefully. She described separating the glommed guts with her gloved hands. "It was unusual," she said. She seemed to remember it all vividly. "The muscles underneath the pelvis were very much inflamed. Red. And they had some blackish spots." What caused the inflammation? Cautious of going beyond her data, Raphael said she didn't know.

Her inspection done, she snipped out tissue samples from virtually every organ: spleen, liver, intestines, heart, lungs, kidneys, brain, lymph nodes. For the SIV-positive cases, she said, lymph nodes were especially important. Yolanda's lymph nodes appeared normal to the eye, but histopathology would later penetrate that illusion. Some of the samples, preserved in RNAlater, went off to Beatrice Hahn. Others, pickled in formalin, were destined for a pathologist in Chicago. When the results came together, this case would challenge prevailing ideas about SIV in chimpanzees. "Previously it was said, they are infected but they don't come down with the disease," Raphael told me. "Yolanda made us to start thinking otherwise."

I followed the pickled samples to Chicago, where the pathologist who had examined them, Karen Terio, welcomed me to a glimpse of the evidence. Terio had trained as a veterinarian, at one of the country's best vet schools, and then did a residency and a doctorate in pathology, specializing in diseases transferred between animals of different species. She worked for the University of Illinois and consulted for the Lincoln Park Zoo, which helps run a health-monitoring project at Gombe. Hence the lymph nodes and other bits of Yolanda came for her expert scrutiny. Terio cut up the tissues, sent them to laboratory technicians for mounting and staining, and sat down for a look at the slides. "It was striking because I couldn't find any lymphocytes," she told me. "When I saw the first lymph node, I thought, 'Hmm, this is weird.'" She asked her boss to have a look through the microscope. He did, and agreed there was something very wrong. She phoned a colleague at the Lincoln Park Zoo, Elizabeth Lonsdorf, who leads the zoo's work on behalf of wild African apes, including the health project at Gombe.

"We have a problem," Terio told Lonsdorf. "She doesn't have any lymphocytes."

"Does that mean what I think it means?"

"Yes. The lesions in this animal look like an end-stage AIDS patient."

Together she and Lonsdorf made a call to Beatrice Hahn. Hahn's first question was, "Are you sure?" Terio was indeed sure, but she quickly emailed images of the slides so that the others could judge for themselves. Brandon Keele was by now in the loop. Terio sent actual slides to another collaborator, an expert on immune-system pathology, to refine the diagnosis. Everyone agreed and, with the sample code broken, everyone knew how these pieces fit together: The chimp Yolanda, dead at age twenty-four, had been SIV-positive and suffering immune deficiency.

Inviting me to a chair at her big double-viewer Olympus microscope, Karen Terio brought out the same slides she had shared with Hahn and Lonsdorf. From her place at the scope she could manipulate a cursor, a little red arrow, moving it over the field to point out what we were seeing. First she showed me a thin-slice section from a lymph node of a normal, SIV-negative chimpanzee. This was for comparison. It looked like a peat bog as viewed on Google Earth, bulging and rife with sphagnum and huckleberry, thick, rich, and riddled just slightly with narrow spaces resembling small sloughs and creeks. The tissue was stained magenta and heavily speckled with darker blue dots. The dots, Terio explained, were lymphocytes in their healthy abundance. In an area where they're especially dense, they pack together into a follicle, like a bag full of jellybeans. She jabbed her red arrow at a follicle.

Then she placed another slide into viewing position. The slide held a slice from one of Yolanda's lymph nodes. Instead of a peat bog, it looked like scrub desert slashed by a large drywash, many days since the last rain.

"Mmmm," I said.

"This is essentially the connective tissue," Terio said. She meant that it was supportive structure only, minus the working innards. Sere and empty. "We've got very, very few lymphocytes left in this animal."

"Yeah."

"And it's collapsed. You see, this whole thing has just sort of collapsed on itself, 'cause there's nothing in there to hold it up." Her little red arrow wandered forlorn through the desert. No sphagnum, no follicles, no little blue dots. I imagined Karen Terio, back in April 2008, examining these slides on her lonesome—and encountering such evidence, before anyone else, at a time when the illusion of nonpathogenic SIV_{cpz} was embraced by researchers everywhere.

"So you sat there, and looked at this . . ."

"And went, 'Oh, no,'" she said.

106

Terio's findings, plus the field data from Gombe, plus the molecular analyses from Hahn's lab—these all came together in a paper published by *Nature* during the summer of 2009. Brandon Keele was first author; Beatrice Hahn was last. "Increased Mortality and AIDS-like Immunopathology in Wild Chimpanzees Infected with SIV_{cpz}" was the catchy title. I think of it—and I'm not alone—as "the Gombe paper." Among the long list of coauthors were Karen Terio, Terio's boss, Elizabeth Lonsdorf, Jane Raphael, two of Hahn's senior colleagues, the expert on primate cell pathology, the chief scientist at Gombe, and Jane Goodall herself.

"Well, I sort of had to be. But I had these long talks with Beatrice first," Jane told me. "She was going to publish it anyway." In the sweep of inevitability and the name of science, Dr. Goodall signed on.

The paper's salient conclusion was that, contrary to Keele's earlier draft abstract, there is indeed a death hazard for SIV-positive chimps at Gombe. Of the eighteen individuals that died during the study period, seven were SIV-positive. Given that less than 20 percent of the population was SIV-positive, and adjusted for normal mortality at a given age, this reflected a risk of death ten times to sixteen times higher for SIV-positive chimps than for SIV-negatives. Repeat: ten to sixteen times higher. The total numbers were small but the margin was significant. Infected animals were falling away. Furthermore, SIV-positive females had lower birth rates and greater infant mortality. Further still, three necropsied individuals (including Yolanda, though her name wasn't mentioned) showed signs of lymphocyte loss and other damage resembling end-stage AIDS.

The authors suggested, cautiously but firmly, "that SIV_{cpz} has a substantial negative impact on the health, reproduction and lifespan of chimpanzees in the wild." So it's not a harmless passenger. It's a hominoid killer, their problem as well as ours.

107

Here's what you have come to understand. That the AIDS pandemic is traceable to a single contingent event. That this event involved a bloody interaction between one chimpanzee and one human. That it occurred in southeastern Cameroon, around the year 1908, give or take. That it led to the proliferation of one strain of virus, now known as HIV-1 group M. That this virus was probably lethal in chimpanzees before the spillover occurred, and that it was certainly lethal in humans afterward. That from southeastern Cameroon it must have traveled downriver, along the Sangha and then the Congo, to Brazzaville and Léopoldville. That from those entrepôts it spread to the world.

Spread how? Once it reached Léopoldville, the group M virus seems to have entered a vortex of circumstances unlike anything at the headwaters of the Sangha. It differed from HIV-2 biologically (having adapted to chimpanzee hosts) and it differed from groups N and O by chance and opportunity (having found itself in an urban environment). Whatever happened to it in Léopoldville during the first half of the twentieth century can only be conjectured. Population density of potential human hosts, a high ratio of males to females, sexual mores different from what prevailed in the villages, and prostitution—these were all parts of the mix. But sex plus crowding may not be a sufficient explanation. A fuller chain of conjecture, and maybe a better one, has been offered by Jacques Pepin, a Canadian professor of microbiology who, during the 1980s, worked for four years at a bush hospital in Zaire. Pepin coauthored several journal papers on the subject and, in 2011, published a book titled *The Origins of AIDS.* Having added some deep historical research to his own field experience and microbiological expertise, he proposed that the crucial factor intermediating between the Cut Hunter and the global pandemic was the hypodermic syringe.

Pepin wasn't referring to recreational drugs and the works shared by addicts at shooting galleries. In a paper titled "Noble Goals, Unforeseen Consequences," and then at greater length in his book, he pointed instead to a series of well-intended campaigns by colonial health authorities, between 1921 and 1959, aimed at treating certain tropical diseases with injectable medicines. There was a massive effort, for instance, against trypanosomiasis (sleeping sickness) in Cameroon. Trypanosomiasis is caused by a persistent little protist (*Trypanosoma brucei*), transmitted in the bite of tsetse flies. The treatment in those years entailed injections of arsenical drugs such as tryparsamide—and a patient didn't get just one shot but a series. In Gabon and Moyen-Congo (the French colonial name for what's now the Republic of the Congo), the regimen for trypanosomiasis sometimes entailed thirty-six injections over three years. And there were similar efforts to control syphilis and yaws. Malaria was treated with injectable forms of quinine. Leprosy patients, in that era before oral antibiotics, underwent a course of injections with extract of chaulmoogra (an Indian medicinal plant), two or three shots per week for a year. In the Belgian Congo, mobile teams of *injecteurs,* people with no formal education but a small bit of technical training, visited trypanosomiasis patients in their villages to give weekly shots. It was a period of mania for the latest medical wonder: needle-delivered cures. Everyone was getting jabbed.

Of course, this was long before the era of the disposable syringe.

Hypodermic syringes, for injecting medicines into muscles or veins, were invented in 1848 and, until after World War I, were handmade of glass and metal by skilled craftsmen. They were expensive, delicate, and meant to be reused like any other precision medical instrument. During the 1920s their manufacture became mechanized, to the point where 2 million syringes were produced globally in 1930, making them more available but not more expendable. To the medical officers working in Central Africa at that time, they seemed invaluable but were in short supply. A famous French colonial doctor named Eugène Jamot, working just east of the upper Sangha River (in a portion of French Equatorial Africa then known as Oubangui-Chari) during 1917–1919, treated 5,347 trypanosomiasis cases using only six syringes. This sort of production-line delivery of injectable medicines didn't allow time for boiling a syringe and needle between uses. It's difficult now, based on skimpy sources and laconic testimony, to know exactly what sort of sanitary precautions were taken. But according to one Belgian doctor, writing in 1953: "The Congo contains various health institutions (maternity centres, hospitals, dispensaries, etc.) where every day local nurses give dozens, even hundreds, of injections in conditions such that sterilisation of the needle or the syringe is impossible." This man was writing about the risk of accidental transmission of hepatitis B during treatment for venereal diseases, but Pepin quoted his report at length, for its potential relevance to AIDS:

> The large number of patients and the small quantity of syringes available to the nursing staff preclude sterilisation by autoclave after each use. Used syringes are simply rinsed, first with water, then with alcohol and ether, and are ready for a new patient. The same type of procedure exists in all health institutions where a small number of nurses have to provide care to a large number of patients, with very scarce supplies. The syringe is used from one patient to the next, occasionally retaining small quantities of infectious blood, which are large enough to transmit the disease.

How much of this went on? Very much. Pepin's diligent search through old colonial archives turned up some big numbers. In the period 1927–1928, Eugène Jamot's team in Cameroon performed 207,089 injections of tryparsamide, plus about 1 million injections of something called atoxyl, another arsenical drug for treating trypanosomiasis. During just the year 1937, throughout French Equatorial Africa, the army of doctors and nurses and semipro jabbers delivered 588,086 injections aimed at trypanosomiasis, not to mention countless more for other diseases. Pepin's arithmetic totaled up 3.9 million injections just against trypanosomiasis, of which 74 percent were intravenous (right into a vein, not just a muscle), the most direct method of drug delivery and also the best for unintentionally transmitting a blood-borne virus.

All those injections, according to Pepin, might account for boosting the incidence of HIV infection beyond a critical threshold. Once the reusable needles and syringes put the virus into enough people—say, several hundred—it wouldn't come to a dead end, it wouldn't burn out, and sexual transmission could do the rest. Some experts, including Michael Worobey and Beatrice Hahn, doubt that needles were necessary in any such way to the establishment of HIV in humans—that is, to its early transmission from one person to another. But even they agree that injection campaigns could have played a role later, spreading the virus in Africa once it was established.

This needle theory didn't originate with Jacques Pepin. It dates back more than a decade to work by an earlier team of researchers, including Preston Marx of the Rockefeller University, who proposed it in 2000 at the same Royal Society meeting on AIDS origins at which Edward Hooper spoke for his oral polio vaccine theory. Marx's group even argued that serial passage of HIV through people, by means of such injection campaigns, might have accelerated the evolution of the virus and its adaptation to humans as a host, just as passaging malarial parasites through 170 syphilis patients (remember the crazed Romanian researcher, Mihai Ciuca?) could increase the virulence of *Plasmodium knowlesi*. Jacques Pepin picked up where Preston Marx left off, though with less emphasis on the evolutionary effect of serial passage. Pepin's main point was simply that dirty needles, used so widely, must have increased the prevalence of the virus among people in Central Africa. Unlike the OPV theory, this one hasn't been discredited by further research, and Pepin's new archival evidence suggests that it's highly plausible, if unprovable.

Most of those injections for trypanosomiasis occurred in the countryside. City dwellers were less exposed to trypanosomiasis, partly because the tsetse fly doesn't thrive in urban jungles as well as it does in green ones. One question that needed answering, therefore, was whether any such mania for injecting had also gripped Léopoldville, where HIV met its most crucial test. Pepin's answer is unexpected, interesting, and persuasive. Never mind trypanosomiasis. He discovered a different but equally aggressive campaign of injections, aimed at limiting syphilis and gonorrhea in the city's population.

In 1929, the Congolese Red Cross established a clinic known as the Dispensaire Antivénérien, open to women and men for the treatment of what we used to call venereal diseases. Located in a neighborhood on the east side of Léopoldville, near the river, it was a private facility providing a public service. Male migrants, arriving to seek work, were required by city regulations to report to the Dispensaire for an exam. Anyone experiencing symptoms could visit the place voluntarily, and there was no charge for treatment. But the bulk of the caseload, according to Pepin, "consisted of thousands of asymptomatic free women who came for screening because they were required to do so by law, in theory every month." The colonial government accepted prostitution as an ineradicable fact but evidently hoped to keep the trade hygienic—so *les femmes libres* were obliged to get checked.

If a person tested positive for syphilis or gonorrhea, he or she would be treated. But the diagnostic testing was imprecise. Any free woman or male migrant who had once been exposed to yaws (caused by a bacterium very similar to the syphilis bacterium, but not sexually transmissible) might flunk the blood test, be classed as syphilitic, and receive a long course of drugs containing arsenic or bismuth. Harmless vaginal flora could be mistaken for gonococcus, the agent of gonorrhea. A woman diagnosed gonorrheic might be injected with typhoid vaccine, or a drug called Gono-yatren, or (even Jacques Pepin seems puzzled by this one) milk. During the 1930s and 1940s, the Dispensaire Antivénérien administered more than forty-seven thousand injections annually. Most were intravenous. Straight into the blood. With increased migration to the city following World War II, the numbers rose. In the early 1950s, the quackier remedies (intravenous milk?) and the metallic poisons gave way to penicillin and streptomycin, which had longer-lasting effects and therefore meant fewer shots. The campaign peaked in 1953, at about 146,800 injections, or roughly 400 per day. Many if not most of those injections were administered to *femme libres,* sex workers, ladies of hospitality, however you want to describe them, who had multiple male clients. They came and went. The syringes were rinsed and reused. This in a city where HIV-1 had arrived.

Six years later came the blood sample that yielded the HIV-1 sequence now known as ZR59. One year after that, DRC60. The virus had spread and diversified. It was at large. No one can say whether either of those two patients had ever visited the Dispensaire Antivénérien for a shot. But if they hadn't, they probably knew someone who had.

108

From this point the story gets huge and various, literally going off in all directions. It explodes out of Léopoldville like an infectious starburst. I won't try to trace those diverging trajectories—a task for ten other books, with purposes different from mine—but I'll sketch the pattern, and then focus briefly on one that's especially notorious.

During its decades of inconspicuous transmission in Léopoldville, the virus continued to mutate (and probably also to recombine, mixing larger sections of genome from one virion to another), and those copying errors drove its diversification. Most mutations are fatal mistakes, bringing the mutant to a dead end, but with so many billions of virions replicating, chance did provide a small, rich supply of viable new variants. The campaigns of injectable drug treatments, at the Dispensaire Antivénérien and elsewhere, may have helped foster this process by transmitting the virus quickly into more human hosts and increasing its total population. The more virions, the more mutations; the more mutations, the more diversity.

The HIV-1 group M lineage became split into nine major subdivisions, now known as subtypes and labeled with letters: A, B, C, D, F, G, H, J, K. (Don't confuse those, if you can help it, with the eight groups of HIV-2, designated A through H. And why are E and I missing? Never mind why. Such edifices of labeling get built piecemeal, like slums of cardboard and tin, not with architectural forethought.) As time passed, as the human population of Léopoldville grew, as travel increased, viruses of those nine subtypes emerged from the city, radiating outward across Africa and the world. Some of them went by airplane and others by more mundane means of transport: bus, boat, train, bicycle, hitchhiking on a transcontinental truck. Foot. Subtype A got to East Africa, probably via the city of Kisangani, halfway between Léopoldville and Nairobi. Subtype C spread to southern Africa, probably via Lubumbashi, way down in the Congolese southeast. Seeping across Zambia, achieving rapid transmission in mining towns full of workers and prostitutes, subtype C proliferated catastrophically throughout South Africa, Mozambique, Lesotho, and Swaziland. It went on to India, which is linked to South Africa by channels of exchange as old as the British empire, and to East Africa. Subtype D established itself alongside subtypes A and C in the countries of East Africa, except for Ethiopia, which for some reason became afflicted early and almost exclusively with subtype C. Subtype G got up into West Africa. Subtypes H, J, and K remained mostly in Central Africa, from Angola to the Central African Republic. In all these places, after the usual lag of years between infection and full-blown AIDS, people began dying. And then there's subtype B.

Sometime around 1966, subtype B crossed from Léopoldville to Haiti.

How it did that is unknown, and can probably never *be* known, but Jacques Pepin's archival burrowing provides new support for one plausible old scenario. When the Belgian government abruptly relinquished its African colony, on June 30, 1960, under the stern encouragement of Patrice Lumumba and his movement, tens of thousands of Belgian expatriates—almost an entire middle class of civil servants, teachers, doctors, nurses, technical experts, and business managers—found themselves unwelcome and uncomfortable in the new republic, and they began flooding homeward. Crowding the planes for Brussels. Their departure created a vacuum, since the Belgian regime had pointedly avoided educating its colonial subjects. There wasn't a single Congolese medical doctor, for instance. Few teachers. The country suddenly needed help. The World Health Organization responded, sending physicians, and the United Nations (through its Educational, Scientific, and Cultural Organization, UNESCO) also began enlisting skilled people to work in the Congo: teachers, lawyers, agronomists, postal administrators, and other bureaucrats, technicians, and professionals. Many of those recruits came from Haiti. It was a natural fit: The Haitians spoke French as did the Congolese; they came from African roots; they had education but very little opportunity at home under the dictatorship of Papa Doc Duvalier.

During the first year of independence, half the teachers sent by UNESCO to the Congo were Haitians. By 1963, according to one estimate, a thousand Haitians were employed in the country. Another estimate says that a total of forty-five hundred Haitians served hitches in the Congo during the 1960s. Evidently there's no surviving, authoritative manifest. Anyway, lots of Haitians, thousands. Some brought families, some came alone. Among the single men, we can assume, few remained celibate. Most of them probably had Congolese girlfriends or visited *femmes libres*. For a few years it may have been a good life. But the Haitians were less needed and less welcome as the Congo began training its own people, especially after Joseph Désiré Mobutu seized power in 1965. Less still when, in the early 1970s, he changed his name to Mobutu Sese Seko, changed his country's to Zaire, and announced a policy of *Zaireanisation*. Many or most of the Haitians, during those years, went home. Their time of being useful and appreciated black brothers from the Americas had passed.

At least one of those returnees, probably among the earliest of them, seems to have carried HIV-1.

More specifically: Someone brought back to Haiti, along with Congolese memories, a dose of HIV-1 group M subtype B.

You can see where this is going, but you might not expect how it gets there. Jacques Pepin's research shed some new light on what may have happened in Haiti during the late 1960s and early 1970s to multiply and forward the virus. One thing that happened was that, from a single HIV-positive person in 1966 or thereabouts, the virus spread fast through the Haitian population. Evidence for that spread came later, from blood samples given by 533 young mothers in a Port-au-Prince slum, who agreed in 1982 to participate in a measles study at a local pediatric clinic. Tested retrospectively, those samples revealed that 7.8 percent of the women had been HIV-positive. That number was startlingly high, for such a newly arrived virus, and caused Pepin to suspect that "there must have been a very effective amplification mechanism" operating in Haiti during the early years—more effective than sex. He found a candidate: the blood plasma trade.

Plasma, the liquid component of blood (minus the cells), is valuable stuff for its antibodies and albumin and clotting factors. Demand for it rose sharply during the period around 1970, and to meet the demand a process called plasmapheresis was developed. Plasmapheresis entails drawing blood from a donor, separating the cells from the plasma by means of filtering or centrifuging, putting the cells back into the donor, and keeping the plasma as a harvested product. One advantage of this process is that it allows donors (who are usually in fact *sellers*, paid for their trouble and needing the money) to be tapped often rather than just a couple times per year. Giving up your plasma, for the good of others or for profit, doesn't leave you anemic. You can go back and give again the following week. One disadvantage of the procedure—and it's a huge one, but wasn't recognized in the early days—is that a plasmapheresis machine, gargling your blood and the blood of many other donors over the course of days, can infect you with a blood-borne virus.

This happened to hundreds of paid plasma donors in Mexico during the mid-1980s. It happened to a quarter million luckless donors in China. Jacques Pepin thinks it happened in Haiti too.

He found reports of a plasmapheresis center in Port-au-Prince, a private business known as Hemo Caribbean, that operated profitably during 1971 and 1972. It was owned by an American investor, a man named Joseph B. Gorinstein, based in Miami, with links to the Haitian Minister of the Interior. Donors received $3 per liter. Their vitals were checked before they could sell plasma, but of course nobody screened them for HIV—which didn't yet exist as an acronym, or an infamous global scourge, only as a quiet little virus that lived in blood. According to an article that ran in *The New York Times* on January 28, 1972, Hemo Caribbean was then exporting between five and six thousand liters of frozen blood plasma to the United States each month. The wholesale customers were American companies, which marketed the product for use in transfusions, tetanus shots, and other medical applications. Mr. Gorinstein wasn't available for comment.

Papa Doc had meanwhile died, in 1971, and been succeeded by his son Jean-Claude (Baby Doc) Duvalier. Annoyed by the *Times* publicity, Baby Doc ordered that Gorinstein's plasmapheresis center be closed. The Haitian Catholic Church condemned the blood trade as exploitation. Beyond that, the story of Hemo Caribbean drew little notice at the time. No one yet realized how devastating blood-product contamination could be. Nor did the CDC's *Morbidity and Mortality Weekly Report* mention it, a decade later, when breaking the news that Haitians seemed especially at risk for the mysterious new immune-deficiency syndrome. Randy Shilts didn't mention it in *And the Band Played On*. The only allusion to Haitian blood plasma that I recall, from the years before Jacques Pepin's book, came during my conversation with Michael Worobey in Tucson.

Shortly before publishing on DRC60 and ZR59, Worobey coauthored another notable paper, dating the emergence of HIV-1 in the Americas. The first author was a postdoc named Tom Gilbert, in Worobey's lab, and in the anchor position was Worobey himself. This was the work, based on analyses of viral fragments from archived blood cells, that placed the arrival of HIV-1 in Haiti to about 1966, give or take a few years. It appeared in the *Proceedings of the National Academy of Sciences.* Soon afterward, Worobey got a peculiar email from a stranger. Not a scientist, just someone who had caught wind of the subject. A reader of newspaper coverage, a listener to radio. "I think he was from Miami," Worobey told me. "He said he worked in an airport that dealt with the blood trade." The man had certain memories. Maybe they haunted him. He wanted to share them. He wanted to tell Worobey about cargo planes arriving full of blood.

109

The next leap of the virus was small in distance and large in consequence. Port-au-Prince is just seven hundred miles from Miami. A ninety-minute flight. Part of the project that Tom Gilbert undertook, in Worobey's lab, was to date when HIV-1 had arrived in the United States. To do that he needed samples of old blood. Whether the blood had reached America in bottles, in bags, or in immigrant Haitians didn't much matter for this purpose.

Worobey, serving as Gilbert's adviser, remembered a study of immunodeficient Haitian immigrants that had been published twenty years earlier. It had been led by a physician named Arthur E. Pitchenik, working at Jackson Memorial Hospital in Miami. Pitchenik was an expert on tuberculosis, and beginning in 1980 he noticed an unusual incidence of that disease, as well as *Pneumocystis* pneumonia, among Haitian patients. He sounded the first alarm about Haitians as a risk group for the new immunodeficiency syndrome, alerting the CDC. In the course of clinical work and research, Pitchenik and his colleagues drew blood from patients and centrifuged it, separating serum from cells, so they could look at certain types of lymphocyte. They also froze some samples, on the assumption those might be useful to other researchers later. They were right. But for a long time no one seemed interested. Then, after two decades, Arthur Pitchenik got a call from Michael Worobey in Tucson. Yes, Pitchenik said, he would be glad to send some material.

Worobey's lab received six tubes of frozen blood cells, and Tom Gilbert managed to amplify viral fragments from five. Those fragments, after genetic sequencing, could be placed into context as limbs on another family tree—just as Worobey himself would later do with DRC60 and ZR59, and as Beatrice Hahn's group was doing with SIV_{cpz}. It was molecular phylogenetics at work. In this case, the tree represented the diversified lineage of HIV-1 group M subtype B. Its major limbs represented the virus as known from Haiti. One of those limbs encompassed a branch from which grew too many small twigs to portray. So in the figure as eventually published, that branch and its twigs were blurred—depicted simply as a solid cone of brown, like a sepia shadow, within which appeared a list of names. The names told where subtype B had gone, after passing through Haiti: the United States, Canada, Argentina, Colombia, Brazil, Ecuador, the Netherlands, France, the United Kingdom, Germany, Estonia, South Korea, Japan, Thailand, and Australia. It had also bounced back to Africa. It was HIV globalized.

This study by Gilbert and Worobey and their colleagues delivered one other piquant finding. Their data and analysis indicated that just a single migration of the virus—one infected person or one container of plasma—accounted for bringing AIDS to America. That sorry advent occurred in 1969, give or take about three years.

So it lurked here for more than a decade before anyone noticed. For more than a decade, it infiltrated networks of contact and exposure. In particular, it followed certain paths of chance and opportunity into certain subcategories of the American populace.

It was no longer a chimpanzee virus. It had found a new host and adapted, succeeding brilliantly, passing far beyond the horizons of its old existence within chimpanzees. It reached hemophiliacs through the blood supply. It reached drug addicts through shared needles. It reached gay men—reached deeply and catastrophically into their circles of love and acquaintance—by sexual transmission, possibly from an initial contact between two males, an American and a Haitian.

For a dozen years it traveled quietly from person to person. Symptoms were slow to arise. Death lagged some distance behind. No one knew. This virus was patient, unlike Ebola, unlike Marburg. More patient even than rabies, but equally lethal. Somebody gave it to Gaëtan Dugas. Somebody gave it to Randy Shilts. Somebody gave it to a thirty-three-year-old Los Angeles man, who eventually fell ill with pneumonia and a weird oral fungus and, in March 1981, walked into the office of Dr. Michael Gottlieb.

IX

IT DEPENDS

110

Finally, let me tell you a little story about caterpillars. This may seem to take us afield from the origins and perils of zoonotic diseases but, trust me, it's very germane.

The caterpillar story begins back in 1993. That year, in the tree-shaded town where I live, it seemed that autumn had comc carly—earlier even than usual for a valley in western Montana, where the cold winds begin blowing in mid-August, the cottonwoods turn color not long after Labor Day, and the first heavy snow often puts a damper on Halloween. This was different. This was June. It seemed like autumn because the leaves were gone from the trees. They had flushed from their buds in May, opening wide and fresh and green; and then, just a month later, they disappeared. They hadn't succumbed to the natural rhythm of season. They hadn't turned yellow, fallen, piled up in the gutters as aromatic autumnal mulch. They had been eaten.

A pestilential abundance of small, hairy larvae had materialized like a plague out of Exodus, stripping the trees of their foliage. The Latinate binomial for these voracious leaf-eaters is *Malacosoma disstria,* though few of us townsfolk knew that at the time. We used another name.

"Tent caterpillars," said the local newspaper, vaguely but not inaccurately. "Tent caterpillars," said the city parks people and the agricultural technicians at the county extension service, who were answering calls from dozens of concerned citizens every day. The radio said "tent caterpillars" too. And so before long we were all out on the sidewalks, saying "tent caterpillars!" back and forth to one another. In the hubbub, we were too occupied to notice that these particular "tent caterpillars" didn't build tents. They just gathered and traveled in dense aggregations, like wildebeests on the Serengeti. Their full common name (their official misnomer?) is the forest tent caterpillar; a closely related insect, the western tent caterpillar (*Malacosoma californicum*) does build tentlike silken shelters. We weren't interested in such entomological subtleties. We wanted to know how we could kill the damned things before they ate all our lovely urban hardwoods down to stumps.

It was awesome, in an ugly way. Not every tree was left naked, but many were, especially among the old towering elms and green ashes that stand along the sidewalks, arching their canopies over the neighborhood lanes. It happened fast. The caterpillars did most of their feeding in full daylight or early evening, but later, on those cool June nights, we could stand beneath a great tree and still hear the gentle crackle, like distant brushfire, of their excrement cascading down through the leaves. In the mornings, we would find the sidewalks heavily sprinkled with those poppy-seed globules of dung. Occasionally a lone caterpillar would rappel down on a filament of silk and dangle there mockingly at eye level. On a day of chilly drizzle, too chilly for caterpillar comfort, we could spot them hunkering sociably, high up on a trunk or in a limb crotch, hundreds of fuzzy gray bodies in each pile, like musk oxen huddled against an Arctic storm. Some of us went away for a weekend, leaving the lawn freshly mowed, all seemingly fine, and came home to find that our trees had been defoliated. We climbed up on ladders and sprayed the caterpillars with soapy dishwater from spritzer bottles. We dosed them with bacterial mists or nasty long-molecule chemicals, as variously prescribed by the local garden-store clerks, who knew little more than we did. We called in SWAT-team strikes by the men from Nitro-Green. All of these measures seemed to be marginally effective at best and, at worst, just poisonous and futile.

The caterpillars continued to chomp. When it appeared that they might move from ravaged trees to healthy ones, in search of more food, we tried to stop them by girdling the tree trunks with barriers of impassable goo. This was pointless (since, as I learned later, a tent caterpillar generally lives out its larval stage in the tree where it hatched) but reflected our desperation. I watched my next-door neighbor, Susan, muster such hopeful defenses for two giant elms in front of her house, each tree banded at waist height with a circular belt of spray-on stickum, and it seemed like a reasonable idea to me too. But the stuff failed to catch a single caterpillar.

They kept coming. They had their way. There were simply too many, and the infestation proceeded along its inexorable course. We stepped on them as they forded the sidewalks. We mooshed them wholesale in the streets. They ate, they grew, they molted their tight old skins and grew further. They marched up and down limbs, all over town, treating our trees like celery.

Eventually they finished eating. They had bulked themselves up to the limits, fulfilled their caterpillar juvenility, and now they were ready for puberty. They spun themselves up inside leaf-wrapped cocoons for a short metamorphic respite, to emerge in a few weeks as little brown moths. The crackling stopped and the treetops, what was left of them, fell silent. The caterpillars, qua caterpillars, were gone. But this vast population of pestiferous insects still lurked over our heads, almost invisible now, like a large gloomy hunch about the future.

Ecologists have a label for such an event. They call it an outbreak.

This use of the word is more general than what's meant by an outbreak of disease. You could think of disease outbreaks as a subset. Outbreak in the broader sense applies to any vast, sudden population increase by a single species. Such outbreaks occur among certain animals but not among others. Lemmings undergo outbreaks; river otters don't. Some kinds of grasshopper do, some kinds of mouse, some kinds of starfish, whereas other kinds of grasshopper, mouse, and starfish do not. An outbreak of woodpeckers is unlikely. An outbreak of wolverines, unlikely. The insect order Lepidoptera (moths and butterflies) contains some notable outbreakers—not just tent caterpillars of several kinds but also gypsy moths, tussock moths, larch budmoths, and others. Those are exceptions, though, to the general rule even for lepidopterans. Among all the forest-dwelling species of butterfly and moth, about 98 percent maintain relatively stable populations at low density through time; no more than 2 percent ever experience outbreaks. What makes a species of insect—or of mammal, or of microbe—capable of the outbreak phenomenon? That's a complicated question that the experts are still trying to answer.

An entomologist named Alan A. Berryman addressed it some years ago in a paper titled "The Theory and Classification of Outbreaks." He began with basics: "From the ecological point of view an outbreak can be defined as an explosive increase in the abundance of a particular species that occurs over a relatively short period of time." Then, in the same bland tone, he noted: "From this perspective, the most serious outbreak on the planet earth is that of the species *Homo sapiens.*" Berryman was alluding, of course, to the rate and the magnitude of human population growth, especially within the last couple centuries. He knew he was being provocative.

But the numbers support him. At the time Berryman wrote, in 1987, the world's human population stood at 5 billion. We had multiplied by a factor of about 333 since the invention of agriculture. We had increased by a factor of 14 since just after the Black Death, by a factor of 5 since the birth of Charles Darwin, and by doubling within the lifetime of Alan Berryman himself. That growth curve, on a coordinate graph, looks like the southwest face of El Capitan. Another way to comprehend it is this: From the time of our beginning as a species (about 200,000 years ago) until the year 1804, human population rose to a billion; between 1804 and 1927, it rose by another billion; we reached 3 billion in 1960; and each net addition of a billion people, since then, has taken only about thirteen years. In October 2011, we came to the 7-billion mark and flashed past like it was a "Welcome to Kansas" sign on the highway. That amounts to a lot of people, and certainly qualifies as an "explosive" increase within Berryman's "relatively short period of time." The rate of growth has declined within recent decades, true, but it's still above 1 percent, meaning that we're adding about 70 million people yearly.

So we're unique in the history of mammals. We're unique in the history of vertebrates. The fossil record shows that no other species of large-bodied beast—above the size of an ant, say, or of an Antarctic krill—has ever achieved anything like such abundance as the abundance of humans on Earth right now. Our total weight amounts to about 750 billion pounds. Ants of all species add up to a greater total mass, krill do too, but not many other groups of organisms. And we are just one species of mammal, not a group. We're big: big in body size, big in numbers, and big in collective weight. We're so big, in fact, that the eminent biologist (and ant expert) Edward O. Wilson felt compelled to do some knowledgeable noodling on the matter. Wilson came up with this: "When *Homo sapiens* passed the six-billion mark we had already exceeded by perhaps as much as 100 times the biomass of any large animal species that ever existed on the land."

Wilson meant wild animals. He omitted consideration of livestock, such as the domestic cow (*Bos taurus*), of which the present global population is about 1.3 billion. We are therefore only five times as numerous as our cattle (and probably less massive in total, since they're each considerably bigger than a human). But of course they wouldn't exist in such excess without us. A trillion pounds of cows, fattening in feedlots and grazing on landscapes that formerly supported wild herbivores, are just another form of human impact. They're a proxy measure of our appetites, and we are hungry. We are prodigious, we are unprecedented. We are phenomenal. No other primate has ever weighed upon the planet to anything like this degree. In ecological terms, we are almost paradoxical: large-bodied and long-lived but grotesquely abundant. We are an outbreak.

111

And here's the thing about outbreaks: They end. In some cases they end after many years, in other cases they end rather soon. In some cases they end gradually, in other cases they end with a crash. In certain cases, even, they end and recur and end again, as though following a regular schedule. Populations of tent caterpillars and several other kinds of forest lepidopterans seem to rise steeply and fall sharply on a cycle of anywhere from five to eleven years. A population of tent caterpillars in British Columbia, for instance, has shown a cycle like that dating back to 1936. The crash endings are especially dramatic and for a long while they seemed mysterious. What could account for such sudden and recurrent collapses? One possible factor is infectious disease. It turns out that viruses, in particular, play that role among outbreak populations of forest insects.

Back in 1993, when the caterpillars hit my town, I got interested in this subject and did some research. It seemed peculiar to me that a critter like the forest tent caterpillar, with a very limited repertoire of behavior, a fixed set of adaptive tactics, should multiply egregiously during one or two summers and then virtually disappear by summer three. The environment hadn't changed drastically, yet the success of one species within that environment had. Why? Variations in weather didn't explain it. Exhaustion of food supplies didn't explain it. I called the county extension service and pestered a fellow there with questions. "I don't think anyone can say why you have the boom and bust," he told me. "It just happens."

Because that reply wasn't satisfactory or convincing, I started reading the entomological literature. Among the experts in the field was one Judith H. Myers, a professor at the University of British Columbia, who had published several papers on tent caterpillars and an overview of insect population outbreaks. Myers offered a solution to the mystery. Although population levels are influenced by many factors, she wrote, the cyclical pattern "seems to imply a dominant force that should be easy to identify and quantify. That driving force, however, has proved surprisingly elusive." But now ecologists had a suspect, she reported. Myers described something called nuclear polyhedrosis viruses, known collectively as NPVs, which "may be the long-sought driving force of population cycles in forest Lepidoptera." Field studies had revealed that NPVs achieve their own outbreaks within outbreaking populations of forest lepidopterans, killing off the insects like the blackest of Black Deaths.

For years I didn't think much about this. The outbreak of tent caterpillars in my town ended quietly but quickly, back in 1993, with no sign of the hairy larvae the following summer. That was a long time ago. But the event came back to mind, during my work on this book, as I sat in the auditorium at a scientific conference on the ecology and evolution of infectious diseases. We were gathered in Athens, Georgia. The agenda was peppered with presentations on zoonoses, to be given by some of the frontline researchers and brainiest theorists in the field, and that's what had attracted me. There would be a talk on Hendra virus and how it emerges from flying foxes; there would be a talk on the spillover dynamics of monkeypox; there would be at least four talks on influenza. But the second morning of the conference began with something different. I sat down politely, and then found myself mesmerized by a smart, puckish fellow named Greg Dwyer, a mathematical ecologist from the University of Chicago, who paced back and forth, speaking quickly, without notes, about population outbreaks and disease among insects.

"You've probably never heard of nucleopolyhedroviruses," Dwyer said to us. The name had changed slightly since 1993 but, thanks to the tent caterpillar episode, and to Judith H. Myers, I had. Dwyer described the devastating effect of NPVs on outbreak populations of forest lepidopterans. He spoke particularly about the gypsy moth (*Lymantria dispar*), another little brown creature, whose outbreaks and crashes he had studied for twenty years. He said that gypsy moth larvae essentially "melt" when infected by NPV. I wasn't taking copious notes, but I did write the word "melt" on my yellow pad. I also wrote, quoting him: "Epizootics tend to occur in very dense populations." After a few other general comments, Greg

Dwyer went on to discuss some mathematical models. At the coffee break, I buttonholed him and asked if we could talk sometime about the fate of moths and the prospect of human pandemic disease. He said sure.

112

Two years passed, but then schedules came into alignment and I called on Greg Dwyer at the University of Chicago. His office, on the ground floor of a biology building just off East 57th Street, was cheerily decorated with the usual posters and cartoons and, along the left wall, a long whiteboard. Dwyer was fifty at the time and seemed young, like an amiable grad student whose beard had gone gray. He wore round tortoiseshell glasses and a black T-shirt printed with a grotesquely complex integral equation. Above and below the equation, the shirt asked in large letters: WHAT PART OF [this gobbledygook] DON'T YOU UNDERSTAND? The shirt was a metajoke, he explained to me. The gobbledygook was one of Maxwell's equations; the joke part, of course, was that no average person would understand the thing at all; the meta part, I think, was that Maxwell's equations are famous but so notoriously abstruse that even a mathematician might not recognize this one. Get it?

We seated ourselves on opposite sides of his desk but then, as soon as the conversation got rolling, Dwyer jumped up and began drawing on the whiteboard. So I stood too, as though being closer to his scribblings would help me comprehend them. He drew a set of coordinate axes, one axis for the number of gypsy moth eggs in a forest, the other axis for time, and explained how scientists measure an outbreak. Between outbreaks, the gypsy moth is so scarce it's undetectable. During an outbreak, in contrast, you find thousands of egg masses per acre. With about 250 eggs in each egg mass, that yields a lot of moths. He drew a graph depicting the rise and fall of a gypsy moth population over successive years. It looked like a Chinese dragon, the line of its back arching way up and then dropping way down, way up again, then again way down. He drew a sketch of NPV particles and described how they package themselves for protection against sunlight and other environmental stresses. Each packet is a solid lump of protein, polyhedral in shape (hence the name) and containing dozens of virions embedded like bits of cherry in a fruitcake. Dwyer drew more graphs and, while drawing, explained to me how this nefarious virus works.

The packets of virus lay besmeared on a leaf, left there after the death of a previous caterpillar victim. A healthy caterpillar comes munching along and swallows packets with the leaf tissue. Once inside the caterpillar, a packet unfolds, sinister and orderly, like a MIRV warhead releasing its little nukes over a city. The virions disperse, attacking cells in the caterpillar's gut. Each virion goes to the cell nucleus (again, hence the name), replicates abundantly, generating new virions that exit the cell and proceed to attack others. "They go from cell to cell, and infect lots and lots of cells," Dwyer said. Before long the caterpillar is essentially just a crawling and eating bag of virus. Still, it doesn't act sick. It doesn't seem to know how sick it is. "If it has eaten a big enough dose," he said, "then it will continue to wander around on leaves and continue to feed—but after maybe ten days, maybe two weeks, sometimes even as long as three weeks, it will melt onto a leaf." There was that word again, the same one he had used in Atlanta, exquisitely vivid: melt.

Other caterpillars meanwhile are suffering the same fate. "The virus has almost completely consumed them before they really stop functioning." Late in this process, as the virions within each caterpillar begin crowding one another, running short of food, they get themselves bundled together again within protective packets. Time to emerge. Time to move on. The caterpillar at this point is filled with virus, consumed by virus, held together only by its skin. But the skin, made of protein and carbohydrates, is tough and flexible. Then the virus releases certain enzymes, which dissolve the skin, and the caterpillar splits open like a water balloon. "They pick up the virus," Dwyer said, and "they go *splat* on a leaf." Each caterpillar disintegrates, leaving little more than a viral smudge—a smudge that, in the crowded conditions of an outbreak population of gypsy moths, is soon gobbled by the next hungry caterpillar. And so on. "Another insect comes along, feeds on that leaf, a week or two later," Dwyer said, then repeated: "It goes *splat*."

There might be five or six generations of *splat* in the course of the summer, five or six waves of transmission, with the virus progressively increasing its prevalence within the caterpillar population. From a starting point of low prevalence—say, 5 percent of the caterpillars are infected—it might grow to 40 percent by the first autumn. After the surviving caterpillars have metamorphosed to moths, and then mated, in a habitat still cluttered with NPV, some packets of the virus are left besmeared not just on foliage but on the egg masses laid by the female moths. So a large portion of the new caterpillars become infected the following spring as they hatch. The prevalence of the infection rises steeply. And that rise, beyond the preceding year's level, "translates into an even higher percentage the following year," Dwyer said. Within two or three years, such ratcheting "basically wipes out the entire population."

The moths disappear and all that remains is the virus. Sometimes there's so much of it, he added, that "you'll see this kind of gray fluid trickling down the bark." Rains come, and the trees weep with a slurry of dissolved caterpillars and virus. I was duly impressed.

It sounds like Ebola, I said.

"Yeah, right." He had sat through some of the same meetings and read some of the same books and papers that I had.

Except not Ebola in reality, I said. The sensationalized Ebola, the popularized nightmare of Ebola, the hyped version of victims "bleeding out" like a sack of liquid guts.

He agreed. And the same distinction between degrees of gruesomeness, the real versus the exaggerated, applies to NPV. "With our virus, people like to say, they'll say, 'Oh, you study that virus that causes the insect to explode!' Like, the virus *doesn't* cause the insect to *explode*," he insisted. "It causes it to *melt*."

Having heard this scenario, and seen his graphs, and appreciated the directness of his language, and admired Maxwell's equation on his T-shirt, I came to the point of my visit: what I called The Analogy. As of last week, I said, we've got 7 billion humans on this planet. It seems like an outbreak population. We live at high densities. Look at Hong Kong, look at Mumbai. We're closely interconnected. We fly around. The 7 million people in Hong Kong are only three hours away from the 12 million people in Beijing. No other large animal has ever been as abundant. And we've also got our share of potentially devastating viruses. Some of those might be as nasty as NPV. So . . . what's the prognosis? Is it valid, The Analogy? Should we expect to crash like a population of gypsy moths?

Dwyer couldn't be rushed into saying yes. Judiciously empirical, wary of easy extrapolations, he wanted to pause and think. He did. And then we found ourselves talking about influenza.

113

I haven't said much about influenza in this book, but not because it isn't important. On the contrary, it's vastly important, vastly complicated, and still potentially devastating in the form of a global influenza pandemic. The Next Big One could very well be flu. Greg Dwyer knew this, which is why he mentioned it. I'm sure you don't need reminding that the 1918–1919 flu killed about 50 million people; and there's still no magical defense, no universal vaccine, no foolproof and widely available treatment, to guarantee that such death and misery don't occur again. Even during an average year, seasonal flu causes at least 3 million cases and more than 250,000 fatalities worldwide. So influenza is hugely dangerous, at best. At worst, it would be apocalyptic. I've left it for now only because it's well suited to suggest some closing thoughts on the whole subject of zoonotic disease.

First, the basics. Influenza is caused by three types of viruses, of which the most worrisome and widespread is influenza A. Viruses of that type all share certain genetic traits: a single-stranded RNA genome, which is partitioned into eight segments, which serve as templates for eleven different proteins. In other words, they have eight discrete stretches of RNA coding, linked together like eight railroad cars, with eleven different deliverable cargoes. The eleven deliverables are the molecules that comprise the structure and functional machinery of the virus. They are what the genes make. Two of those molecules become spiky protuberances from the outer surface of the viral envelope: hemagglutinin and neuraminidase. Those two, recognizable by an immune system, and crucial for penetrating and exiting cells of a host, give the various subtypes of influenza A their definitive labels: H5N1, H1N1, and so on. The term "H5N1" indicates a virus featuring subtype 5 of the hemagglutinin protein combined with subtype 1 of the neuraminidase protein. Sixteen different kinds of hemagglutinin, plus nine kinds of neuraminidase, have been detected in the natural world. Hemagglutinin is the key that unlocks a cell membrane so that the virus can get in, and neuraminidase is the key for getting back out. Okay so far? Having absorbed this simple paragraph, you understand more about influenza than 99.9 percent of the people on Earth. Pat yourself on the back and get a flu shot in November.

At the time of the 1918–1919 pandemic, no one knew what was causing it (though there were plenty of guesses). No one could find the guilty bug, no one could see it, no one could name it or comprehend it, because virology itself had scarcely begun to exist. Techniques of viral isolation hadn't yet been developed. Electron microscopes hadn't yet been invented. The virus responsible, which turned out to be a variant of H1N1, wasn't precisely identified until . . . 2005! During the intervening decades there were other flu pandemics, including one in 1957, which killed roughly 2 million people, and another in 1968, which became known as the Hong Kong flu (for where it began) and killed a million. By the end of the 1950s, scientists had recognized the influenza viruses as a somewhat mystifying group, highly diverse and variously capable of infecting pigs, horses, ferrets, cats, domestic ducks, and chickens as well as people. But no one knew where these things lived in the wild.

Were they zoonoses? Did they have reservoir hosts? One hint appeared in 1961, when a number of common terns (*Sterna hirundo*, a kind of seabird) died in South Africa and were found to contain influenza. If the flu virus had killed them, then by definition the terns weren't its reservoir; but maybe their life histories put them *in contact* with the reservoir. Soon after that, a young biologist from New Zealand went for a walk along the coast of New South Wales with a young Australian biochemist. They saw some dead birds.

These two men were great pals, sharing a love for the outdoors. Their beach walk, in fact, was part of a fishing trip. The New Zealander was Robert G. Webster, transplanted to Australia to do his PhD, and the Australian was William Graeme Laver, educated in Melbourne and London, inspired to a research career by Macfarlane Burnet. Laver was such an adventurous soul that, when he finished his doctoral work in London, he and his wife *drove* home to Australia rather than fly. Several years later he and Webster took their historic stroll, found the beach littered with carcasses of wedge-tailed shearwaters (another seabird, of the species *Puffinus pacificus*), and wondered—with the South African terns in mind—whether these birds too might have been killed by influenza. Laver suggested, almost as a lark, that it would be good to go up to the Great Barrier Reef and sample some birds there for influenza. The Great Barrier Reef is not generally perceived as a hardship venue. They might get a bit of fishing, bake in the sun, enjoy the clear blue-green waters, and do the science. Laver asked his boss at the Australian National University, in Canberra, to fund Webster and him for such a study. You must be hallucinating, said the boss. Not on my dollar, you're not. So they appealed to the World Health Organization, in Geneva, where a trusting officer gave them $500, a substantial bit of money at the time. Laver and Webster went to a place called Tryon Island, fifty miles off the coast of Queensland, and found influenza virus in wedge-tailed shearwaters.

"So we have flu, related to human flu, in the wild migratory birds of the world," Robert Webster told me, forty years later. In the scientific literature he had been rather unassuming about this work but in conversation he laid it out: Sure, Graeme Laver made the discovery that waterfowl are the reservoirs of influenza, with my help. Laver by now was dead, but fondly remembered by Dr. Webster.

Robert Webster today is arguably the most eminent influenza scientist in the world. He grew up on a New Zealand farm, studied microbiology, did his doctorate at Canberra, worked and cavorted with Laver, then moved to the United States in 1969, taking a post at St. Jude Children's Research Hospital in Memphis, and has been there (apart from his frequent travels) ever since. He was almost eighty when I met him but still on the job, still robust, and still at the forefront of influenza research as it responds daily to viral news from all over the world. We spoke in his office, upstairs in a sleek building at St. Jude's, after he had bought me a cup of strong coffee in the hospital cafeteria. On the office wall hung two mounted fish—a large green grouper and a handsome red snapper—as though in tribute to Graeme Laver. One of the things that makes influenza so problematic, Webster said, is its propensity to change.

He explained. First of all there's the high rate of mutation, as in any RNA virus. No quality control as it replicates, he said, echoing what I'd heard from Eddie Holmes. Continual copying errors at the level of individual letters of code. But that's not the half of it. Even more important is the reassortment. ("Reassortment" means the accidental swapping of entire genomic segments between virions of two different subtypes. It's similar to recombination, as occurs sometimes between crossed chromosomes in dividing cells, except that reassortment is somewhat more facile and orderly. It happens often among influenza viruses because the segmentation allows their RNA to snap apart neatly at the points of demarcation between genes: those eight railroad cars in a switching yard.) Sixteen available kinds of hemagglutinin, Webster reminded me. Nine kinds of neuraminidase. "You can do the arithmetic," he said. (I did: 144 possible pairings.) The changes are random and most yield bad combinations, making the virus less viable. But random changes do constitute variation, and variation is the exploration of possibilities. It's the raw material of natural selection, adaptation, evolution. That's why influenza is such a protean sort of bug, always full of surprises, full of newness, full of menace: so much mutation and reassortment.

The steady incidence of mutations yields incremental change in how the virus looks and behaves. Ergo you need another flu shot every autumn: This year's version of flu is different enough from last year's. Reassortment yields big changes. Such major innovations by reassortment, introducing new subtypes, which may be infectious but unfamiliar to the human population, are what generally lead to pandemics.

But it's not all about human disease. Different subtypes, Webster noted, have their affinities for different species of host. H7N7 does well among horses. The dead terns in South Africa, back in 1961, had been infected with H5N3. Only subtypes bearing H1, H2, or H3 as their hemagglutinin cause human flu epidemics, because only those spread readily from person to person. Pigs offer conditions intermediate between what a flu virus finds in people and what it finds in birds; therefore pigs get infected with both human subtypes and bird subtypes. When an individual pig is infected simultaneously with two viruses—one adapted to humans, one adapted to birds—the opportunity exists for reassortment between those two. Although wild aquatic birds are now known to be the ultimate origin of all influenzas, the viruses reassort themselves in pigs and elsewhere (quail also serve as mixing bowls), and by the time they get into humans, they have generally been assembled from H1, H2, or H3 plus the ten other necessary proteins, some of those in forms borrowed from this or that bird flu or pig flu virus. Other subtypes, featuring H7 and H5, have occasionally "tried on" the prospect of targeting people, Webster said. And in all cases so far, the fit has been bad.

"They infect humans," he said, "but they haven't acquired transmissibility." They don't pass from person to person. They may kill a lot of poultry, spreading through entire flocks, but they don't travel on human sneezes. (Influenza among birds is primarily an infection of the gastrointestinal tract, with transmission occurring by the fecal-oral route; a sick bird shits the virus onto the floor of its coop, or onto the ground of a barnyard, or into the water of a lake or an estuary, and another bird picks it up while pecking or dabbling for food. That's presumably how those South African terns and Australian shearwaters encountered the virus.) So you've got to handle a hen, or butcher a duck, to get infected. Still, with such a variable group of viruses, always mutating, continually reassorting, the next "try on" could be different. Consequently there's "not a hope in hell, at this time," Webster said, of predicting just what the next pandemic will be.

But some things bear watching. Case in point: H5N1, more familiar to you and me as bird flu.

Webster himself played a crucial role in responding to that scary subtype when it first emerged. A three-year-old boy died in Hong Kong of influenza, in May 1997, and a swab sample from his windpipe yielded virus. The lab scientists in Hong Kong didn't recognize that virus. Some of the boy's sample went to the CDC, but no one there got around to characterizing it. Then a Dutch scientist on a visit to Hong Kong was given a bit of the virus, and he went home and worked on it immediately. *Hmm, mijn God.* The Dutchman informed his international colleagues that it looked like an H5. A bird flu. "And we all said, 'no, impossible,'" Webster recalled. "Since H5 doesn't affect humans. We thought it was a mistake." It wasn't. What seemed so alarming is that this was the first documented case of a purely avian influenza virus—containing no human-flu genes brought in by reassortment—to cause killer respiratory illness in a person. Three more cases turned up in November, at which point Webster himself jumped on a plane for Hong Kong.

It was bad timing for a medical emergency, 1997, that being the year of Hong Kong's big political transition from a British colony into a special administrative region of China. Public institutions were unsettled, management and staffing were in flux, and Robert Webster found the University of Hong Kong depleted of influenza experts. Then still more human cases appeared, for a total of eighteen by the end of the year, with a case fatality rate of 33 percent. The bird subtype was highly virulent. But how transmissible? No one had traced its origin, let alone learned whether it might spread quickly among humans. "So I whistled up all the postdocs that I had trained around the Pacific," Webster said, "and told them to get to Hong Kong. And within three days, we located the virus in the live poultry markets."

It was a crucial start. Hong Kong officials ordered the culling of all domestic poultry (1.5 million birds) and closed the bird markets, which solved the immediate problem. For a while there were no further cases, not in Hong Kong, not anywhere. But the nasty new virus hadn't been eradicated. It continued to circulate quietly among domestic ducks in the coastal provinces of China, where many rural people kept small flocks of quackers and led them out daily to feed in the rice paddies. The virus was hard to trace in such circumstances, harder still to eliminate, because infected ducks showed no symptoms. "The duck is the Trojan horse," Webster told me. That's where the danger lurked secretly, he meant. Wild ducks might land on your flooded paddy, carrying the virus, fouling the water, and infecting your domestic ducks. Your ducks would appear fine, but when your son brought them home to their coop for the night, they could infect your chickens. Before long your chickens—and your son too—might be dead of bird flu.

"The duck is the Trojan horse," he repeated. It was a good line, vivid and clear, and I had seen it also in some of his published work. But today he was even more specific: mallards and pintails. The pathogenicity of this virus differs starkly for different kinds of birds. "It depends on the species," Webster said. "Some duck species die. The bar-headed goose dies. The swans die. But the mallard, and the pintail in particular, carry. And spread."

Six years after its first outbreak in Hong Kong, H5N1 returned, infecting three members of a family and killing two. As I've described earlier, this occurred during the first alarms over what came to be known as SARS, complicating efforts to identify that very different bug. Around the same time, H5N1 started turning up among domestic poultry in South Korea, Vietnam, Japan, Indonesia, and elsewhere throughout the region, killing many chickens and at least a couple more people. It also traveled in wild birds—traveled pretty far. Qinghai Lake, in western China, thirteen hundred miles northwest of Hong Kong, became the site of one ominous event, to which Webster had alluded with his mention of bar-headed geese.

Qinghai Lake is an important breeding site for migratory waterfowl, whose flyways lead variously from there to India, Siberia, and Southeast Asia. In April and May 2005, six thousand birds died at Qinghai of H5N1 influenza. The first animal affected was the bar-headed goose, but the disease also struck ruddy shelducks, great cormorants, and two kinds of gull. Bar-headed geese, with large wing areas relative to their weight, are well adapted to flying high and far. They nest on the Tibetan plateau. They migrate over the Himalayas. They shed H5N1.

"And then presumably," Webster told me, "the wild birds carried it westward to India, Africa, Europe, and so on." It got to Egypt in 2006, for instance, and has been especially problematic for that country. "The virus is *everywhere* in Egypt. Through the commercial poultry, through the duck populations." Egyptian health authorities tried vaccinating their poultry, with vaccine imported from Asia, but the vaccine efforts didn't work. "It's surprising there are not more human cases." The toll in Egypt is high enough: 151 confirmed as of August 2011, of which 52 were fatal. Those numbers represent more than a quarter of all the world's known human cases of bird flu, and more than a third of all fatalities, since H5N1 emerged in 1997. But here's a critical fact: Few if any of the Egyptian cases resulted from human-to-human transmission. Those unfortunate Egyptian patients all seem to have acquired the virus directly from birds. This indicates that the virus hasn't yet found an efficient way to pass from one person to another.

Two aspects of the situation are dangerous, according to Robert Webster. The first is that Egypt, given its recent political upheavals and the uncertainty about where those will lead, may be unable to stanch an outbreak of transmissible avian flu, if one occurs. His second point of concern is shared by influenza researchers and public health officials around the globe: With all that mutating, with all that contact between people and their infected birds, the virus *could* hit upon a genetic configuration making it highly transmissible among people.

"As long as H5N1 is out there in the world," Webster said, "there is the possibility of disaster. That's really the bottom line with H5N1. So long as it's out there in the human population, there is the theoretical possibility that it can acquire the ability to transmit human-to-human." He paused. "And then God help us."

114

This whole subject, like an airborne virus, is at large on the breezes of discourse. Most people aren't familiar with the word "zoonotic," but they have heard of SARS, they have heard of West Nile virus, they have heard of bird flu. They know someone who has suffered through Lyme disease and someone else who has died of AIDS. They have heard of Ebola, and they know that it's a terrifying thing (though they may confuse it with *E. coli*, the bacterium that can kill you if you eat the wrong spinach). They are concerned. They are vaguely aware. But they don't have the time or the interest to consider a lot of scientific detail. I can say from experience that some people, if they hear you're writing a book about such things—about scary emerging diseases, about killer viruses, about pandemics—want you to cut to the chase. So they ask: "Are we all gonna die?" I have made it my little policy to say yes.

Yes, we are all gonna die. Yes. We are all gonna pay taxes and we are all gonna die. Most of us, though, will probably die of something more mundane than a new virus lately emerged from a duck or a chimpanzee or a bat.

The dangers presented by zoonoses are real and severe but the degree of uncertainties is also high. There's not a hope in hell, as Robert Webster pungently told me, of predicting the nature and timing of the next influenza pandemic. Too many factors vary randomly, or almost randomly, in that system. Prediction, in general, so far as all these diseases are concerned, is a tenuous proposition, more likely to yield false confidence than actionable intelligence. I have asked not just Webster but also many other eminent disease scientists, including some of the world's experts on Ebola, on SARS, on bat-borne viruses generally, on the HIVs, and on viral evolution, the same two-part question: (1) Will a new disease emerge, in the near future, sufficiently virulent and transmissible to cause a pandemic on the scale of AIDS or the 1918 flu, killing tens of millions of people? and (2) If so, what does it look like and whence does it come? Their answers to the first part have ranged from Maybe to Probably. Their answers to the second have focused on RNA viruses, especially those for which the reservoir host is some kind of primate. None of them has disputed the premise, by the way, that if there *is* a Next Big One it will be zoonotic.

In the scientific literature, you find roughly the same kind of cautious, informed speculation. A highly regarded infectious-disease epidemiologist named Donald S. Burke, presently dean of the Graduate School of Public Health at the University of Pittsburgh, gave a lecture (later published) back in 1997 in which he listed the criteria that might implicate certain kinds of viruses as likeliest candidates to cause a new pandemic. "The first criterion is the most obvious: recent pandemics in human history," Burke told his audience. That would point to the orthomyxoviruses (including the influenzas) and the retroviruses (including the HIVs), among others. "The second criterion is proven ability to cause major epidemics in non-human animal populations." This would again spotlight the orthomyxoviruses, but also the family of paramyxoviruses, such as Hendra and Nipah, and the coronaviruses, such as that virus later known as SARS-CoV. Burke's third criterion was "intrinsic evolvability," meaning readiness to mutate and to recombine (or reassort), which "confers on a virus the potential to emerge into and to cause pandemics in human populations." As examples he returned to retroviruses, orthomyxoviruses, and coronaviruses. "Some of these viruses," he warned, citing coronaviruses in particular, "should be considered as serious threats to human health. These are viruses with high evolvability and proven ability to cause epidemics in animal populations." It's interesting in retrospect to note that he had augured the SARS epidemic six years before it occurred.

Much more recently, Burke told me: "I made a lucky guess." He laughed a self-deprecating hoot and then added that "prediction is too strong a word" for what he had been doing.

Donald Burke can be trusted on this as much as anyone alive. But the difficulty of predicting precisely doesn't oblige us to remain blind, unprepared, and fatalistic about emerging and re-emerging zoonotic diseases. No. The practical alternative to soothsaying, as Burke put it, is "improving the scientific basis to improve readiness." By "the scientific basis" he meant the understanding of which virus groups to watch, the field capabilities to detect spillovers in remote places before they become regional outbreaks, the organizational capacities to control outbreaks before they become pandemics, plus the laboratory tools and skills to recognize known viruses speedily, to characterize new viruses almost as fast, and to create vaccines and therapies without much delay. If we can't predict a forthcoming influenza pandemic or any other newly emergent virus, we can at least be vigilant; we can be well-prepared and quick to respond; we can be ingenious and scientifically sophisticated in the forms of our response.

To a considerable degree, such things are already being done on our behalf by some foresighted institutions and individuals in the realm of disease science and public health. Ambitious networks and programs have been created, by the World Health Organization, the Centers for Disease Control and Prevention, the United States Agency for International Development, the European Center for Disease Prevention and Control, the World Organization for Animal Health, and other national and international agencies, to address the danger of emerging zoonotic diseases. Because of concern over the potential of "bioterrorism," even the US Department of Homeland Security and the Defense Advanced Research Projects Agency (aka Darkest DARPA, whose motto is "Creating & Preventing Strategic Surprise") of the US Department of Defense have their hands in the mix. (Since the United States foreswore offensive bioweapons research back in 1969, presumably DARPA's disease program is now aimed at preventing, not creating, strategic surprise of the epidemiological sort.) These efforts carry names and acronyms such as the Global Outbreak Alert and Response Network (GOARN, of WHO), Prophecy (of DARPA), the Emerging Pandemic Threats program (EPT, of USAID), and the Special Pathogens Branch (SPB, of the CDC), all of which sound like programmatic boilerplate but which harbor some dedicated people working in field sites where spillovers happen and secure labs where new pathogens can be quickly studied. Private organizations, such as EcoHealth Alliance (led by a former parasitologist named Peter Daszak and now employing Jon Epstein for his Nipah work in Bangladesh and elsewhere, Aleksei Chmura for his bat research in China, Billy Karesh for his continuing wildlife-health studies around the world, and others), have also tackled the problem. There is an intriguing effort called the Global Viral Forecasting Initiative (GVFI), financed in part by Google and created by a bright, enterprising scientist named Nathan Wolfe, one of whose mentors was Don Burke. GVFI gathers blood samples on small patches of filter paper from bushmeat hunters and other people across tropical Africa and Asia, and screens those samples for new viruses, in a systematic effort to detect spillovers and stop the next pandemic before it begins to spread. Wolfe learned the filter-paper technique from Balbir Singh and Janet Cox-Singh (the malaria researchers who study *Plasmodium knowlesi* in humans, remember?), during field time he spent with them as a graduate student in the 1990s. At the Mailman School of Public Health, part of Columbia University, Ian Lipkin's laboratory is a whiz-bang center of efforts to develop new molecular diagnostic tools. Lipkin, trained as a physician as well as a molecular biologist, calls his métier "pathogen discovery" and uses techniques such as high-throughput sequencing (which can sequence thousands of DNA samples quickly and cheaply), MassTag PCR (identifying amplified genome segments by mass spectrometry), and the GreeneChip diagnostic system, which can simultaneously screen for thousands of different pathogens. When Jon Epstein takes serum from flying foxes in Bangladesh, when Aleksei Chmura bleeds bats in southern China, some of those samples go straight to Ian Lipkin.

These scientists are on alert. They are our sentries. They watch the boundaries across which pathogens spill. And they are productively interconnected with one another. When the next novel virus makes its way from a chimpanzee, a bat, a mouse, a duck, or a macaque into a human, and maybe from that human into another human, and thereupon begins causing a small cluster of lethal illnesses, they will see it—we hope they will, anyway—and raise the alarm.

Whatever happens after that will depend on science, politics, social mores, public opinion, public will, and other forms of human behavior. It will depend on how we citizens respond.

So before we respond, either calmly or hysterically, either intelligently or doltishly, we should understand in some measure the basic outlines and dynamics of the situation. We should appreciate that these recent outbreaks of new zoonotic diseases, as well as the recurrence and spread of old ones, are part of a larger pattern, and that humanity is responsible for generating that pattern. We should recognize that they reflect things that we're *doing*, not just things that are *happening* to us. We should understand that, although some of the human-caused factors may seem virtually inexorable, others are within our control.

The experts have alerted us to these factors and it's easy enough to make a list. We have increased our population to the level of 7 billion and beyond. We are well on our way toward 9 billion before our growth trend is likely to flatten. We live at high densities in many cities. We have penetrated, and we continue to penetrate, the last great forests and other wild ecosystems of the planet, disrupting the physical structures and the ecological communities of such places. We cut our way through the Congo. We cut our way through the Amazon. We cut our way through Borneo. We cut our way through Madagascar. We cut our way through New Guinea and northeastern Australia. We shake the trees, figuratively and literally, and things fall out. We kill and butcher and eat many of the wild animals found there. We settle in those places, creating villages, work camps, towns, extractive industries, new cities. We bring in our domesticated animals, replacing the wild herbivores with livestock. We multiply our livestock as we've multiplied ourselves, operating huge factory-scale operations involving thousands of cattle, pigs, chickens, ducks, sheep, and goats, not to mention hundreds of bamboo rats and palm civets, all confined en masse within pens and corrals, under conditions that allow those domestics and semidomestics to acquire infectious pathogens from external sources (such as bats roosting over the pig pens), to share those infections with one another, and to provide abundant opportunities for the pathogens to evolve new forms, some of which are capable of infecting a human as well as a cow or a duck. We treat many of those stock animals with prophylactic doses of antibiotics and other drugs, intended not to cure them but to foster their weight gain and maintain their health just sufficiently for profitable sale and slaughter, and in doing that we encourage the evolution of resistant bacteria. We export and import livestock across great distances and at high speeds. We export and import other live animals, especially primates, for medical research. We export and import wild animals as exotic pets. We export and import animal skins, contraband bushmeat, and plants, some of which carry secret microbial passengers. We travel, moving between cities and continents even more quickly than our transported livestock. We stay in hotels where strangers sneeze and vomit. We eat in restaurants where the cook may have butchered a porcupine before working on our scallops. We visit monkey temples in Asia, live markets in India, picturesque villages in South America, dusty archeological sites in New Mexico, dairy towns in the Netherlands, bat caves in East Africa, racetracks in Australia—breathing the air, feeding the animals, touching things, shaking hands with the friendly locals—and then we jump on our planes and fly home. We get bitten by mosquitoes and ticks. We alter the global climate with our carbon emissions, which may in turn alter the latitudinal ranges within which those mosquitoes and ticks live. We provide an irresistible opportunity for enterprising microbes by the ubiquity and abundance of our human bodies.

Everything I've just mentioned is encompassed within this rubric: the ecology and evolutionary biology of zoonotic diseases. Ecological circumstance provides opportunity for spillover. Evolution seizes opportunity, explores possibilities, and helps convert spillovers to pandemics.

It's a neat but sterile historical coincidence that the germ theories of disease came to scientific prominence at about the same time, in the late nineteenth century, as the Darwinian theory of evolution—neat because these were two great bodies of insight with much to offer each other, and sterile because their synergy was long delayed, as germ theories remained for another sixty years largely uninformed by evolutionary thinking. Ecological thinking, in its modern form, arose even later and was equally slow to be absorbed by disease science. The other absent science, until the second half of the twentieth century, was molecular biology. Medical people of the earlier eras might guess that bubonic plague was somehow related to rodents, yes, but they didn't know how or why until Alexandre Yersin, during an 1894 epidemic in Hong Kong, found the plague bacterium in rats. Even that didn't illuminate the path to human infection until Paul-Louis Simond, several years later, showed that the bacterium is transmitted by rat fleas. Anthrax, caused by another bacterium, was known to kill cows and people but seemed to arise by spontaneous generation until Koch proved otherwise in 1876. Rabies was even more obviously associated with transmission to humans from animals—notably, mad dogs—and Pasteur introduced a rabies vaccine in 1885, injecting a bitten boy, who survived. But rabies virus itself, so much smaller than a bacterium, couldn't be directly detected nor traced to wild carnivores until much later. During the early twentieth century, disease scientists from the Rockefeller Foundation and other institutions conceived the ambitious goal of eradicating some infectious diseases entirely. They tried hard with yellow fever, spending millions of dollars and many years of effort, and failed. They tried with malaria, and failed. They tried later with smallpox, and succeeded. Why? The differences among those three diseases are many and complex, but probably the most crucial one is that smallpox resided neither in a reservoir host nor in a vector. Its ecology was simple. It existed in humans—in humans only—and was therefore much easier to eradicate. The campaign to eradicate polio, begun in 1988 by WHO and other institutions, is a realistic effort for the same reason: Polio isn't zoonotic. And malaria is now targeted again. The Bill and Melinda Gates Foundation announced, in 2007, a new long-term initiative to eradicate that disease. It's an admirable goal, a generously imaginative dream, but a person is left to wonder how Mr. and Mrs. Gates and their scientific advisers propose to deal with *Plasmodium knowlesi*. Do you exterminate the parasite by killing off its reservoir hosts, or do you somehow apply your therapeutics to those hosts, curing every macaque in the forests of Borneo?

That's the salubrious thing about zoonotic diseases: They remind us, as St. Francis did, that we humans are inseparable from the natural world. In fact, there *is* no "natural world," it's a bad and artificial phrase. There is only the world. Humankind is part of that world, as are the ebolaviruses, as are the influenzas and the HIVs, as are Nipah and Hendra and SARS, as are chimpanzees and bats and palm civets and bar-headed geese, as is the next murderous virus—the one we haven't yet detected.

I don't say these things about the ineradicability of zoonoses to render you hopeless and depressed. Nor am I trying to be scary for the sake of scariness. The purpose of this book is not to make you more worried. The purpose of this book is to make you more smart. That's what most distinguishes humans from, say, tent caterpillars and gypsy moths. Unlike them, we can be pretty smart.

Greg Dwyer came around to this point during our talk in Chicago. He had studied all the famous mathematical models proposed to explain disease outbreaks in humans—Anderson and May, Kermack and McKendrick, George MacDonald, John Brownlee, and the others. He had noted the crucial effect of individual behavior on rate of transmission. He had recognized that what people do as individuals, what moths do as individuals, has a large effect on R_0. The transmission of HIV, for instance, Dwyer said, "depends on human behavior." Who could argue? It has been proven. Consult the changes in rate of transmission among American gay men, among the general populace of Uganda, or among sex workers in Thailand. The transmission of SARS, Dwyer said, seems to depend much on superspreaders—and their behavior, not to mention the behavior of people around them, can be various. The mathematical ecologist's term for variousness of behavior is "heterogeneity," and Dwyer's models have shown that heterogeneity of behavior, even among forest insects, let alone among humans, can be very important in damping the spread of infectious disease.

"If you hold mean transmission rate constant," he told me, "just adding heterogeneity by itself will tend to reduce the overall infection rate." That sounds dry. What it means is that individual effort, individual discernment, individual choice can have huge effects in averting the catastrophes that might otherwise sweep through a herd. An individual gypsy moth may inherit a slightly superior ability to avoid smears of NPV as it grazes on a leaf. An individual human may choose not to drink the palm sap, not to eat the chimpanzee, not to pen the pig beneath mango trees, not to clear the horse's windpipc with his bare hand, not to have unprotected sex with the prostitute, not to share the needle in a shooting gallery, not to cough without covering her mouth, not to board a plane while feeling ill, or not to coop his chickens along with his ducks. "Any tiny little thing that people do," Dwyer said, if it makes them different from one another, from the idealized standard of herd behavior, "is going to reduce infection rates." This was after I had asked him to consider The Analogy and he had pushed his brain against it for half an hour.

"There's only so many ways gypsy moths can differ," he said finally. "But the number of ways that humans can differ is really, really huge. And especially in their behavior. Right. Which gets back to your question, which is, How much does it matter that humans are smart? And so, I guess I'm actually going to say that it matters a whole lot. Now that I stop to think about it carefully. I think it will matter a great deal."

Then he took me into the basement of the building and gave me a glimpse of the experimental side of his work. He unlocked a door to what he called "the dirty room," opened an incubator, took out a plastic container, and showed me gypsy moth caterpillars infected with NPV. I saw what it looks like to go *splat* on a leaf.

115

Of the two giant elm trees that stood before my neighbor Susan's house, only one remains. The other died about four years ago, senescent, drought stricken, and harried by aphids. A contract arborist came with his crew and his truck and took it down, limb by limb, section by section. That was a sad day for Susan—for me too, having lived in the shade of that majestic hardwood for almost three decades. Then even the stump, big enough to serve as a coffee table, vanished. It had been ground down with a stump grinder and covered with grass. The tree is now gone but not forgotten. The neighborhood is less graceful for its loss. But there was no choice.

The other big elm is still here, arching grandly over our little street. Circling the tree's grayish brown bark, at waist level, is a stain—a dark band of discoloration, evidently indelible against weather and time, marking where it was defended with toxic goo against the tent caterpillars, twenty years ago. The caterpillars are long departed, just another outbreak population that crashed, but this mark is like their fossil record.

When I'm home in Montana, I walk past that tree every day. Usually I notice the dark band. Usually I remember the caterpillars, how they came in such numbers and then disappeared. Conditions had been good for them. But something happened. Maybe luck was the crucial element. Maybe circumstance. Maybe their sheer density. Maybe genetics. Maybe behavior. Often nowadays, when I see the mark on the tree, I recall what Greg Dwyer told me: It all depends.

NOTES

I. Pale Horse

24. "*Viruses have no locomotion*": Morse (1993), ix.
28. "*He remained deeply unconscious*": O'Sullivan et al. (1997), 93.
29. "*It seems," McCormack's group concluded, "that very close contact*": McCormack et al. (1999), 23.
35. "*Economically, it is the most important*": Brown (2001), 239.
41. "*If you look at the world from the point of view*": William H. McNeill, in Morse (1993), 33–34.
44. "*Furthermore, 71.8% of these zoonotic*": Jones-Engel et al. (2008), 990.

II. Thirteen Gorillas

54. "*The chimpanzee seems to have been the index case*": Georges et al. (1999), S70.
70. "*Only limited ecological investigations*": Johnson et al. (1978), 272.
71. "*No more dramatic or potentially explosive epidemic*": Johnson et al. (1978), 288.
72. "*No evidence of Ebola virus infection*": Breman et al. (1999), S139.
77. "*Contact with nature is intimate*": Heymann et al. (1980), 372–73.
84. "*Viruses of each species have genomes that*": Towner et al. (2008), 1.
87. "*bad human-like spirits that cause illness*": Hewlett and Hewlett (2008), 6.

88. *a final "love touch" of the deceased*: Hewlett and Amola (2003), 1245.
90. *"This illness is killing everyone"*: Hewlett and Hewlett (2008), 75.
91. *"Sorcery does not kill without reason"*: Hewlett and Hewlett (2008), 75.
92. *"jumped from bed to bed, killing patients left and right"*: Preston (1994), 68.
92. *"transforms virtually every part of the body"*: Preston (1994), 72.
92. *"suddenly deteriorates," its internal organs deliquescing*: Preston (1994), 75.
92. *"essentially melts down with Marburg"*: Preston (1994), 293.
92. *comatose, motionless, and "bleeding out"*: Preston (1994), 184.
93. *"Droplets of blood stand out on the eyelids"*: Preston (1994), 73.
99. *"It is difficult to describe working with a horse infected with Ebola"*: *Yaderny Kontrol* (Nuclear Control) *Digest*, No. 11, Center for Policy Studies in Russia, Summer 1999.
119. *"Taken together, our results clearly point"*: Walsh et al. (2005), 1950.
120. *"Thus, Ebola outbreaks probably do not occur as"*: Leroy et al. (2004), 390.

III. Everything Comes from Somewhere

132. *some interesting points about "smouldering" epidemics*: Hamer (1906), 733–35.
132. *This idea became known as the "mass action principle"*: Fine (1979), 348.
133. *"the acquisition by an organism of a high grade of infectivity"*: Brownlee (1907), 516.
133. *"the condition of the germ"*: Brownlee (1907), 517.
133. *"extirpated once and forever"*: Ross (1910), 313.
133. *a "theory of happenings"*: Ross (1916), 206.
134. *"so little mathematical work should have been done"*: Ross (1916), 204–5.
141. *"This indicates," they wrote confidently, "that human* P. falciparum*"*: Liu et al. (2010), 424.
141. *"a monophyletic lineage within the gorilla* P. falciparum *radiation"*: Liu et al. (2010), 423.
143. *"One of the most important problems in epidemiology"*: Kermack and McKendrick (1927), 701.
144. *"Small increases of the infectivity rate"*: Kermack and McKendrick (1927), 721.
146. *"very small changes in the essential transmission factors"*: MacDonald (1953), 880.

146. *"the number of infections distributed in a community"*: MacDonald (1956), 375.
147. *"It all but destroyed malariology"*: Harrison (1978), 258.
151. *"The effect was remarkable"*: Desowitz (1993), 129.
152. *"This occurrence," wrote a quartet of the doctors involved*: Chin et al. (1965), 865.
161. *"it is possible that we are setting the stage for a switch"*: Cox-Singh and Singh (2008), 408.

IV. Dinner at the Rat Farm

169. *"hospitalized for treatment of severe, acute respiratory syndrome"*: World Health Organization (2006), 257.
169. *"During the past week," it said, "WHO has received reports"*: World Health Organization (2006), 259–60.
171. *described simply as "a local government official"*: Abraham (2007), 30.
171. *labeling it "atypical pneumonia"*: Abraham (2007), 34.
172. *"Population estimates of* R_0 *can obscure"*: Lloyd-Smith et al. (2005), 355.
173. *"Each time they began to insert the tube"*: Abraham (2007), 37.
182. *alarming rumors about "a strange contagious disease"*: World Health Organization (2006), 5.
184. *"The first thing going through our minds"*: Normile (2003), 886.
185. *announcing this new coronavirus as "a possible cause"*: Peiris (2003), 1319.
186. *"We were too cautious," one of them said later*: Enserink (2003), 294.
187. *"Southern Chinese have always noshed more widely"*: Greenfeld (2006), 10.
189. *"The animals are packed in tiny spaces"*: Lee et al. (2004), 12.
191. *"from another, as yet unknown, animal source"*: Guan et al. (2003), 278.
195. *"An infectious consignment of bats"*: Li et al. (2005), 678.
206. *"humankind has had a lucky escape"*: Weiss and McLean (2004), 1139.

V. The Deer, the Parrot, and the Kid Next Door

211. *known initially as "abattoir fever"*: Sexton (1991), 93.
212. *an example of "public hysteria" commensurate with flagellation*: *The Washington Post*, January 26, 1930, 1.
214. *"three died in agony"*: Van Rooyen (1955), 4.

214. *"The year 1929 marked a turning point"*: Van Rooyen (1955), 5.
215. *"tall with a gnarled Lincolnian face"*: De Kruif (1932), 178.
218. *"If the young cockatoo, after capture"*: Burnet and MacNamara (1936), 88.
219. *"a distinct clinical entity"*: Derrick (1937), 281.
219. *"a filterable virus," meaning an agent so small*: Burnet and Freeman (1937), 299.
220. *"Most significant discoveries just grow on one"*: Burnet (1967), 1067.
220. *"From that moment, there was no doubt"*: Burnet (1967), 1068.
220. *"Problems of nomenclature arose"*: Burnet (1967), 1068.
221. *"the Nine Mile agent"*: McDade (1990), 12.
221. *"sharp pains in the eyeballs"*: McDade (1990), 16.
221. *"There is no disease to match Q fever"*: Burnet (1967), 1068.
222. *"One of the more bizarre episodes"*: Burnet (1967), 1068.
223. *"there was no drop of rain"*: Karagiannis et al. (2009), 1289.
226. *The other was a "hobby farm"*: Karagiannis et al. (2009), 1286, 1288.
228. *"windborne transmission" as the most likely source*: Karagiannis et al. (2009), 1292.
231. *"a filterable virus," a microbe so tiny*: Burnet (1940), 19.
233. *"I just don't know if I can watch it"*: Enserink (2010), 266.
234. *"were on the whole too busy to think of anything but"*: Burnet (1940), 2–3.
235. *"Other workers with an appreciation of modern developments"*: Burnet (1940), 3.
235. *"The parasitic mode of life is essentially similar"*: Burnet (1940), 8.
236. *"It will be clear, however," Burnet wrote*: Burnet (1940), 12.
236. *"Like many other infectious diseases, psittacosis"*: Burnet (1940), 19.
237. *"those cockatoos, left to a natural life in the wild"*: Burnet (1940), 23.
237. *"It is a conflict between man and his parasites"*: Burnet (1940), 23.
238. *such a thing as "chronic Lyme disease"*: Feder et al. (2007), 1422.
238. *"No convincing biologic evidence exists"*: *IDSA News,* Vol. 16, No. 3, Fall 2006, 2.
239. *"post-Lyme disease syndrome" was another matter*: *IDSA News,* Vol. 16, No. 3, Fall 2006, 1.
239. *"by allowing individuals with financial interests"*: Quoted in press release, Office of the Attorney General of Connecticut, May 1, 2008, 2.
239. *"no convincing evidence for the existence"*: Quoted in press release, IDSA (Infectious Diseases Society of America), April 22, 2010, 2.
241. *began calling the syndrome "Lyme arthritis"*: Steere et al. (1977a), 7.
242. *were now calling "Lyme disease"*: Steere and Malawista (1979), 730.

243. *"a disease of the past," no longer justifying*: Burgdorfer (1986), 934.
244. *"No longer did we hear, 'get out'"*: Burgdorfer (1986), 936.
244. *later jovially called the "lymelight"*: Burgdorfer (1986), 936.
245. *"Dammin's northeastern deer ixodid"*: Ostfeld (2011), 26.
246. *"The notion that Lyme disease risk is closely tied"*: Ostfeld (2011), 22.
246. *One journal article had called white-tailed deer*: Both this article and the next, quoted in Ostfeld (2011), 22.
247. *"The higher the number of deer in an area"*: *The Dover-Sherborn Press*, January 12, 2011.
247. *"Any infectious disease is inherently an ecological system"*: Ostfeld (2011), 4.
248. *"Thus began my interest in Lyme disease ecology"*: Ostfeld (2011), x.
249. *"a messy and challenging task"*: Ostfeld (2011), 48.
250. *"exquisitely sensitive" to chemical and physical signals*: Ostfeld (2011), 23.
250. *the word is "questing"*: Ostfeld (2011), 23.
251. *Ostfeld and others call "reservoir competence"*: Ostfeld (2011), 12.
258. *"We know that walking into a small woodlot"*: Ostfeld (2011), 9.
258. *Some people take "All life is connected" to be*: Ostfeld (2011), 6–7.
258. *a sort of cystlike stage known as a "round body"*: Margulis et al. (2009), 52.

VI. Going Viral

265. *"the sap of leaves infected with tobacco mosaic disease"*: Levine (1992), 2.
267. *"encouraged by the study of the so-called 'filterable virus' agents"*: Zinsser (1934), 63.
267. *"Here, as in bacterial disease"*: Zinsser (1934), 64.
268. *"a piece of bad news wrapped up in a protein"*: Quoted in Crawford (2000), 6.
273. *"pain, redness, and slight swelling" around the bite*: Sabin and Wright (1934), 116.
273. *They called it simply "the B virus"*: Sabin and Wright (1934), 133.
278. *"no case" of human infection with the virus*: Engel et al. (2002), 792.
288. *"a virus in search of a disease"*: Weiss (1988), 497.
295. *the most "efficient" parasite, in Pasteur's view*: Pasteur's view as summarized and reaffirmed by Rene Dubos, quoted in Ewald (1994), 188–89.
295. *"a more perfect mutual tolerance"*: Zinsser (1934), 61.
295. *"In general terms, where two organisms have developed"*: Burnet (1940), 37.

296. *"A disease organism that kills its host quickly"*: McNeill (1976), 9.
297. *"started jumping up and down, biting other animals"*: Quoted in ProMED-mail post, April 22, 2011.
297. *"He barked like a dog," his wife recalled later*: Quoted in ProMED-mail post, April 1, 2011.
298. *Austin was an "ardent acclimatizer"*: Fenner and Ratcliffe (1965), 17.
299. *causing what was called a "spectacular epizootic"*: Fenner and Ratcliffe (1965), 276.
301. *"Laboratory experiments showed that all field strains"*: Fenner (1983), 265.
304. *"weave together" the two approaches*: Anderson and May (1979), 361.
304. *"unsupported statements" in medical and ecological textbooks*: Anderson and May (1982), 411.
306. *"Our major conclusion," wrote Anderson and May*: Anderson and May (1982), 424.

VII. Celestial Hosts

315. *"Pigs are a common host for the virus"*: *New Straits Times,* January 7, 1999.
316. *"It became known as a one-mile barking cough"*: Hume Field was the expert, quoted in a *60 Minutes* (of Australia) television interview.
327. *"touching dead animals" looked like it might be important*: Montgomery et al. (2008), 1529, Table 2.
328. *"increases the risk for wider spread"*: Gurley et al. (2007), 1036.
331. *"Owners viewed the fruit bats as a nuisance"*: Luby et al. (2006), 1892.
344. *"the revenge of the rain forest"*: Preston (1994), 289.
350. *Do bats have a different "set point"*: Calisher et al. (2006), 536.
351. *"Emphasis, sometimes complete emphasis, on nucleotide sequence"*: Calisher et al. (2006), 541.
351. *"we are simply waiting for the next"*: Calisher et al. (2006), 540.
351. *"The natural reservoir hosts of these viruses have not yet been identified"*: Calisher et al. (2006), 539.
356. *"is only one of many such cave populations"*: Towner et al. (2009), 2.
372. *"Patient C was the father of a 4-year-old girl"*: Leroy et al. (2009), 5.
372. *"Thus, virus transmission may have occurred"*: Leroy et al. (2009), 6.
373. *"In fact, it is highly likely that several other persons"*: Leroy et al. (2009), 5.

VIII. The Chimp and the River

385. *"profoundly depressed" in number*: Gottlieb et al. (1981), 251.
387. *"strikingly similar to the syndrome of immunodeficiency"*: Pitchenik et al. (1983), 277.
387. *written about as the man who "carried the virus out of Africa"*: e.g., Wikipedia, "Gaëtan Dugas," citing Auerbach et al. (1984), although Auerbach et al. do not make that assertion.
387. *vain but charming, even "gorgeous" in some eyes:* Shilts (1987), 47.
388. *"I've got gay cancer"*: Shilts (1987), 165.
388. *"Although the cause of AIDS is unknown"*: Auerbach et al. (1984), 490.
388. *to the more resonant "Patient Zero" of his book*: Shilts (1987), 23.
389. *"I'd better go home to die"*: Shilts (1987), 6.
391. *"AIDS could not be caused by a conventional bacterium"*: Montagnier (2000), 42.
393. *"more than 4000 individuals in the world"*: Levy et al. (1984), 840.
393. *"Our data cannot reflect a contamination"*: Levy et al. (1984), 842.
396. *"In 1985, the highest rates of HIV were reported"*: Essex and Kanki (1988), 68.
396. *"must have evolved mechanisms"*: Essex and Kanki (1988), 68.
396. *"not close enough to make it likely that SIV"*: Essex and Kanki (1988), 69.
399. *HUMAN AIDS VIRUS NOT FROM MONKEYS*: Mulder (1988), 396.
399. *sampled by the Japanese team, because it was "of Kenyan origin"*: Fukasawa et al. (1988), 457.
401. *revealed that the virus was "endemic" among them*: Murphey-Corb et al. (1986), 437.
402. *"These results suggest that SIV_{sm} has infected macaques"*: Hirsch et al. (1989), 389.
414. *with material direct from a "vaccinal sore"*: Willrich (2011), 181.
415. *"The origin of the AIDS virus is of no importance"*: Quoted in Curtis (1992), 21.
415. *"It's distracting, it's nonproductive, it's confusing"*: Quoted in Curtis (1992), 21.
416. *"The controversy surrounding the source of the Nile"*: Hooper (1999), 4.
421. *"Our estimation of divergence times"*: Worobey et al. (2008), 663.
423. *"the most persuasive evidence yet"*: Weiss and Wrangham (1999), 385.
428. *"We show here that the SIV_{cpzPtt} strain that gave rise"*: Keele at al. (2006), 526.
428. *"In humans, direct exposure to animal blood"*: Hahn et al. (2000), 611.

428. *"The likeliest route of chimpanzee-to-human transmission"*: Sharp and Hahn (2010), 2492.
429. *"a hard mission field," according to one Swedish missionary*: Quoted in Martin (2002), 25.
430. *"a low-risk type of prostitution"*: Pepin (2011), 90.
437. *"Until recently, the Bakweles have been using chimps"*: From the typewritten, unpublished report of my anonymous source in Yokadouma.
464. *"survived their own AIDS-like pandemic"*: Cohen (2002), 15.
477. *"that* SIV_{cpz} *has a substantial negative impact"*: Keele et al. (2009), 515.
479. *"The Congo contains various health institutions"*: Beheyt (1953), quoted in Pepin (2011), 164.
479. *"The large number of patients and the small quantity of syringes"*: Beheyt (1953), quoted in Pepin (2011), 164.
481. *"consisted of thousands of asymptomatic free women"*: Pepin (2011), 161.
485. *"there must have been a very effective amplification mechanism"*: Pepin (2011), 196.

IX. It Depends

496. *"From the ecological point of view an outbreak"*: Berryman (1987), 3.
497. *"When* Homo sapiens *passed the six-billion mark"*: Wilson (2002), 86.
498. *"seems to imply a dominant force"*: Myers (1993), 240.
512. *"The first criterion is the most obvious"*: Burke (1998), 7.

BIBLIOGRAPHY

Abraham, Thomas. 2007. *Twenty-First Century Plague: The Story of SARS*. Baltimore: The Johns Hopkins University Press.

AbuBakar, Sazaly, Li-Yen Chang, A. R. Mohd Ali, S. H. Sharifah, Khatijah Yusoff, and Zulkeflie Zamrod. 2004. "Isolation and Molecular Identification of Nipah Virus from Pigs." *Emerging Infectious Diseases*, 10 (12).

Aguirre, A. Alonso, Richard S. Ostfeld, Gary M. Tabor, Carol House, and Mary C. Pearl, eds. 2002. *Conservation Medicine: Ecological Health in Practice*. Oxford: Oxford University Press.

Alibek, Ken. 1999. *Biohazard: The Chilling True Story of the Largest Covert Biological Weapons Program in the World—Told from the Inside by the Man Who Ran It*. With Stephen Handelman. New York: Delta/Dell Publishing.

Anderson, Roy M., and Robert M. May. 1978. "Regulation and Stability of Host-Parasite Population Interactions." *Journal of Animal Ecology*, 47.

———. 1979. "Population Biology of Infectious Diseases: Part I." *Nature*, 280.

———. 1980. "Infectious Diseases and Populations of Forest Insects." *Science*, 210.

———. 1982. "Coevolution of Hosts and Parasites." *Parasitology*, 85.

———. 1992. *Infectious Diseases of Humans: Dynamics and Control*. Oxford: Oxford University Press.

Arricau-Bouvery, Nathalie, and Annie Rodolakis. 2005. "Is Q Fever an Emerging or Re-emerging Zoonosis?" *Veterinary Research*, 36.

Auerbach, D. M., W. W. Darrow, H. W. Jaffe, and J. W. Curran. 1984. "Cluster of Cases of the Acquired Immune Deficiency Syndrome. Patients Linked by Sexual Contact." *The American Journal of Medicine,* 76 (3).

Bacon, Rendi Murphree, Kiersten J. Kugeler, and Paul S. Mead. 2008. "Surveillance for Lyme Disease—United States, 1992–2006." *Morbidity and Mortality Weekly Report,* 57.

Bailes, Elizabeth, Feng Gao, Frederic Biboilet-Ruche, Valerie Courgnaud, Martine Peeters, Preston A. Marx, Beatrice H. Hahn, and Paul M. Sharp. 2003. "Hybrid Origin of SIV in Chimpanzees." *Science,* 300.

Baize, S., E. M. Leroy, M. C. Georges-Courbot, J. Lansoud-Soukate, P. Debré, S. P. Fisher-Hoch, J. B. McCormick, and A. J. Georges. 1999. "Defective Humoral Responses and Extensive Intravascular Apoptosis are Associated with Fatal Outcome in Ebola Virus-Infected Patients." *Nature Medicine,* 5 (4).

Barbosa, Pedro, and Jack C. Schultz, eds. 1987. *Insect Outbreaks.* San Diego: Academic Press.

Barin, F., S. M'Boup, F. Denis, P. Kanki, J. S. Allan, T. H. Lee, and M. Essex. 1985. "Serological Evidence for Virus Related to Simian T-Lymphotropic Retrovirus III in Residents of West Africa." *The Lancet,* 2.

Barré-Sinoussi, F., J. C. Chermann, F. Rey, M. T. Nugeyre, S. Chamaret, J. Gruest, C. Dauguet, et al. 1983. "Isolation of a T-Lymphotropic Retrovirus from a Patient at Risk for Acquired Immune Deficiency Syndrome (AIDS)." *Science,* 220.

Barré-Sinoussi, Françoise. 2003a. "The Early Years of HIV Research: Integrating Clinical and Basic Research." *Nature Medicine,* 9 (7).

———. 2003b. "Barré-Sinoussi Replies." *Nature Medicine,* 9 (7).

Barry, John M. 2005. *The Great Influenza: The Epic Story of the Deadliest Plague in History.* New York: Penguin Books.

Beaudette, F. R., ed. 1955. *Psittacosis: Diagnosis, Epidemiology and Control.* New Brunswick, NJ: Rutgers University Press.

Beheyt, P. 1953. "*Contribution à l'étude des hepatites en Afrique. L'hépatite épidémique et l'hépatite par inoculation.*" *Annales de la Société Belge de Médicine Tropicale.*

Bermejo, Magdalena, José Domingo Rodríguez-Teijeiro, Germán Illera, Alex Barroso, Carles Vilà, and Peter D. Walsh. 2006. "Ebola Outbreak Killed 5000 Gorillas." *Science,* 314.

Bernoulli, Daniel. 2004. "An Attempt at a New Analysis of the Mortality Caused by Smallpox and of the Advantages of Inoculation to Prevent It." Reprinted in *Reviews in Medical Virology,* 14.

Berryman, Alan A. 1987. "The Theory and Classification of Outbreaks." In *Insect Outbreaks,* ed. P. Barbosa and J. C. Schultz. San Diego: Academic Press.

Biek, Roman, Peter D. Walsh, Eric M. Leroy, and Leslie A. Real. 2006. "Recent Common Ancestry of Ebola Zaire Virus Found in a Bat Reservoir." *PLoS Pathogens,* 2 (10).

Blum, L. S., R. Khan, N. Nahar, and R. F. Breiman. 2009. "In-Depth Assessment of an Outbreak of Nipah Encephalitis with Person-to-Person Transmission in Bangladesh: Implications for Prevention and Control Strategies." *American Journal of Tropical Medicine and Hygiene,* 80 (1).

Boaz, Noel T. 2002. *Evolving Health: The Origins of Illness and How the Modern World Is Making Us Sick.* New York: John Wiley and Sons.

Boulos, R., N. A. Halsey, E. Holt, A. Ruff, J. R. Brutus, T. C. Quin, M. Adrien, and C. Boulos. 1990. "HIV-1 in Haitian Women 1982–1988." *Journal of Acquired Immune Deficiency Syndromes,* 3.

Breman, Joel G., Karl M. Johnson, Guido van der Groen, C. Brian Robbins, Mark V. Szczeniowski, Kalisa Ruti, Patrician A. Webb, et al. 1999. "A Search for Ebola Virus in Animals in the Democratic Republic of the Congo and Cameroon: Ecologic, Virologic, and Serologic Surveys, 1979–1980." In *Ebola: The Virus and the Disease,* ed. C. J. Peters and J. W. LeDuc. Special issue of *The Journal of Infectious Diseases,* 179 (S1).

Brown, Corrie. 2001. "Update on Foot-and-Mouth Disease in Swine." *Journal of Swine and Health Production,* 9 (5).

Brownlee, John. 1907. "Statistical Studies in Immunity: The Theory of an Epidemic." *Proceedings of the Royal Society of Edinburgh,* 26.

Burgdorfer, W., A. G. Barbour, S. F. Hayes, J. L. Benach, E. Grunwaldt, and J. P. Davis. 1982. "Lyme Disease—A Tick-Borne *Spirochetosis*?" *Science,* 216.

Burgdorfer, Willy. 1986. "The Enlarging Spectrum of Tick-Borne *Spirochetoses*: R. R. Parker Memorial Address." *Reviews of Infectious Diseases,* 8 (6).

Burke, Donald S. 1998. "Evolvability of Emerging Viruses." In *Pathology of Emerging Infections 2,* ed. A. M. Nelson and C. Robert Horsburgh, Jr. Washington: ASM Press.

Burnet, F. M. 1934. "*Psittacosis* in Australian Parrots." *The Medical Journal of Australia,* 2.

———. 1940. *Biological Aspects of Infectious Disease.* Cambridge: Cambridge University Press.

Burnet, F. M., and Mavis Freeman. 1937. "Experimental Studies on the Virus of 'Q' Fever." *The Medical Journal of Australia,* 2.

Burnet, F. M., and Jean MacNamara. 1936. "Human *Psittacosis* in Australia." *The Medical Journal of Australia,* 2.

Burnet, MacFarlane. 1967. "Derrick and the Story of Q Fever." *The Medical Journal of Australia,* 2 (24).

Bwaka, M. A., M. J. Bonnet, P. Calain, R. Colebunders, A. De Roo, Y. Guimard, K. R. Katwiki, et al. 1999. "Ebola Hemorrhagic Fever in Kikwit, Democratic Republic of the Congo: Clinical Observations in 103 Patients." In *Ebola: The Virus and the Disease,* ed. C. J. Peters and J. W. LeDuc. Special issue of *The Journal of Infectious Diseases,* 179 (S1).

Bygbjerg, I. C. 1983. "AIDS in a Danish Surgeon (Zaire, 1976)." *The Lancet,* 1 (2).

Caillaud, D., F. Levréro, R. Cristescu, S. Gatti, M. Dewas, M. Douadi, A. Gautier-Hion, et al. 2006. "Gorilla Susceptibility to Ebola Virus: The Cost of Sociality." *Current Biology,* 16 (13).

Calisher, Charles H., James E. Childs, Hume E. Field, Kathryn V. Holmes, and Tony Schountz. 2006. "Bats: Important Reservoir Hosts of Emerging Viruses." *Clinical Microbiology Reviews,* 19 (3).

Chen, Hualan, Yanbing Li, Zejun Li, Jianzhong Shi, Kyoko Shinya, Guohua Deng, Qiaoling Qi, et al. 2006. "Properties and Dissemination of H5N1 Viruses Isolated during an Influenza Outbreak in Migratory Waterfowl in Western China." *Journal of Virology,* 80 (12).

Chin, William, Peter G. Contacos, G. Robert Coatney, and Harry R. Kimball. 1965. "A Naturally Acquired Quotidian-Type Malaria in Man Transferable to Monkeys." *Science,* 149.

Chitnis, Amit, Diana Rawls, and Jim Moore. 2000. "Origin of HIV Type 1 in Colonial French Equatorial Africa?" *AIDS Research and Human Retroviruses,* 16 (1).

Chua, K. B., W. J. Bellini, P. A. Rota, B. H. Harcourt, A. Tamin, S. K. Lam, T. G. Ksiazek, et al. 2000. "Nipah Virus: A Recently Emergent Deadly Paramyxovirus." *Science,* 288.

Chua, K. B., B. H. Chua, and C. W. Wang. 2002. "Anthropogenic Deforestation, El Niño and the Emergence of Nipah Virus in Malaysia." *Malaysian Journal of Pathology,* 24 (1).

Chua, K. B., K. J. Goh, K. T. Wong, A. Kamarulzaman, P. S. Tan, T. G. Ksiazek, S. R. Zaki, et al. 1999. "Fatal Encephalitis due to Nipah among Pig-Farmers." *The Lancet,* 354.

Chua, K. B., C. L. Koh, P. S. Hooi, K. F. Wee, J. H. Khong, B. H.

Chua, Y. P. Chan, et al. 2002. "Isolation of Nipah Virus from Malaysian Island Flying-Foxes." *Microbes and Infection*, 4.

Chua, Kaw Bing. 2002. "Nipah Virus Outbreak in Malaysia." *Journal of Clinical Virology*, 26.

———. 2010. "Risk Factors, Prevention and Communication Strategy During Nipah Virus Outbreak in Malaysia." *Malaysian Journal of Pathology*, 32 (2).

Chua, Kaw Bing, Gary Crameri, Alex Hyatt, Meng Yu, Mohd Rosli Tompang, Juliana Rosli, Jennifer McEachern, et al. 2007. "A Previously Unknown Reovirus of Bat Origin Is Associated with an Acute Respiratory Disease in Humans." *Proceedings of the National Academy of Sciences*, 104 (27).

Churchill, Sue. 1998. *Australian Bats*. Sydney: New Holland Publishers.

Clavel, F., D. Guétard, F. Brun-Vézinet, S. Chamaret, M. A. Rey, M. O. Santos-Ferreira, A. G. Laurent, et al. 1986. "Isolation of a New Human Retrovirus from West African Patients with AIDS." *Science*, 233.

Coatney, G. Robert, William E. Collins, and Peter G. Contacos. 1971. "The Primate Malarias." Bethesda, Maryland: National Institutes of Health.

Cohen, Philip. 2002. "Chimps Have Already Conquered AIDS." *New Scientist*, August 24.

Cohn, Samuel K., Jr. 2003. *The Black Death Transformed: Disease and Culture in Early Renaissance Europe*. London: Arnold.

Cornejo, Omar E., and Ananias A. Escalante. 2006. "The Origin and Age of *Plasmodium vivax*." *Trends in Parasitology*, 22 (12).

Cory, Jenny S., and Judith H. Myers. 2003. "The Ecology and Evolution of Insect Baculoviruses." *Annual Review of Ecology, Evolution, and Systematics*, 34.

———. 2009. "Within and Between Population Variation in Disease Resistance in Cyclic Populations of Western Tent Caterpillars: A Test of the Disease Defence Hypothesis." *Journal of Animal Ecology*, 78.

Cox-Singh, J., T. M. Davis, K. S. Lee, S. S. Shamsul, A. Matusop, S. Ratnam, H. A. Rahman, et al. 2008. "*Plasmodium knowlesi* Malaria in Humans Is Widely Distributed and Potentially Life Threatening." *Clinical Infectious Diseases*, 46.

Cox-Singh, Janet, and Balbir Singh. 2008. "Knowlesi Malaria: Newly Emergent and of Public Health Importance?" *Trends in Parasitology*, 24 (9).

Crawford, Dorothy H. 2000. *The Invisible Enemy: A Natural History of Viruses*. Oxford: Oxford University Press.

Crewdson, John. 2002. *Science Fictions: A Scientific Mystery, a Massive Coverup, and the Dark Legacy of Robert Gallo.* Boston: Little, Brown.

Crosby, Alfred W. 1989. *America's Forgotten Pandemic: The Influenza of 1918.* Cambridge: Cambridge University Press.

Curtis, Tom. 1992. "The Origin of AIDS." *Rolling Stone,* March 19.

Daniel, M. D., N. L. Letvin, N. W. King, M. Kannagi, P. K. Sehgal, R. D. Hunt, P. J. Kanki, et al. 1985. "Isolation of T-Cell Tropic HTLV-III-like Retrovirus from Macaques." *Science,* 228.

Daszak, P., A. A. Cunningham, and A. D. Hyatt. 2001. "Anthropogenic Environmental Change and the Emergence of Infectious Diseases in Wildlife." *Acta Tropica,* 78.

Daszak, Peter, Andrew H. Cunningham, and Alex D. Hyatt. 2000. "Emerging Infectious Diseases of Wildlife–Threats to Biodiversity and Human Health." *Science,* 287.

Davis, Gordon E., and Herald R. Cox. 1938. "A Filter-Passing Infectious Agent Isolated from Ticks." *Public Health Reports,* 53 (52).

De Groot, N. G., N. Otting, G. G. Doxiadis, S. S. Balla-Jhagjoorsingh, J. L. Heeney, J. J. van Rood, P. Gagneux, et al. 2002. "Evidence for an Ancient Selective Sweep in the MHC Class I Gene Repertoire of Chimpanzees." *Proceedings of the National Academy of Sciences,* 99 (18).

De Kruif, Paul. 1932. *Men Against Death.* New York: Harcourt, Brace and Company.

Derrick, E. H. 1937. "Q Fever, A New Fever Entity: Clinical Features, Diagnosis and Laboratory Investigation." *The Medical Journal of Australia,* 2 (8).

Desowitz, Robert S. 1993. *The Malaria Capers: More Tales of Parasites, People, Research and Reality.* New York: W. W. Norton.

Diamond, Jared. 1997. *Guns, Germs, and Steel: The Fates of Human Societies.* New York: W. W. Norton.

Dobson, Andrew P., and E. Robin Carper. 1996. "Infectious Diseases and Human Population History." *BioScience,* 46 (2).

Dowdle, W. R., and D. R. Hopkins, eds. 1998. *The Eradication of Infectious Diseases.* New York: John Wiley and Sons.

Drosten, C., S. Günter, W. Preiser, S. van der Werf, H. R. Brodt, S. Becker, H. Rabenau, et al. 2003. "Identification of a Novel Coronavirus in Patients with Severe Acute Respiratory Syndrome." *New England Journal of Medicine,* 348 (20).

Drucker, Ernest, Phillip C. Alcabes, and Preston A. Marx. 2001. "The Injection Century: Massive Unsterile Injections and the Emergence of Human Pathogens." *The Lancet,* 358.

Duesberg, Peter. 1996. *Inventing the AIDS Virus.* Washington, D.C.: Regnery Publishing.

Dwyer, Greg. 1991. "The Roles of Density, Stage, and Patchiness in the Transmission of an Insect Virus." *Ecology,* 72 (2).

Dwyer, Greg, and Joseph S. Elkinton. 1993. "Using Simple Models to Predict Virus Epizootics in Gypsy Moth Populations." *Journal of Animal Ecology,* 62.

Eaton, Bryan T. 2001. "Introduction to Current Focus on Hendra and Nipah Viruses." *Microbes and Infection,* 3.

Edlow, Jonathan A. 2003. *Bull's-Eye: Unraveling the Medical Mystery of Lyme Disease.* New Haven: Yale University Press.

Elderd, B. D., J. Dushoff, and G. Dwyer. 2008. "Host-Pathogen Interactions, Insect Outbreaks, and Natural Selection for Disease Resistance." *The American Naturalist,* 172 (6).

Elderd, Bret D., Vanja M. Dukic, and Greg Dwyer. 2006. "Uncertainty in Predictions of Disease Spread and Public Health Responses to Bioterrorism and Emerging Diseases." *Proceedings of the National Academy of Sciences,* 103 (42).

Elkinton, J. S. 1990. "Populations Dynamics of Gypsy Moth in North America." *Annual Reviews of Entomology,* 35.

Emmerson, A. M., P. M. Hawkey, and S. H. Gillespie. 1997. *Principles and Practice of Clinical Bacteriology.* Chichester and New York: John Wiley and Sons.

Emond, R. T., B. Evans, E. T. Bowen, and G. Lloyd. 1977. "A Case of Ebola Virus Infection." *British Medical Journal,* 2.

Engel, Gregory A., Lisa Jones-Engel, Michael A. Schillaci, Komang Gde Suaryana, Artha Putra, Agustin Fuentes, and Richard Henkel. 2002. "Human Exposure to Herpesvirus B-Seropositive Macaques, Bali, Indonesia." *Emerging Infectious Diseases,* 8 (8).

Engel, Jonathan. 2006. *The Epidemic: A Global History of AIDS.* New York: Smithsonian Books/HarperCollins.

Enserink, Martin. 2003. "China's Missed Chance." *Science,* 301.

———. 2010. "Questions Abound in Q-Fever Explosion in The Netherlands." *Science,* 327.

Epstein, Helen. 2007. *The Invisible Cure: Why We Are Losing the Fight against AIDS in Africa.* New York: Picador.

Epstein, Jonathan H., Vibhu Prakash, Craig S. Smith, Peter Daszak, Amanda B. McLaughlin, Greer Meehan, Hume E. Field, and Andrew A. Cunningham. 2008. "*Henipavirus* Infection in Fruit Bats (*Pteropus giganteus*), India." *Emerging Infectious Diseases,* 14 (8).

Escalante, Ananias A., Omar E. Cornejo, Denise E. Freeland, Amanda

C. Poe, Ester Durego, William E. Collins, and Altaf A. Lal. 2005. "A Monkey's Tale: The Origin of *Plasmodium vivax* as a Human Malaria Parasite." *Proceedings of the National Academy of Sciences,* 102 (6).

Essex, Max, and Phyllis J. Kanki. 1988. "The Origins of the AIDS Virus." *Scientific American,* 259 (4).

Essex, Max, Souleymane Mboup, Phyllis J. Kanki, Richard G. Marlink, and Sheila D. Tlou, eds. 2002. *AIDS in Africa.* 2nd ed. New York: Kluwer Academic/Plenum Publishers.

Ewald, Paul W. 1994. *Evolution of Infectious Disease.* Oxford: Oxford University Press.

Feder, Henry M., Jr., Barbara J. B. Johnson, Susan O'Connell, Eugene D. Shapiro, Allen C. Steere, Gary P. Wormser, and the Ad Hoc International Lyme Disease Group. 2007. "A Critical Appraisal of Chronic Lyme Disease." *New England Journal of Medicine,* 357 (14).

Fenner, F. 1983. "Biological Control, as Exemplified by Smallpox Eradication and Myxomatosis." *Proceedings of the Royal Society,* B, 218.

Fenner, Frank, and F. N. Ratcliffe. 1965. *Myxomatosis.* Cambridge: Cambridge University Press.

Field, Hume. 2001. "The Natural History of Hendra and Nipha Viruses." *Microbes and Infection,* 3.

Fields, Bernard N., David M. Knipe, and Peter M. Howley, eds. 1996. *Fundamental Virology.* 3rd ed. Philadelphia: Lippincott Williams & Wilkins.

Figtree, M., R. Lee, L. Bain, T. Kennedy, S. Mackertich, M. Urban, Q. Cheng, and B. J. Hudson. 2010. "*Plasmodium knowlesi* in Human, Indonesian Borneo." *Emerging Infectious Diseases,* 16 (4).

Fine, Paul E. M. 1979. "John Brownlee and the Measurement of Infectiousness: An Historical Study in Epidemic Theory." *Journal of the Royal Statistical Society,* A, 142 (P3).

Formenty, P., C. Boesch, M. Wyers, C. Steiner, F. Donati, F. Dind, F. Walker, and B. Le Guenno. 1999. "Ebola Virus Outbreak among Wild Chimpanzees Living in a Rain Forest of Côte d'Ivoire." In *Ebola: The Virus and the Disease,* ed. C. J. Peters and J. W. LeDuc. Special issue of *The Journal of Infectious Diseases,* 179 (S1).

Freifeld, A. G., J. Hilliard, J. Southers, M. Murray, B. Savarese, J. M. Schmitt, S. E. Strauss. 1995. "A Controlled Seroprevalence Survey of Primate Handlers for Evidence of Asymptomatic Herpes B Virus Infection." *The Journal of Infectious Diseases,* 171.

Friedman-Kein, Alvin E. 1981. "Disseminated Kaposi's Sarcoma Syndrome in Young Homosexual Men." *Journal of the American Academy of Dermatology,* 5.

Fukasawa, M., T. Miura, A. Hasegawa, S. Morikawa, H. Tsujimoto, K. Miki, T. Kitamura, and M. Hayami. 1988. "Sequence of Simian Immunodeficiency Virus from African Green Monkey, A New Member of the HIV/SIV Group." *Nature,* 333.

Gallo, R. C., S. Z. Salahuddin, M. Popovic, G. M. Shearer, M. Kaplan, B. F. Haynes, T. J. Palker, et al. 1984. "Frequent Detection and Isolation of Cytopathic Retroviruses (HTLV-III) from Patients with AIDS and at Risk for AIDS." *Science,* 224.

Gallo, R. C., P. S. Sarin, E. P. Gelmann, M. Robert-Guroff, E. Richardson, V. S. Kalyanaraman, D. Mann, et al. 1983. "Isolation of Human T-Cell Leukemia Virus in Acquired Immune Deficiency Syndrome (AIDS)." *Science,* 220.

Gallo, Robert. 1991. *Virus Hunting: AIDS, Cancer, and the Human Retrovirus: A Story of Scientific Discovery.* New York: Basic Books.

Gallo, Robert C., and Luc Montagnier. 1988. "AIDS in 1988." *Scientific American,* 259 (4).

Galvani, Alison P., and Robert M. May. 2005. "Dimensions of Superspreading." *Nature,* 438.

Gao, F., E. Bailes, D. L. Robertson, Y. Chen, C. M. Rodenburg, S. F. Michael, L. B. Cummins, et al. 1999. "Origin of HIV-1 in the Chimpanzee *Pan troglodytes troglodytes.*" *Nature,* 397.

Garrett, Laurie. 1994. *The Coming Plague: Newly Emerging Diseases in a World Out of Balance.* New York: Farrar, Straus and Giroux.

Georges, A. J., E. M. Leroy, A. A. Renaut, C. T. Benissan, R. J. Nabias, M. T. Ngoc, P. I. Obiang, et al. 1999. "Ebola Hemorrhagic Fever Outbreaks in Gabon, 1994–1997: Epidemiologic and Health Control Issues." In *Ebola: The Virus and the Disease,* ed. C. J. Peters and J. W. LeDuc. Special issue of *The Journal of Infectious Diseases,* 179 (S1).

Gilbert, M. Thomas P., Andrew Rambaud, Gabriela Wlasiuk, Thomas J. Spira, Arthur E. Pitchenik, and Michael Worobey. 2007. "The Emergence of HIV/AIDS in the Americas and Beyond." *Proceedings of the National Academy of Sciences,* 104 (47).

Giles-Vernick, Tamara. 2002. *Cutting the Vines of the Past: Environmental Histories of the Central African Rain Forest.* Charlottesville: University Press of Virginia.

Gopalakrishna, G., P. Choo, Y. S. Leo, B. K. Tay, Y. T. Lim, A. S. Khan, and C. C. Tan. 2004. "SARS Transmission and Hospital Containment." *Emerging Infectious Diseases,* 10 (3).

Gormus, Bobby J., Louis N. Martin, and Gary B. Baskin. 2004. "A Brief History of the Discovery of Natural Simian Immunodeficiency

Virus (SIV) Infections in Captive Sooty Mangabey Monkeys." *Frontiers in Bioscience*, 9.

Gottlieb, M. S., H. M. Shankar, P. T. Fan, A. Saxon, J. D. Weisman, and I. Pozalski. 1981. "*Pneumocystic* Pneumonia—Los Angeles." *Morbidity and Mortality Weekly Report*, June 5.

Greenfeld, Karl Taro. 2006. *China Syndrome: The True Story of the 21st Century's First Great Epidemic.* New York: HarperCollins Publishers.

Guan, Y., B. J. Zheng, Y. Q. He, X. L. Liu, Z. X. Zhuang, C. L. Cheung, S. W. Luo, et al. 2003. "Isolation and Characterization of Viruses Related to the SARS Coronavirus from Animals in Southern China." *Science*, 302.

Gurley, Emily S., Joel M. Montgomery, M. Jahangir Hossain, Michael Bell, Abul Kalam Azad, Mohammad Rafiqul Islam, Mohammad Abdur Rahim Molla, et al. 2007. "Person-to-Person Transmission of Nipah Virus in a Bangladeshi Community." *Emerging Infectious Diseases*, 13 (7).

Hahn, Beatrice H., George M. Shaw, Kevin M. De Cock, and Paul M. Sharp. 2000. "AIDS as a Zoonosis: Scientific and Public Health Implications." *Science*, 287.

Halpin, K., P. L. Young, H. E. Field, and J. S. Mackenzie. 2000. "Isolation of Hendra Virus from Pteropid Bats: A Natural Reservoir of Hendra Virus." *Journal of General Virology*, 81.

Hamer, W. H. 1906. "Epidemic Disease in England—The Evidence of Variability and of Persistency of Type." *The Lancet*, March 17.

Harcourt, Brian H., Azaibi Tamin, Thomas G. Ksiazek, Pierre E. Rollin, Larry J. Anderson, William J. Bellini, and Paul A. Rota. 2000. "Molecular Characterization of Nipah Virus, a Newly Emergent Paramyxovirus." *Virology*, 271.

Harms, Robert W. 1981. *River of Wealth, River of Sorrow: The Central Zaire Basin in the Era of the Slave and Ivory Trade, 1500–1891.* New Haven: Yale University Press.

Harris, Richard L., and Temple W. Williams, Jr. 1985. "Contribution to the Question of Pneumotyphus: A Discussion of the Original Article by J. Ritter in 1880." *Review of Infectious Diseases*, 7 (1).

Harrison, Gordon. 1978. *Mosquitoes, Malaria and Man: A History of the Hostilities Since 1880.* New York: E. P. Dutton.

Hawgood, Barbara J. 2008. "Alexandre Yersin (1864–1943): Discoverer of the Plague Bacillus, Explorer and Agronomist." *Journal of Medical Biography*, 16.

Hay, Simon I. 2004. "The Global Distribution and Population at Risk of Malaria: Past, Present, and Future." *Lancet Infectious Disease*, 4 (6).

Haydon, D. T., S. Cleaveland, L. H. Taylor, and M. K. Laurenson. 2002. "Identifying Reservoirs of Infection: A Conceptual and Practical Challenge." *Emerging Infectious Diseases,* 8 (12).

Hemelaar, J., E. Gouws, P. D. Ghys, and S. Osmanov. 2006. "Global and Regional Distribution of HIV-1 Genetic Subtypes and Recombinants in 2004." *AIDS,* 20 (16).

Hennessey, A. Bennett, and Jessica Rogers. 2008. "A Study of the Bushmeat Trade in Ouesso, Republic of Congo." *Conservation and Society,* 6 (2).

Henig, Robin Marantz. 1993. *A Dancing Matrix: Voyages along the Viral Frontier.* New York: Alfred A. Knopf.

Hewlett, B. S., A. Epelboin, B. L. Hewlett, and P. Formenty. 2005. "Medical Anthropology and Ebola in Congo: Cultural Models and Humanistic Care." *Bulletin de la Société Pathologie Exotique,* 98 (3).

Hewlett, Barry S., and Richard P. Amola. 2003. "Cultural Contexts of Ebola in Northern Uganda." *Emerging Infectious Diseases,* 9 (10).

Hewlett, Barry S., and Bonnie L. Hewlett. 2008. *Ebola, Culture, and Politics: The Anthropology of an Emerging Disease.* Belmont, CA: Thomson Wadsworth.

Heymann, D. L., J. S. Weisfeld, P. A. Webb, K. M. Johnson, T. Cairns, and H. Berquist. 1980. "Ebola Hemorrhagic Fever: Tandala, Zaire, 1977–1978." *The Journal of Infectious Diseases,* 142 (3).

Hirsch, V. M., R. A. Olmsted, M. Murphy-Corb, R. H. Purcell, and P. R. Johnson. 1989. "An African Primate Lentivirus (SIV_{sm}) Closely Related to HIV-2." *Nature,* 339.

Holmes, Edward C. 2009. *The Evolution and Emergence of RNA Viruses.* Oxford: Oxford University Press.

Hoong, Chua Mui. 2004. *A Defining Moment: How Singapore Beat SARS.* Singapore: Institute of Policy Studies.

Hooper, Ed. 1990. *Slim: A Reporter's Own Story of AIDS in East Africa.* London: The Bodley Head.

Hooper, Edward. 1999. *The River: A Journey to the Source of HIV and AIDS.* Boston: Little, Brown.

———. 2001. "Experimental Oral Polio Vaccines and Acquired Immune Deficiency Syndrome." *Philosophical Transactions of the Royal Society of London,* 356.

Huff, Jennifer L., and Peter A. Barry. 2003. "B-Virus (*Cercopithecine herpesvirus* 1) Infection in Humans and Macaques: Potential for Zoonotic Disease." *Emerging Infectious Diseases,* 9 (2).

Huijbregts, Bas, Pawel De Wachter, Louis Sosthene Ndong Obiang, and Marc Ella Akou. 2003. "Ebola and the Decline of Gorilla *Gorilla*

gorilla and Chimpanzee *Pan troglodytes* Populations in Minkebe Forest, North-eastern Gabon." *Oryx,* 37 (4).

Hsu, Vincent P., Mohammed Jahangir Hossain, Umesh D. Parashar, Mohammed Monsur Ali, Thomas G. Ksiazek, Ivan Kuzmin, Michael Niezgoda, et al. 2004. "Nipah Virus Encephalitis Reemergence, Bangladesh." *Emerging Infectious Diseases,* 10 (12).

Jiang, Ning, Qiaocheng Chang, Xiaodong Sun, Huijun Lu, Jigang Yin, Zaixing Zhang, Mats Wahlgren, and Qijun Chen. 2010. "Co-Infections with *Plasmodium knowlesi* and Other Malaria Parasites, Myanmar." *Emerging Infectious Diseases,* 16 (9).

Johara, Mohd Yob, Hume Field, Azmin Mohd Rashdi, Christopher Morrissy, Brenda van der Heide, Paul Rota, Azri bin Adzhar, et al. 2001. "Nipah Virus Infection in Bats (Order *Chiroptera*) in Peninsular Malaysia." *Emerging Infectious Diseases,* 7 (3).

Johnson, K. M., and Members of the International Commission. 1978. "Ebola Haemorrhagic Fever in Zaire, 1976." *Bulletin of the World Health Organization,* 56.

Johnson, Karl M. 1999. "Gleanings from the Harvest: Suggestions for Priority Actions against Ebola Virus Epidemics." In *Ebola: The Virus and the Disease,* ed. C. J. Peters and J. W. LeDuc. Special issue of *The Journal of Infectious Diseases,* 179 (S1).

Johnson, Russell C., George P. Schmid, Fred W. Hyde, A. G. Steigerwalt, and Don J. Brenner. 1984. "*Borrelia burgdorferi* sp. no.: Etiologic Agent of Lyme Disease." *International Journal of Systematic Bacteriology,* 34 (4).

Jones-Engel, L., G. A. Engel, M. A. Schillaci, A. Rompis, A. Putra, K. G. Suaryana, A. Fuentes, et al. 2005. "Primate-to-Human Retroviral Transmission in Asia." *Emerging Infectious Diseases,* 11 (7).

Jones-Engel, Lisa, Cynthia C. May, Gregory A. Engel, Katherine A. Steinkraus, Michael A. Schillaci, Agustin Fuentes, Aida Rompis, et al. 2008. "Diverse Contexts of Zoonotic Transmission of Simian Foamy Viruses in Asia." *Emerging Infectious Diseases,* 14 (8).

Jones-Engel, Lisa, Katherine A. Steinkraus, Shannon M. Murray, Gregory A. Engel, Richard Grant, Nantiya Aggimarangsee, Benjamin P. Y.-H. Lee, et al. 2007. "Sensitive Assays for Simian Foamy Viruses Reveal a High Prevalence of Infection in Commensal, Free-Ranging Asian Monkeys." *Journal of Virology,* 81 (14).

Jongwutiwes, Somchai, Chaturong Putaporntip, Takuya Iwasaki, Tetsutaro Sata, and Hiroji Kanbara. 2004. "Naturally Acquired *Plasmodium knowlesi* Malaria in Human, Thailand." *Emerging Infectious Diseases,* 10 (12).

Kanki, P. J., J. Alroy, and M. Essex. 1985. "Isolation of T-Lymphotropic

Retrovirus Related to HTLV-III/LAV from Wild-Caught African Green Monkeys." *Science*, 230.

Kanki, P. J., F. Barin, S. M'Boup, J. S. Allan, J. L. Romet-Lemonne, R. Marlink, M. F. Maclane, et al. 1986. "New Human T-Lymphotropic Retrovirus Related to Simian T-Lymphotropic Virus Type III (STVL-III_{AGM})." *Science*, 232.

Kanki, P. J., M. F. MacLane, N. W. King, Jr., N. L. Letvin, R. D. Hunt, P. Sehgal, M. D. Daniel, et al. 1985. "Serologic Identification and Characterization of a Macaque T-Lymphotropic Retrovirus Closely Related to HTLV-III." *Science*, 228.

Kantele, Anu, Hanspeter Marti, Ingrid Felger, Dania Müller, and T. Sakari Jokiranta, et al. 2008. "Monkey Malaria in a European Traveler Returning from Malaysia." *Emerging Infectious Diseases*, 14 (9).

Kappe, Stefan H. I., Ashley M. Vaughan, Justin A. Boddey, and Alan F. Cowman. 2010. "That Was Then But This Is Now: Malaria Research in the Time of an Eradication Agenda." *Science*, 328.

Karagiannis, I., G. Morroy, A. Rietveld, A. M. Horrevorts, M. Hamans, P. Francken, and B. Schimmer. 2007. "Q Fever Outbreak in The Netherlands: A Preliminary Report." *Eurosurveillance*, 12 (32).

Karagiannis, I., B. Schimmer, A. Van Lier, A. Timen, P. Schneeberger, B. Van Rotterdam, A. De Bruin, et al. 2009. "Investigation of a Q Fever Outbreak in a Rural Area of The Netherlands." *Epidemiology and Infection*, 137.

Karesh, William B. 1999. *Appointment at the Ends of the World: Memoirs of a Wildlife Veterinarian*. New York: Warner Books.

Karesh, William B., and Robert A. Cook. 2005. "The Animal-Human Link." *Foreign Affairs*, 84 (4).

Keele, Brandon F., Fran Van Heuverswyn, Yingying Li, Elizabeth Bailes, Jun Takehisa, Mario L. Santiago, Frederic Bibollet-Ruche, et al. 2006. "Chimpanzee Reservoirs of Pandemic and Nonpandemic HIV-1." *Science*, 313.

Keele, Brandon F., James Holland Jones, Karen A. Terio, Jacob D. Estes, Rebecca S. Rudicell, Michael L. Wilson, Yingying Li, et al. 2009. "Increased Mortality and AIDS-like Immunopathology in Wild Chimpanzees Infected with SIVcpz." *Nature*, 460.

Kermack, W. O., and A. G. McKendrick. 1927. "A Contribution to the Mathematical Theory of Epidemics." *Proceedings of the Royal Society*, A, 115.

Kestler, H. W., III, Y. Li, Y. M. Naidu, C. V. Butler, M. F. Ochs, G. Jaenel, N. W. King, et al. 1988. "Comparison of Simian Immunodeficiency Virus Isolates." *Nature*, 331.

Khan, Naveed Ahmed. 2008. *Microbial Pathogens and Human Disease.* Enfield, New Hampshire: Science Publishers.

Klenk, H.-D., M. N. Matrosovich, and J. Stech, eds. 2008. *Avian Influenza.* Basel: Karger.

Knowles, R., and B. M. Das Gupta. 1932. "A Study of Monkey-Malaria and its Experimental Transmission to Man." *The Indian Medical Gazette,* June.

Koene, R. P. M., B. Schimmer, H. Rensen, M. Biesheuvel, A. De Bruin, A. Lohuis, A. Horrevorts, et al. 2010. "A Q Fever Outbreak in a Psychiatric Care Institution in The Netherlands." *Epidemiology and Infection*, 139 (1).

Kolata, Gina. 2005. *Flu: The Story of the Great Influenza Pandemic of 1918 and the Search for the Virus that Caused It.* New York: Touchstone/Simon & Schuster.

Koprowski, Hilary. 2001. "Hypothesis and Facts." *Philosophical Transactions of the Royal Society of London,* 356.

Korber, B., M. Muldoon, J. Theiler, F. Gao, R. Gupta, A. Lapedes, B. H. Hahn, et al. 2000. "Timing the Ancestor of the HIV-1 Pandemic Strains." *Science,* 288.

Krief, Sabrina, Ananias A. Escalante, M. Andreina Pacheco, Lawrence Mugisha, Claudine André, Michel Halbwax, Anne Fischer, et al. 2010. "On the Diversity of Malaria Parasites in African Apes and the Origin of *Plasmodium falciparum* from Bonobos." *PLoS Pathogens,* 6 (2).

Ksiazek, T. G., D. Erdman, C. S. Goldsmith, S. R. Zaki, T. Peret, S. Emery, S. Tong, et al. 2003. "A Novel Coronavirus Associated with Severe Acute Respiratory Syndrome." *New England Journal of Medicine,* 348 (20).

Kuhn, Jens. 2008. *Filoviruses: A Compendium of 40 Years of Epidemiological, Clinical, and Laboratory Studies.* C. H. Calisher, ed. New York: Springer-Verlag.

Lahm, S. A., M. Kobila, R. Swanepoel, and R. F. Barnes. 2006. "Morbidity and Mortality of Wild Animals in Relation to Outbreaks of Ebola Haemorrhagic Fever in Gabon, 1994–2003." *Transactions of the Royal Society of Tropical Medicine and Hygiene,* 101 (1).

Lau, Susanna K. P., Patrick C. Y. Woo, Kenneth S. M. Li, Yi Huang, Hoi-Wah Tsoi, Beatrice H. L. Wong, Samson S. Y. Wong, et al. 2005. "Severe Acute Respiratory Syndrome Coronavirus-like Virus in Chinese Horseshoe Bats." *Proceedings of the National Academy of Sciences,* 102 (39).

Lee, K. S., M.W. N. Lau, and B.P.L. Chan. 2004. "Wild Animal Trade

Monitoring at Selected Markets in Guangzhou and Shenzhen, South China, 2000–2003." *Kadoorie Farm & Botanic Garden Technical Report* (2).

Le Guenno, B., P. Formenty, M. Wyers, P. Gounon, F. Walker, and C. Boesch. 1995. "Isolation and Partial Characterisation of a New Strain of Ebola." *The Lancet,* 345 (8960).

Lepore, Jill. 2009. "It's Spreading." *The New Yorker,* June 1.

Leroy, E. M., A. Epelboin, V. Mondonge, X. Pourrut, J. P. Gonzalez, J. J. Muyembe-Tamfun, P. Formenty, et al. 2009. "Human Ebola Outbreak Resulting from Direct Exposure to Fruit Bats in Luebo, Democratic Republic of Congo, 2007." *Vector-Borne and Zoonotic Diseases,* 9 (6).

Leroy, Eric M., Brice Kumulungui, Xavier Pourrut, Pierre Rouquet, Alexandre Hassanin, Philippe Yaba, André Délicat, et al. 2005. "Fruit Bats as Reservoirs of Ebola Virus." *Nature,* 438.

Leroy, Eric M., Pierre Rouquet, Pierre Formenty, Sandrine Souquière, Annelisa Kilbourne, Jean-Marc Froment, Magdalena Bermejo, et al. 2004. "Multiple Ebola Virus Transmission Events and Rapid Decline of Central African Wildlife." *Science,* 303.

Letvin, Norman L., Kathryn A. Eaton, Wayne R. Aldrich, Prabhat K. Sehgal, Beverly J. Blake, Stuart F. Schlossman, Norval W. King, and Ronald D. Hunt. 1983. "Acquired Immunodeficiency Syndrome in a Colony of Macaque Monkeys." *Proceedings of the National Academy of Sciences,* 80.

Levine, Arnold J. 1992. *Viruses.* New York: Scientific American Library.

Levy, J. A., A. D. Hoffman, S. M. Kramer, J. A. Landis, J. M. Shimabukuro, and L. S. Oshiro. 1984. "Isolation of Lymphocytopathic Retroviruses from San Francisco Patients with AIDS." *Science,* 225.

Li, Wendong, Zhengli Shi, Meng Yu, Wuze Ren, Craig Smith, Jonathan H. Epstein, Hanzhong Wang, et al. 2005. "Bats Are Natural Reservoirs of SARS-like Coronavirus." *Science,* 310.

Liang, W., Z. Zhu, J. Guo, Z. Liu, W. Zhou, D. P. Chin, A. Schuchat, et al. 2004. "Severe Acute Respiratory Syndrome, Beijing, 2003." *Emerging Infectious Diseases,* 10 (1).

Lillie, R. D. 1930. "*Psittacosis:* Rickettsia-like Inclusions in Man and in Experimental Animals." *Public Health Reports,* 45 (15).

Liu, Weimin, Yingying Li, Gerald H. Learn, Rebecca S. Rudicell, Joel D. Robertson, Brandon F. Keele, Jean-Bosco N. Ndjango, et al. 2010. "Origin of the Human Malaria Parasite *Plasmodium falciparum* in Gorillas." *Nature,* 467.

Lloyd-Smith, J. O., S. J. Schreiber, P. E. Kopp, and W. M. Getz. 2005.

"Superspreading and the Effect of Individual Variation on Disease Emergence." *Nature,* 438.

LoGiudice, Kathleen, Richard S. Ostfeld, Kenneth A. Schmidt, and Felicia Keesing. 2003. "The Ecology of Infectious Disease: Effects of Host Diversity and Community Composition on Lyme Disease Risk." *Proceedings of the National Academy of Sciences,* 100 (2).

Luby, Stephen P., M. Jahangir Hossain, Emily S. Gurley, Be-Nazir Ahmed, Shakila Banu, Salah Uddin Khan, Nusrat Homaira, et al. 2009. "Recurrent Zoonotic Transmission of Nipah Virus into Humans, Bangladesh, 2001–2007." *Emerging Infectious Diseases,* 15 (8).

Luby, Stephen P., Mahmudur Rahman, M. Jahangir Hossain, Lauren S. Blum, M. Mustaq Husain, Emily Gurley, Rasheda Khan, et al. 2006. "Foodborne Transmission of Nipah Virus, Bangladesh." *Emerging Infectious Diseases,* 12 (12).

Luchavez, J., F. Espino, P. Curameng, R. Espina, D. Bell, P. Chiodini, D. Nolder, et al. 2008. "Human Infections with *Plasmodium knowlesi,* the Philippines." *Emerging Infectious Diseases,* 14 (5).

MacDonald, G. 1956. "Theory of the Eradication of Malaria." *Bulletin of the World Health Organization,* 15.

MacDonald, George. 1953. "The Analysis of Malaria Epidemics." *Tropical Diseases Bulletin,* 50 (10).

Margulis, Lynn, Andrew Maniotis, James MacAllister, John Scythes, Oystein Brorson, John Hall, Wolfgang E. Krumbein, and Michael J. Chapman. 2009. "Spirochete Round Bodies. Syphilis, Lyme Disease & AIDS: Resurgence of 'The Great Imitator?' " *Symbiosis,* 47.

Marrie, Thomas J., ed. 1990. *Q Fever. Vol. I: The Disease.* Boca Raton: CRC Press.

Martin, Phyllis M. 2002. *Leisure and Society in Colonial Brazzaville.* Cambridge: Cambridge University Press.

Martinsen, Ellen S., Susan L. Perkins, and Jos J. Schall. 2008. "A Three-Genome Phylogeny of Malaria Parasites (*Plasmodium* and Closely Related Genera): Evolution of Life-History Traits and Host Switches." *Molecular Phylogenetics and Evolution,* 47.

Marx, Jean L. 1983. "Human T-Cell Leukemia Virus Linked to AIDS." *Science,* 220.

Marx, P. A., P. G. Alcabes, and E. Drucker. 2001. "Serial Human Passage of Simian Immunodeficiency Virus by Unsterile Injections and the Emergence of Epidemic Human Immunodeficiency Virus in Africa." *Philosophical Transactions of the Royal Society of London,* 356.

May, Robert. 2001. "Memorial to Bill Hamilton." *Philosophical Transactions of the Royal Society of London,* 356.

McCormack, J. G., A. M. Allworth, L. A. Selvey, and P. W. Selleck. 1999. "Transmissibility from Horses to Humans of a Novel Paramyxovius, Equine Morbillivirus (EMV)." *Journal of Infection,* 38.

McCormick, Joseph B., and Susan Fisher-Hoch. 1996. *Level 4: Virus Hunters of the CDC.* With Leslie Alan Horvitz. Atlanta: Turner Publishing.

McCoy, G. W. 1930. "Accidental *Psittacosis* Infection Among the Personnel of the Hygienic Laboratory." *Public Health Reports,* 45 (16).

McDade, Joseph E. 1990. "Historical Aspects of Q Fever." In *Q Fever. Vol. I: The Disease*, ed. T. Marrie. Boca Raton: CRC Press.

McKenzie, F. Ellis, and Ebrahim M. Samba. 2004. "The Role of Mathematical Modeling in Evidence-Based Malaria Control." *American Journal of Tropical Medicine and Hygiene,* 71.

McLean, Angela, Robert May, John Pattison, and Robin Weiss, eds. 2005. *SARS: A Case Study in Emerging Infections.* Oxford: Oxford University Press.

McNeill, William H. 1976. *Plagues and Peoples.* New York: Anchor Books.

Meiering, Christopher D., and Maxine L. Linial. 2001. "Historical Perspective of Foamy Virus Epidemiology and Infection." *Clinical Microbiology Reviews,* 14 (1).

Meyer, K. F., and B. Eddie. 1934. "*Psittacosis* in the Native Australian Budgerigars." *Proceedings of the Society for Experimental Biology & Medicine,* 31.

Miranda, M. E. 1999. "Epidemiology of Ebola (Subtype Reston) Virus in the Philippines, 1996." In *Ebola: The Virus and the Disease,* ed. C. J. Peters and J. W. LeDuc. Special issue of *The Journal of Infectious Diseases,* 179 (S1).

Monath, Thomas P. 1999. "Ecology of Marburg and Ebola Viruses: Speculations and Directions for Future Research." In *Ebola: The Virus and the Disease,* ed. C. J. Peters and J. W. LeDuc. Special issue of *The Journal of Infectious Diseases,* 179 (S1).

Montagnier, Luc. 2000. *Virus: The Co-Discoverer of HIV Tracks Its Rampage and Charts the Future.* Translated from the French by Stephen Sartelli. New York: W. W. Norton.

———. 2003. "Historical Accuracy of HIV Isolation." *Nature Medicine,* 9 (10).

Montgomery, Joel M., Mohammed J. Hossain, E. Gurley, D. S. Carroll, A. Croisier, E. Bertherat, N. Asgari, et al. 2008. "Risk Factors for Nipah Virus Encephalitis in Bangladesh." *Emerging Infectious Diseases,* 14 (10).

Moore, Janice. 2002. *Parasites and the Behavior of Animals.* Oxford: Oxford University Press.

Morse, Stephen S., ed. 1993. *Emerging Virsues.* New York: Oxford University Press.

Mulder, Carel. 1988. "Human AIDS Virus Not from Monkeys." *Nature,* 333.

Murphey-Corb, M., L. N. Martin, S. R. Rangan, G. B. Baskin, B. J. Gormus, R. H. Wolf, W. A. Andres, et al. 1986. "Isolation of an HTLV-III-related Retrovirus from Macaques with Simian AIDS and Its Possible Origin in Asymptomatic Mangabeys." *Nature,* 321.

Murray, K., R. Rogers, L. Selvey, P. Selleck, A. Hyatt, A. Gould, L. Gleeson, et al. 1995. "A Novel Morbillivirus Pneumonia of Horses and its Transmission to Humans." *Emerging Infectious Diseases,* 1 (1).

Murray, K., P. Selleck, P. Hooper, A. Hyatt, A. Gould, L. Gleeson, H. Westbury, et al. 1995. "A Morbillivirus that Caused Fatal Disease in Horses and Humans." *Science,* 268.

Myers, Judith H. 1990. "Population Cycles of Western Tent Caterpillars: Experimental Introductions and Synchrony of Fluctuations." *Ecology,* 71 (3).

———. 1993. "Population Outbreaks in Forest Lepidoptera." *American Scientist,* 81.

———. 2000. "Population Fluctuations of the Western Tent Caterpillar in Southwestern British Columbia." *Population Ecology,* 42.

Nahmias, A. J., J. Weiss, X. Yao, F. Lee, R. Kodsi, M. Schanfield, T. Matthews, et al. 1986. "Evidence for Human Infection with an HTLV III/LAV-like Virus in Central Africa, 1959." *The Lancet,* 1 (8492).

Nathanson, Neal, and Rafi Ahmed. 2007. *Viral Pathogenesis and Immunity.* London: Elsevier.

Neghina, Raul, A. M. Neghina, I. Marincu, and I. Iacobiciu. 2011. "Malaria and the Campaigns Toward its Eradication in Romania, 1923–1963." *Vector-Borne and Zoonotic Diseases,* 11 (2).

Nelson, Anne Marie, and C. Robert Horsburgh, Jr., eds. 1998. *Pathology of Emerging Infections 2.* Washington: ASM Press.

Ng, Lee Ching, Eng Eong Ooi, Cheng Chuan Lee, Piao Jarrod Lee, Oong Tek Ng, Sze Wong Pei, Tian Ming Tu, et al. 2008. "Naturally Acquired Human *Plasmodium knowlesi* Infection, Singapore." *Emerging Infectious Diseases,* 14 (5).

Normile, Dennis. 2003. "Up Close and Personal with SARS." *Science,* 300.

———. 2005. "Researchers Tie Deadly SARS Virus to Bats." *Science,* 309.

Normile, Dennis, and Martin Enserink. 2003. "Tracking the Roots of a Killer." *Science,* 301.

Novembre, F. J., M. Saucier, D. C. Anderson, S. A. Klumpp, S. P. O'Neil, C. R. Brown II, C. E. Hart, et al. 1997. "Development of AIDS in a Chimpanzee Infected with Human Immunodeficiency Virus Type 1." *Journal of Virology,* 71 (5).

Nye, Edwin R., and Mary E. Gibson. 1997. *Ronald Ross: Malariologist and Polymath.* New York: St. Martin's Press.

Oldstone, Michael B. A. 1998. *Viruses, Plagues, and History.* New York: Oxford University Press.

Olsen, S. J., H. L. Chang, T. Y. Cheung, A. F. Tang, T. L. Fisk, S. P. Ooi, H. W. Kuo, et al. 2003. "Transmission of the Severe Acute Respiratory Syndrome on Aircraft." *New England Journal of Medicine,* 349 (25).

Oshinsky, David M. 2006. *Polio: An American Story.* Oxford: Oxford University Press.

Ostfeld, Richard S. 2011. *Lyme Disease: The Ecology of a Complex System.* Oxford: Oxford University Press.

Ostfeld, Richard S., Felicia Keesing, and Valerie T. Eviner, eds. 2008. *Infectious Disease Ecology: The Effects of Ecosystems on Disease and of Disease on Ecosystems.* Princeton: Princeton University Press.

O'Sullivan, J. D., A. M. Allworth, D. L. Paterson, T. M. Snow, R. Boots, L. J. Gleeson, A. R. Gould, et al. 1997. "Fatal Encephalitis Due to Novel Paramyxovirus Transmitted from Horses." *The Lancet,* 349 (9045).

Palmer, Amos E. 1987. "B Virus, *Herpesvirus simiae:* Historical Perspective." *Journal of Medical Primatology,* 16.

Parashar, U. D., L. M. Sunn, F. Ong, A. W. Mounts, M. T. Arif, T. G. Ksiazek, M. A. Kamaluddin, et al. 2000. "Case-Control Study of Risk Factors for Human Infection with a New Zoonotic Paramyxovirus, Nipah Virus, during a 1998–1999 Outbreak of Severe Encephalitis in Malaysia." *The Journal of Infectious Diseases,* 181.

Paton, N. I., Y. S. Leo, S. R. Zaki, A. P. Auchus, K. E. Lee, A. E. Ling, S. K. Chew, et al. 1999. "Outbreak of Nipah-virus Infection among Abattoir Workers in Singapore." *The Lancet,* 354 (9186).

Pattyn, S. R., ed. 1978. *Ebola Virus Haemorrhagic Fever.* Proceedings of an International Colloquium on Ebola Virus Infection and Other Haemorrhagic Fevers held in Antwerp, Belgium, December 6–8, 1977. Amsterdam: Elsevier/North-Holland Biomedical Press.

Peeters, M., K. Fransen, E. Delaporte, M. Van den Haesevelde, G. M. Gershy-Damet, L. Kestens, G. van der Groen, and P. Piot. 1992. "Isolation and Characterization of a New Chimpanzee Lentivirus (Simian Immunodeficiency Virus Isolate cpz-ant) from a Wild-Captured Chimpanzee." *AIDS,* 6 (5).

Peeters, M., C. Honoré, T. Huet, L. Bedjabaga, S. Ossari, P. Bussi, R. W. Cooper, and E. Delaporte. 1989. "Isolation and Partial Characterization of an HIV-related Virus Occurring Naturally in Chimpanzees in Gabon." *AIDS,* 3 (10).

Peiris, J. S., Y. Guan, and K. Y. Yuen. 2004. "Severe Acute Respiratory Syndrome." *Nature Medicine Supplement,* 10 (12).

Peiris, J. S., W. C. Yu, C. W. Leung, C. Y. Cheung, W. F. Ng, J. M. Nicholls, T. K. Ng, et al. 2004. "Re-emergence of Fatal Human Influenza A Subtype H5N1 Disease." *The Lancet,* 363 (9409).

Peiris, J. S. M., S. T. Lai, L. L. M. Poon, Y. Guan, L. Y. C. Yam, W. Lim, J. Nicholls, et al. 2003. "Coronavirus as a Possible Cause of Severe Acute Respiratory Syndrome." *The Lancet,* 361 (9366).

Peiris, J. S. Malik, Menno D. de Jong, and Yi Guan. 2007. "Avian Influenza Virus (H5N1): A Threat to Human Health." *Clinical Microbiology Reviews,* 20 (2).

Pepin, Jacques. 2011. *The Origins of AIDS.* Cambridge: Cambridge University Press.

Pepin, Jacques, and Eric H. Frost. 2011. "Reply to Marx et al." *Clinical Infectious Diseases*, Correspondence 52.

Pepin, Jacques, and Annie-Claude Labbé. 2008. "Noble Goals, Unforeseen Consequences: Control of Tropical Diseases in Colonial Central Africa and the Iatrogenic Transmission of Blood-borne Diseases." *Tropical Medicine and International Health,* 13 (6).

Pepin, Jacques, Annie-Claude Labbé, Fleurie Mamadou-Yaya, Pascal Mbélesso, Sylvestre Mbadingaï, Sylvie Deslandes, Marie-Claude Locas, and Eric Frost. 2010. "Iatrogenic Transmission of Human T Cell Lymphotropic Virus Type 1 and Hepatitis C Virus through Parenteral Treatment and Chemoprophylaxis of Sleeping Sickness in Colonial Equatorial Africa." *Clinical Infectious Diseases,* 51.

Pepin, K. M., S. Lass, J. R. Pulliam, A. F. Read, and J. O. Lloyd-Smith. 2010. "Identifying Genetic Markers of Adaptation for Surveillance of Viral Host Jumps." *Nature,* 8.

Peters, C. J., and James W. LeDuc, eds. 1999. *Ebola: The Virus and the Disease.* Special issue of *The Journal of Infectious Diseases*, 179 (S1).

Peters, C. J., and Mark Olshaker. 1997. *Virus Hunter: Thirty Years of Battling Hot Viruses around the World.* New York: Anchor Books.

Peterson, Dale. 2003. *Eating Apes.* With an afterword and photographs by Karl Ammann. Berkeley: University of California Press.

Pisani, Elizabeth. 2009. *The Wisdom of Whores: Bureaucrats, Brothels, and the Business of AIDS*. New York: W. W. Norton.

Pitchenik, Arthur E., Margaret A. Fischl, Gordon M. Dickinson, Daniel M. Becker, Arthur M. Fournier, Mark T. O'Connell, Robert D. Colton, and Thomas J. Spira. 1983. "Opportunistic Infections and Kaposi's Syndrome among Haitians: Evidence of a New Acquired Immunodeficiency State." *Annals of Internal Medicine,* 98 (3).

Plantier, J. C., M. Leoz, J. E. Dickerson, F. De Oliveira, F. Cordonnier, V. Lemée, F. Damond, et al. 2009. "A New Human Immunodeficiency Virus Derived from Gorillas." *Nature Medicine,* 15.

Plotkin, Stanley A. 2001. "Untruths and Consequences: The False Hypothesis Linking CHAT Type 1 Polio Vaccination to the Origin of Human Immunodeficiency Virus." *Philosophical Transactions of the Royal Society of London,* 356.

Plowright, R. K., H. E. Field, C. Smith, A. Divljan, C. Palmer, G. Tabor, P. Daszak, and J. E. Foley. 2008. "Reproduction and Nutritional Stress Are Risk Factors for Hendra Virus Infection in Little Red Flying Foxes (*Pteropus scapulatus*)." *Proceedings of the Royal Society,* B, 275.

Plowright, Raina K., P. Foley, H. E. Field, A. P. Dobson, J. E. Foley, P. Eby, and P. Daszak. 2011. "Urban Habituation, Ecological Connectivity and Epidemic Dampening: The Emergence of Hendra Virus from Flying Foxes (*Pteropus spp.*)." *Proceedings of the Royal Society,* B, 278.

Popovic, M., M. G. Sarngadharan, E. Read, and R. C. Gallo. 1984. "Detection, Isolation, and Continuous Production of Cytopathic Retroviruses (HTLV-III) from Patients with AIDS and Pre-AIDS." *Science,* 224.

Poon, L. L. M., D. K. W. Chu, K. H. Chan, O. K. Wong, T. M. Ellis, Y. H. C. Leung, S. K. P. Lau, et al. 2005. "Identification of a Novel Coronavirus in Bats." *Journal of Virology,* 79 (4).

Pourrut, X., B. Kumulungui, T. Wittmann, G. Moussavou, A. Délicat, P. Yaba, D. Nkoghe, et al. 2005. "The Natural History of Ebola Virus in Africa." *Microbes and Infection,* 7.

Poutanen, S. M., D. E. Low, B. Henry, S. Finkelstein, D. Rose, K. Green, R. Tellier, et al. 2003. "Identification of Severe Acute Respiratory Syndrome in Canada." *New England Journal of Medicine,* 348 (20).

Preston, Richard. 1994. *The Hot Zone.* New York: Random House.

Price-Smith, Andrew T. 2009. *Contagion and Chaos: Disease, Ecology, and National Security in the Era of Globalization.* Cambridge, MA: The MIT Press.

Read, Andrew F. 1994. "The Evolution of Virulence." *Trends in Microbiology,* 2 (3).

Reeves, Jacqueline D., and Robert W. Doms. 2002. "Human Immunodeficiency Virus Type 2." *Journal of General Virology,* 83.

Reynes, J. M., D. Counor, S. Ong, C. Faure, V. Seng, S. Molia, J. Walston, et al. 2005. "Nipah Virus in Lyle's Flying Foxes, Cambodia." *Emerging Infectious Diseases,* 11 (7).

Rich, Stephen M., Fabian H. Leendertz, Guang Xu, Matthew LeBreton, Cyrille F. Djoko, Makoah N. Aminake, Eric E. Takang, et al. 2009. "The Origin of Malignant Malaria." *Proceedings of the National Academy of Sciences,* 106 (35).

Richter, D., A. Spielman, N. Komar, and F. R. Matuschka. 2000. "Competence of American Robins as Reservoir Hosts for Lyme Disease *Spirochetes.*" *Emerging Infectious Diseases,* 6 (2).

Roest, H. I., J. J. Tilburg, W. van der Hoek, P. Vellema, F. G. van Zijdervelde, C. H. Klaassen, and D. Raoult. 2010. "The Q Fever Epidemic in The Netherlands: History, Onset, Response and Reflection." *Epidemiology and Infection,* 139 (1).

Roest, H. I., R. C. Ruuls, J. J. Tilburg, M. H. Nabuurs-Franssen, C. H. Klaassen, P. Vellema, R. van den Brom, et al. 2011. "Molecular Epidemiology of *Coxiella burnetii* from Ruminants in Q Fever Outbreak, The Netherlands." *Emerging Infectious Diseases,* 17 (4).

Ross, Ronald. 1910. *The Prevention of Malaria.* New York: E. P. Dutton.

———. 1916. "An Application of the Theory of Probabilities to the Study of *a priori* Pathometry." *Proceedings of the Royal Society,* A, 92 (638).

———. 1923. *Memoirs.* London: John Murray.

Rothman, Kenneth J., and Sander Greenland, eds. 1998. *Modern Epidemiology.* Philadelphia: Lippincott Williams & Wilkins.

Sabin, Albert B., and Arthur M. Wright. 1934. "Acute Ascending Myelitis Following a Monkey Bite, with the Isolation of a Virus Capable of Reproducing the Disease." *Journal of Experimental Medicine,* 59.

Salomon, Rachelle, and Robert G. Webster. 2009. "The Influenza Virus Enigma." *Cell,* 136.

Santiago, Mario L., Friederike Range, Brandon F. Keele, Yingying Li, Elizabeth Bailes, Frederic Bibollet-Ruche, Cecile Fruteau, et al. 2005. "Simian Immunodeficiency Virus Infection in Free-Ranging Sooty Mangabeys (*Cercocebus atys atys)* from the Taï Forest, Côte d'Ivoire: Implications for the Origin of Epidemic Human Immunodeficiency Virus Type 2." *Journal of Virology,* 79 (19).

Santiago, Mario L., Cynthia M. Rodenburg, Shadrack Kamenya, Frederic Bibollet-Ruche, Feng Gao, Elizabeth Bailes, Sreelatha Meleth, et al. 2002. "SIVcpz in Wild Chimpanzees." *Science*, 295.

Scrimenti, Rudolph J. 1970. "Erythema Chronicum Migrans." *Archives of Dermatology*, 102.

Sellers, R. F., and A. J. Forman. 1973. "The Hampshire Epidemic of Foot-and-Mouth Disease, 1967." *Journal of Hygiene*, 71.

Sellers, R. F., and J. Parker. 1969. "Airborne Excretion of Foot-and-Mouth Disease Virus." *Journal of Hygiene*, 67.

Selvey, L. A., R. M. Wells, J. G. McCormack, A. J. Ansford, K. Murray, R. J. Rogers, P. S. Lavercombe, et al. 1995. "Infection of Humans and Horses by a Newly Described Morbillivirus." *Medical Journal of Australia*, 162.

Selvey, Linda, Roscoe Taylor, Antony Arklay, and John Gerrard. 1996. "Screening of Bat Carers for Antibodies to Equine Morbillivirus." *Communicable Diseases*, 20 (22).

Severo, Richard. 1972. "Impoverished Haitians Sell Plasma for Use in the U.S." *The New York Times*, January 28.

Sexton, Christopher. 1991. *The Seeds of Time: The Life of Sir Macfarlane Burnet*. Oxford: Oxford University Press.

Shah, Keerti V. 2004. "Simian Virus 40 and Human Disease." *The Journal of Infectious Diseases*, 190.

Shah, Kccrti, and Ncal Nathanson. 1976. "Human Exposurc to SV40: Review and Comment." *American Journal of Epidemiology*, 103 (1).

Sharp, Paul M., and Beatrice H. Hahn. 2010. "The Evolution of HIV-1 and the Origin of AIDS." *Philosophical Transactions of the Royal Society of London*, 365.

Shilts, Randy. 1987. *And the Band Played On: Politics, People, and the AIDS Epidemic*. New York: St Martin's Griffin.

Simpson, D. I. H., and the Members of the WHO/International Study Team. 1978. "Ebola Haemorrhagic Fever in Sudan, 1976." *Bulletin of the World Health Organization*, 56 (2).

Singh, Balbir, Lee Kim Sung, Asmad Matusop, Anand Radhakrishnan, Sunita S. G. Shamsul, Janet Cox-Singh, Alan Thomas, and David J. Conway. 2004. "A Large Focus of Naturally Acquired *Plasmodium knowlesi* Infections in Human Beings." *The Lancet*, 363 (9414).

Smith, Davey, and Diana Kuh. 2001. "Commentary: William Ogilvy Kermack and the Childhood Origins of Adult Health and Disease." *International Journal of Epidemiology*, 30.

Snow, John 1855. *On the Mode of Communication of Cholera*. London: John Churchill.

Sompayrac, Lauren. 2002. *How Pathogenic Viruses Work*. Sudbury, MA: Jones and Bartlett Publishers.

Sorensen, J. H., D. K. Mackay, C. O. Jensen, and A. I. Donaldson. 2000. "An Integrated Model to Predict the Atmospheric Spread of Foot-and-Mouth Disease Virus." *Epidemiology and Infection,* 124.

Stearns, Jason K. 2011. *Dancing in the Glory of Monsters: The Collapse of the Congo and the Great War of Africa*. New York: PublicAffairs.

Steere, Allen C. 2001. "Lyme Disease." *New England Journal of Medicine,* 345 (2).

Steere, Allen C., and Stephen E. Malawista. 1979. "Cases of Lyme Disease in the United States: Locations Correlated with Distribution of *Ixodes dammini*." *Annals of Internal Medicine,* 91.

Steere, Allen C., Stephen E. Malawista, John A. Hardin, Shaun Ruddy, Philip W. Askenase, and Warren A. Andiman. 1977a. "Erythema Chronicum Migrans and Lyme Arthritis, The Enlarging Clinical Spectrum." *Annals of Internal Medicine,* 86 (6).

Steere, Allen C., Stephen E. Malawista, David R. Snydman, Robert E. Shope, Warren A. Andiman, Martin R. Ross, and Francis M. Steele. 1977b. "Lyme Arthritis. An Epidemic of Oligoarticular Arthritis in Children and Adults in Three Connecticut Communities." *Arthritis and Rheumatism,* 20 (1).

Stepan, Nancy Leys. 2011. *Eradication: Ridding the World of Diseases Forever?* London: Reaktion Books.

Strauss, James H., and Ellen G. Strauss. 2002. *Viruses and Human Disease*. San Diego: Academic Press.

Sureau, Pierre H. 1989. "Firsthand Clinical Observations of Hemorrhagic Manifestations in Ebola Hemorrhagic Fever in Zaire." *Reviews of Infectious Diseases,* 11 (S4).

Switzer, William M. 2005. "Ancient Co-Speciation of Simian Foamy Viruses and Primates." *Nature,* 434.

Taylor, Barbara S., Magdalena E. Sobieszczyk, Francine E. McCutchan, and Scott M. Hammer. 2008. "The Challenge of HIV-1 Subtype Diversity." *New England Journal of Medicine,* 358 (15).

Timen, Aura, Marion P. G. Koopmans, Ann C. T. M. Vossen, Gerard J. J. van Doornum, Stephan Gunther, Franchette Van den Berkmortel, Kees M. Verduin, et al. 2009. "Response to Imported Case of Marburg Hemorrhagic Fever, The Netherlands." *Emerging Infectious Diseases,* 15 (8).

Towner, Jonathan S., Brian S. Amman, Tara K. Sealy, Serena A. Reeder Carroll, James A. Comer, Alan Kemp, Robert Swanepoel, et al. 2009.

"Isolation of Genetically Diverse Marburg Viruses from Egyptian Fruit Bats." *PLoS Pathogens,* 5 (7).

Towner, Jonathan S., Tara K. Sealy, Marina L. Khristova, César G. Albariño, Sean Conlan, Serena A. Reeder, Phenix-Lan Quan, et al. 2008. "Newly Discovered Ebola Virus Associated with Hemorrhagic Fever Outbreak in Uganda." *PLoS Pathogens,* 4 (11).

Tu, Changchun, Gary Crameri, Xiangang Kong, Jinding Chen, Yanwei Sun, Meng Yu, Hua Xiang, et al. 2004. "Antibodies to SARS Coronavirus in Civets." *Emerging Infectious Diseases,* 10 (12).

Tutin, C. E. G., and M. Fernandez. 1984. "Nationwide Census of Gorilla (*Gorilla g. gorilla*) and Chimpanzee (*Pan t. troglodytes*) Populations in Gabon." *American Journal of Primatology,* 6.

Van den Brom, R., and P. Vellema. 2009. "Q Fever Outbreaks in Small Ruminants and People in The Netherlands." *Small Ruminant Research,* 86.

Van der Hoek, W., F. Dijkstra, B. Schimmer, P. M. Schneeberger, P. Vellema, C. Wijkmans, R. ter Schegget, et al. "Q Fever in The Netherlands: An Update on the Epidemiology and Control Measures." *Eurosurveillance,* 15.

Van Rooyen, G. E. 1955. "The Early History of Psittacosis." In *Psittacosis: Diagnosis, Epidemiology and Control,* ed. F. R. Beaudette. New Brunswick, NJ: Rutgers University Press.

Uppal, P. K. 2000. "Emergence of Nipah Virus in Malaysia." *Annals of the New York Academy of Sciences,* 916.

Varia, Monali, Samantha Wilson, Shelly Sarwal, Allison McGeer, Effie Gournis, Elena Galanis, Bonnie Henry, et al. 2003. "Investigation of a Nosocomial Outbreak of Severe Acute Respiratory Syndrome (SARS) in Toronto, Canada." *Canadian Medical Association Journal,* 169 (4).

Volberding, Paul A., Merle A. Sande, Joep Lange, Warner C. Greene, and Joel E. Gallant, eds. 2008. *Global HIV/AIDS Medicine.* Philadelphia: Saunders Elsevier.

Voyles, Bruce A. 2002. *The Biology of Viruses.* Boston: McGraw-Hill.

Wacharapluesadee, Supaporn, Boonlert Lumlertdacha, Kalyanee Boongird, Sawai Wanghongsa, Lawan Chanhome, Pierrie Rollin, Patrick Stockton, et al. 2005. "Bat Nipah Virus, Thailand." *Emerging Infectious Diseases,* 11 (12).

Walsh, Peter D., Roman Biek, and Leslie A. Real. 2005. "Wave-Like Spread of Ebola Zaire." *PLoS Biology,* 3 (11).

Walsh, Peter D., Thomas Breuer, Crickette Sanz, David Morgan, and Diane Doran-Sheehy. 2007. "Potential for Ebola Transmission

Between Gorilla and Chimpanzee Social Groups." *The American Naturalist,* 169 (5).

Walters, Marc Jerome. 2003. *Six Modern Plagues: And How We Are Causing Them.* Washington: Island Press/Shearwater Books.

Wamala, Joseph F., Luswa Lukwago, Mugagga Malimbo, Patrick Nguku, Zabulon Yoti, Monica Musenero, Jackson Amone, et al. 2010. "Ebola Hemorrhagic Fever Associated with Novel Virus Strain, Uganda, 2007–2008." *Emerging Infectious Diseases,* 16 (7).

Waters, A. P., D. G. Higgins, and T. F. McCutchan. 1991. "*Plasmodium falciparum* Appears to Have Arisen as a Result of Lateral Transfer Between Avian and Human Hosts." *Proceedings of the National Academy of Sciences,* 88.

Webster, Robert G. 1998. "Influenza: An Emerging Disease." *Emerging Infectious Diseases,* 4 (3).

———. 2004. "Wet Markets—a Continuing Source of Severe Acute Respiratory Syndrome and Influenza?" *The Lancet,* 363 (9404).

———. 2010. "William Graeme Laver, 3 June 1929–26 September 2008." *Biographical Memoirs of the Fellows of the Royal Society,* 56.

Weeks, Benjamin S., and I. Edward Alcamo. 2006. *AIDS: The Biological Basis.* Sudbury, MA: Jones and Bartlett.

Weigler, Benjamin J. 1992. "Biology of B Virus in Macaque and Human Hosts: A Review." *Clinical Infectious Diseases,* 14.

Weiss, Robin A. 1988. "A Virus in Search of a Disease." *Nature,* 333.

———. 2001. "The Leeuwenhoek Lecture 2001. Animal Origins of Human Infectious Disease." *Philosophical Transactions of the Royal Society of London,* B, 356.

Weiss, Robin A., and Jonathan L. Heeney. 2009. "An Ill Wind for Wild Chimps?" *Nature,* 460.

Weiss, Robin A., and Angela R. McLean. 2004. "What Have We Learnt from SARS?" *Philosophical Transactions of the Royal Society of London,* B, 359.

Weiss, Robin A., and Richard W. Wrangham. 1999. "From *PAN* to Pandemic." *Nature,* 397.

Wertheim, Joel O., and Michael Worobey. 2009. "Dating the Age of the SIV Lineages that Gave Rise to HIV-1 and HIV-2." *PLoS Computational Biology,* 5 (5).

White, N. J. 2008. "*Plasmodium knowlesi*: The Fifth Human Malaria Parasite." *Clinical Infectious Diseases,* 46.

Williams, Jim C., and Herbert A. Thompson. 1991. *Q Fever: The Biology of* Coxiella burnetii. Boca Raton: CRC Press.

Willrich, Michael. 2011. *Pox: An American History.* New York: Penguin.

Wills, Christopher. 1996. *Yellow Fever, Black Goddess: The Coevolution of People and Plagues.* New York: Basic Books.

Wilson, Edward O. 2002. "The Bottleneck." *Scientific American,* February.

Wolf, R. H., B. J. Gormus, L. N. Martin, G. B. Baskin, G. P. Walsh, W. M. Meyers, and C. H. Binford. 1985. "Experimental Leprosy in Three Species of Monkeys." *Science,* 227.

Wolfe, Nathan. 2011. *The Viral Storm: The Dawn of a New Pandemic Age.* New York: Times Books/Henry Holt.

Wolfe, Nathan D., Claire Panosian Dunavan, and Jared Diamond. 2004. "Origins of Major Human Infectious Diseases." *Nature,* 447.

Wolfe, Nathan D., William M. Switzer, Jean K. Carr, Vinod B. Bhullar, Vedapuri Shanmugam, Ubald Tamoufe, A. Tassy Prosser, et al. 2004. "Naturally Acquired Simian Retrovirus Infections in Central African Hunters." *The Lancet,* 363 (9413).

Woolhouse, Mark E. J. 2002. "Population Biology of Emerging and Re-emerging Pathogens." *Trends in Microbiology,* 10 (10, Suppl.).

Worboys, Michael. 2000. *Spreading Germs: Disease Theories and Medical Practice in Britain, 1865–1900.* Cambridge: Cambridge University Press.

World Health Organization. 2006. *SARS: How a Global Pandemic Was Stopped.* Geneva: World Health Organization.

Worobey, Michael. 2008. "The Origins and Diversification of HIV." In *Global HIV/AIDS Medicine,* ed. P. A. Volberding, M. A. Sande, J. Lange, W. C. Greene, and J. E. Gallant. Philadelphia: Saunders Elsevier.

Worobey, Michael, Marlea Gemmel, Dirk E. Teuwen, Tamara Haselkorn, Kevin Kuntsman, Michael Bunce, Jean-Jacques Muyembe, et al. 2008. "Direct Evidence of Extensive Diversity of HIV-1 in Kinshasa by 1960." *Nature,* 455.

Wrong, Michela. 2001. *In the Footsteps of Mr. Kurtz: Living on the Brink of Disaster in Mobutu's Congo.* New York: HarperCollins.

Xu, Rui-Heng, Jian-Feng He, Guo-Wen Peng, De-Wen Yu, Hui-Min Luo, Wei-Sheng Lin, Peng Lin, et al. 2004. "Epidemiologic Clues to SARS Origin in China." *Emerging Infectious Diseases,* 10 (6).

Yates, Terry L., James N. Mills, Cheryl A. Parmenter, Thomas G. Ksiazek, Robert R. Parmenter, John R. Vande Castle, Charles H. Calisher, et al. 2002. "The Ecology and Evolutionary History of an Emergent Disease: Hantavirus Pulmonary Syndrome." *BioScience,* 52 (11).

Young, P., H. Field, and K. Halpin. 1996. "Identification of Likely Natural Hosts for Equine Morbillivirus." *Communicable Diseases Intelligence*, 20 (22).

Zhong, N. S., B. J. Zheng, Y. M. Li, L. L. M. Poon, Z. H. Xie, K. H. Chan, P. H. Li, et al. 2003. "Epidemiology and Cause of Severe Acute Respiratory Syndrome (SARS) in Guangdong, People's Republic of China, in February, 2003." *The Lancet*, 362 (9393).

Zhu, Tuofu, and David D. Ho. 1995. "Was HIV Present in 1959?" *Nature*, 374.

Zhu, Tuofu, Bette T. Korber, Andre J. Nahmias, Edward Hooper, Paul M. Sharp, and David D. Ho. 1998. "An African HIV-1 Sequence from 1959 and Implications for the Origin of the Epidemic." *Nature*, 391.

Zimmer, Carl. 2011. *A Planet of Viruses*. Chicago: The University of Chicago Press.

Zinsser, Hans. 1934. *Rats, Lice and History*. Reprint edition (undated), New York: Black Dog & Leventhal Publishers.

ACKNOWLEDGMENTS

This book had its earliest origin around a campfire in a Central African forest, in July 2000, when two Gabonese men spoke to me about the Ebola outbreak that had struck their village, Mayibout 2, and the thirteen dead gorillas they had seen nearby in the forest around the same time their family members and friends were dying. My thanks must therefore go first to those two men: Thony M'Both and Sophiano Etouck. I'm also indebted to the people who put me at that campfire: Bill Allen, Oliver Payne, Kathy Moran, and their colleagues at *National Geographic* magazine; Nick Nichols, my photographic partner on that assignment (and on many others since); Tomo Nishihara and John Brown, logisticians; Neeld Messler, field assistant to Nick (and asset to us all); the Bantu and Pygmy crewmen who, serving as porters and much more, made the expedition through that Gabonese forest possible, including not just Thony and Sophiano but also Jean-Paul, Jacques, Celestin, Kar, Alfred, Mayombo, Boba, Yeye, and the point man with the machete, tireless Bebe; and most of all, J. Michael Fay, the mad dreamer of African conservation, whose dedication to preserving wild ecosystems and their fauna and flora is exceeded only, if at all, by his physical and intellectual toughness. Walking for weeks through Congolese and Gabonese forests with Mike Fay has been one of the great privileges of my life.

And because *National Geographic* has continued to sustain me with other work and privileged field experiences in the years since—including the assignment that led to "Deadly Contact," a feature story on zoonotic diseases, published in the October 2007 issue—I also declare here my ongoing gratitude to Chris Johns (editor in chief, having succeeded

Bill Allen), Carolyn White, Victoria Pope, again my longtime editor Oliver Payne, and all the other people involved in producing that great magazine. Lynn Johnson did a brilliant job on the photographic side of "Deadly Contact." Billy Karesh and Peter Daszak helped brainstorm the coverage plan for the article. Billy also provided fine company and veterinary insights on three continents. Peter Reid opened a crucial line into the subject when, in a former paddock near Brisbane, amid newly built houses and dark memories, he said: "That's it. That's the bloody tree."

Jens Kuhn, Charlie Calisher, and Mike Gilpin read the complete book in draft and gave me many invaluable corrections, suggestions, and remonstrations. Their expertise, thoroughness, and generosity made the book much better, but don't hold them responsible for any of its failings. Karl Johnson, from a very early stage, shared his thoughts and memories as an expert and as a friend, and allowed me to read his own book-in-progress on the Machupo story. Les Real counseled me on disease ecology and on the historical development of mathematical disease theory, from Bernoulli to Anderson and May. Karl Johnson, Les Real, and these other scientists and informants also found time to read and correct various sections in draft: Sazaly AbuBakr, Brian Amman, Brenda Ang, Michelle Barnes, Donald Burke, Aleksei Chmura, Jenny Cory, Janet Cox-Singh, Greg Dwyer, Gregory Engel, Jonathan Epstein, Kylie Forster, Emily Gurley, Beatrice Hahn, Barry Hewlett, Eddie Holmes, Lisa Jones-Engel, Jean-Marie Kabongo, Phyllis Kanki, Billy Karesh again, Brandon Keele, Eric Leroy, Steve Luby, Martin Muller, Judith Myers, Rick Ostfeld, Martine Peeters, Raina Plowright, Peter Reid, Hendrik-Jan Roest, Linda Selvey, Balbir Singh, Jaap Taal, Karen Terio, Dirk Teuwen, Jonathan Towner, Kelly Warfield, Robert Webster, and Michael Worobey. Lin-fa Wang gave me a day-long tour of the BSL-4 and other facilities at AAHL, in Geelong. Kelly Warfield likewise gave me a day, poured out her story, and got me into (and back out of) the Slammer. Ian Lipkin opened his lab and his people to me as well. Quite a few other scientists, mentioned below, trusted me with the opportunity to accompany them during fieldwork. Larry Madoff provided me inestimable assistance, without knowing it, through his ProMED-mail alerts on disease incidents around the world. And there were many others, so many, in so many places, who aided my research efforts so variously—as interviewees or expert consultants or traveling companions or providers of leads—that my further thanks are best organized geographically and alphabetically.

In Australia: Natalie Beohm, Jennifer Crane, Bart Cummings, Rebekah Day, Carol de Jong, Hume Field, Kylie Forster, Kim Halpin,

Peter Hulbert, Brenton Lawrence, David Lovell, Deb Middleton, Nigel Perkins, Raina Plowright, Stephen Prowse, Peter Reid, Linda Selvey, Neil Slater, Craig Smith, Gary Tabor, Barry Trail, Ray Unwin, Craig Walker, Lin-fa Wang, Emma Wilkins, and Dick Wright.

In Africa: Patrick Atimnedi, Bruno Baert, Prosper Balo, Paul Bates, Roman Biek, Ken Cameron, Anton Collins, Zacharie Dongmo, Bob Downing, Ofir Drori, Clelia Gasquet, Jane Goodall, Barry Hewlett, Naftali Honig, Jean-Marie Kabongo, Winyi Kaboyo, Glady Kalema-Zikusoka, Shadrack Kamenya, Billy Karesh, John Kayiwa, Sally Lahm, Eric Leroy, Iddi Lipende, Julius Lutwama, Pegue Manga, Neville Mbah, Apollonaire Mbala, Alastair McNeilage, Achille Mengamenya, Jean Vivien Mombouli, Albert Munga, J. J. Muyembe, Max Mviri, Cécile Neel, Hanson Njiforti, Alain Ondzie, Cindy Padilla, Andrew Plumptre, Xavier Pourrut, Jane Raphael, Trish Reed, Paul Roddy, Innocent Rwego, Jordan Tappero, Moïse Tchuialeu, Peter Walsh, Joe Walston, Nadia Wauquier, Beryl West, and Lee White.

In Asia: Sazaly AbuBakar, Brenda Ang, Mohammad Aziz, Aleksei Chmura, Janet Cox-Singh, Jim Desmond, Gregory Engel, Jonathan Epstein, Mustafa Feeroz, Martin Gilbert, Emily Gurley, Johangir Hossain, Arif Islam, Yang Jian, Lisa Jones-Engel, Rasheda Khan, Salah Uddin Khan, Steve Luby, Sue Meng, Joe Meyer, Nazmun Nahar, Malik Peiris, Leo Poon, Mahmudur Rahman, Muhammad Rahman, Sohayati Rahman, Sorn San, Balbir Singh, Gavin Smith, Juliet Tseng, and Guangjian Zhu.

In Europe: Rob Besselink, Arnout de Bruin, Pierre Formenty, Fabian Leendertz, Viktor Molnar, Martine Peeters, Hendrik-Jan Roest, Barbara Schimmer, Jaap Taal, Dirk Teuwen, Wim van der Hoek, Yvonne van Duynhoven, Jim van Steenbergen, and Ineke Weers.

In the United States: Brian Amman, Kevin Anderson, Mike Antolin, Jesse Brunner, Charlie Calisher, Deborah Cannon, Darin Carroll, David Daigle, Inger Damon, Peter Daszak, Andy Dobson, Tony Dolan, Rick Douglass, Shannon Duerr, Ginny Emerson, Eileen Farnon, Robert Gallo, Tom Gillespie, Barney Graham, Beatrice Hahn, Barbara Harkins, Eddie Homes, Pete Hudson, Vivek Kapur, Kevin Karem, Billy Karesh, Brandon Keele, Ali Khan, Marm Kilpatrick, Lonnie King, Tom Ksiazek, Amy Kuenzi, Jens Kuhn, Edith Lederman, Julie Ledgerwood, Jill Lepore, Ian Lipkin, Andrew Lloyd-Smith, Elizabeth Lonsdorf, Adam MacNeil, Jennifer McQuiston, Nina Marano, Jim Mills, Russ Mittermeier, Jennifer Morcone, Stephen Morse, Martin Muller, Stuart Nichol, Rick Ostfeld, Mary Pearl, Mary Poss, Andrew Price-Smith, Juliet Pulliam, Anne Pusey, Andrew Read, Les Real, Zach Reed, Russ Regnery,

Anne Rimoin, Pierre Rollin, Charles Rupprecht, Anthony Sanchez, Tony Schountz, Nancy Sullivan, Karen Terio, Jonathan Towner, Giliane Trindade, Murray Trostle, Abbigail Tumpey, Sally and Robert Uhlmann, Caree Vander Linden, Kelly Warfield, Robert Webster, Nathan Wolfe, and Michael Worobey.

There were others who helped too, omitted here only because my memory is bad and my notebooks and journals, just slightly more orderly than a Congolese forest, still hold some secrets even from me. Apologies for the omission, and thank you,

Maria Guarnaschelli, of W. W. Norton, my editor through many years and half a dozen books, has played her usual keen-eyed, penetrating, structurally astute, and deeply supportive role with this one. Her contributions are no less precious to me for having continued so reliably over decades. Amanda Urban of ICM, my agent, helped shape the project from the stage of a first-draft proposal and has blessed it with her ferocious advocacy ever since. These two formidable women make it possible for me to write the sort of books (requiring a bit of time and travel) that I want to write. A third, Renée Wayne Golden, played that role in earlier times and without her too this book wouldn't exist. Melanie Tortoroli, Maria's assistant, and their colleagues at Norton have given this project the focus, support, and professionalism for which an author always wishes. Daphne Gillam, creator of the maps (www.handcraftedmaps.com), put the artistry of human touch to the lineaments of geography. Chip Kidd's jacket reminded us all what a spooky subject this is. Emily Krieger combined assiduous research with a reader's sense of flow, both crucial attributes, in serving as my fact-checker. Gloria Thiede, faithful Gloria, again helped me immensely with secretarial tasks, including the transcription of interviews recorded while air conditioners, coffee grinders, street traffic, and cockatoos screeched in the background. Jodi Solomon, my lecture agent, has brokered the way to live audiences. Dan Smith, Dan Krza, Danny Schotthoefer (my three Daniels), and Don Killian assisted me greatly in the digital dimension, handling tasks of Web site design, computer repair and data rescue, and social media wrangling, most of which are even more mysterious to me than the mathematics of Anderson and May. The late Chuck West will be very much missed. Betsy my amazing wife, and Harry and Kevin and Skipper (and Nelson, now departed), our dependents, warmed the home in which this book was written.

INDEX

A NOTE ABOUT THE AUTHOR

DAVID QUAMMEN is the author of four books of fiction and seven acclaimed nonfiction titles, including *The Reluctant Mr. Darwin* and *The Song of the Dodo*, which was awarded the John Burroughs Medal for natural history writing. He has been honored with an Academy Award in Literature by the American Academy of Arts and Letters and is a three-time recipient of the National Magazine Award. Quammen holds honorary doctorates from Montana State University, where he served as the Wallace Stegner Chair of Western American Studies from 2007 to 2009, and Colorado College. He is a contributing writer for *National Geographic* magazine and lives with his wife, Betsy Gaines, in Bozeman, Montana.